MICROARRAY BIOINFORMATICS

This book is a comprehensive guide to all of the mathematics, statistics and computing you will need to successfully operate DNA microarray experiments. It is written for researchers, clinicians, laboratory heads and managers, from both biology and bioinformatics backgrounds, who work with or who intend to work with microarrays. The book covers all aspects of microarray bioinformatics, giving you the tools to design arrays and experiments, to analyze your data, and to share your results with your organisation or with the international community. There are chapters covering sequence databases, oligonucleotide design, experimental design, image processing, normalisation, identifying differentially expressed genes, clustering, classification and data standards. The book is based on the highly successful Microarray Bioinformatics course at Oxford University and, therefore, is ideally suited for teaching the subject at the postgraduate or professional level.

Dov Stekel is Director of the Microarray Bioinformatics professional course at Oxford University and is a visiting academic at the Department of Biochemistry there. He also operates a bioinformatics consultancy, Bius, providing services to customers in the biotechnology industry and academia. Previously, he was the Manager of Bioinformatics at Ed Southern's microarray company, Oxford Gene Technology, and has worked as a Bioinformatics Scientist at Glaxo Wellcome. Dov lives in London and is an avid chocolate maker.

Microarray Bioinformatics

DOV STEKEL

Oxford University and Bius

PUBLISHED BY THE PRESS SYNDICATE OF THE UNIVERSITY OF CAMBRIDGE
The Pitt Building, Trumpington Street, Cambridge, United Kingdom

CAMBRIDGE UNIVERSITY PRESS
The Edinburgh Building, Cambridge CB2 2RU, UK
40 West 20th Street, New York, NY 10011-4211, USA
477 Williamstown Road, Port Melbourne, VIC 3207, Australia
Ruiz de Alarcón 13, 28014 Madrid, Spain
Dock House, The Waterfront, Cape Town 8001, South Africa

http://www.cambridge.org

First published 2003

Printed in the United States of America

Typeface Utopia 9.5/13.5 pt. and ITC Symbol *System* LaTeX 2_ε [TB]

A catalog record for this book is available from the British Library.

Library of Congress Cataloging in Publication Data
Stekel, Dov, 1971–
Microarray bioinformatics / Dov Stekel.
p. cm.
Includes bibliographical references and index.
ISBN 0-521-81982-2 (hbk.) – ISBN 0-521-52587-X (pbk.)
1. DNA microarrays – Mathematics. 2. DNA microarrays – Data processing. 3. DNA microarrays – Statistical methods. 4. Bioinformatics. I. Title.
QP624.5.D726S74 2003
572.8′636 – dc21 2003043959

ISBN 0 521 81982 2 hardback
ISBN 0 521 52587 X paperback

To my parents, Zsuzsi and Ronnie Stekel

In memory of Stephen Zatman, 1971–2002

Contents

Foreword

DNA array technology is almost fifteen years old, and still rapidly evolving. It is one of very few platforms capable of matching the scale of sequence data produced by genome sequencing. Applications range from analysing single base changes, SNPs, to detecting deletion or amplification of large segments of the genome, CGH. At present, its most widespread use is in the analysis of gene expression levels. When carried out globally on all the genes of an organism, this analysis exposes its molecular anatomy with unprecedented clarity. In basic research, it reveals gene activities associated with biological processes and groups genes into networks of interconnected activities. There have been practical outcomes, too. Most notably, large-scale expression analysis has revealed genes associated with disease states, such as cancer, informed the design of new methods of diagnosis, and provided molecular targets for drug development.

At face value, the method is appealingly simple. An array is no more than a set of DNA reagents for measuring the amount of sequence counterparts among the mRNAs of a sample. However, the quality of the result is affected by several factors, including the quality of the array and the sample, the uniformity of hybridisation process, and the method of reading signals. Errors, inevitable at each stage, must be taken into account in the design of the experiment and in the interpretation of results. It is here that the scientist needs the help of advanced statistical tools.

Dr. Stekel is a mathematician with several years of experience in the microarray field. He has used his expertise in a company setting, developing advanced methods for probe design and for the analysis of large, complex data sets. This book is based on this practical experience and, more particularly, on experience gained in designing and running a course on Bioinformatics at the University of Oxford. The demand for this course showed that there are hundreds of biologists wanting to learn how to get the most from their microarray experiments. This book will help them to understand the nature of the data and the likely sources of error. It provides them with practical guidance and tools for handling large data sets and the statistical methods that can deal with them.

Ed Southern

Preface

DNA microarrays are devices that measure the expression of many thousands of genes in parallel. They have revolutionised molecular biology, and in the past five years their use has grown rapidly in academia, medicine, and the pharmaceutical, biotechnology, agrochemical and food industries.

One of the principal features of microarrays is the volume of quantitative data that they generate. As a result, the major challenge in the field is how to handle, interpret and make use of this data. The field of bioinformatics has come to mean the applications of mathematics, statistics and information technology in the biological sciences, and the bioinformatics of microarrays is the answer to that challenge.

This book is a comprehensive guide to all of the bioinformatics you will need to successfully operate DNA microarray experiments. It is written for researchers, clinicians, laboratory heads and managers, from both biology and bioinformatics backgrounds, who work with or who intend to work with microarrays. The book covers all aspects of microarray bioinformatics, giving you the tools to design arrays and experiments, to analyze your data, and to share you results with your organisation or with the international community. It has been inspired by the Microarray Bioinformatics professional course at Oxford University, and thus would also be suitable for teaching the subject at postgraduate or professional level.

The book assumes a minimum knowledge of molecular biology, computer use and statistics. On the biology front, readers will find it helpful if they have an understanding of the basic principles of molecular biology, i.e., DNA, RNA, transcription and translation, as well as the notions of genome sequencing and the existence of sequence databases. On the computing side, it is assumed that readers are familiar with the World Wide Web, and how to obtain data and software from it. No programming experience is needed to understand the book, although some of the ideas in the book would require programming skills to implement. It is also assumed that readers are familiar with the basic ideas of statistical descriptions of populations, such as means, standard deviations, histograms and scatterplots.

Where possible, the chapters include worked examples using real microarray data from published experiments. Each chapter ends with references to the data used, to a small selection of specialized research papers and textbooks relevant for further study, and to Internet resources and software relevant for the implementation of the methods described in it. Unpublished data sets and errata are available at the web site for this book, http://www.microarraybioinformatics.com.

Outline of Contents

The book is split into eleven chapters:

Chapter 1: Microarrays: Making Them and Using Them, gives an introduction to microarray technologies, the different platforms by which microarrays are manufactured, and the laboratory process involved in microarray use.

Chapter 2: Sequence Databases for Microarrays, is a description of the international sequence databases that are used for microarray design and annotation.

Chapter 3: Computer Design of Oligonucleotide Probes, describes the algorithmic methods by which oligonucleotide probes for microarrays can be designed.

Chapter 4: Image Processing, looks at the computational algorithms used to convert microarray images into quantitative data.

Chapter 5: Normalisation, describes methods that are used to eliminate systematic bias introduced by the microarray platform from microarray data.

Chapter 6: Measuring and Quantifying Microarray Variability, describes methods for measuring and quantifying the stratified variability that is a feature of microarray data.

Chapter 7: Analysis of Differentially Expressed Genes, looks at the analysis of microarray data where the microarray is being used to identify genes that may be up-regulated or down-regulated in different tissues or conditions.

Chapter 8: Analysis of Relationships Between Genes, Tissues or Treatments, describes methods that are used to explore the relationships between different genes or samples, including clustering and other related methods.

Chapter 9: Classification of Tissues and Samples, discusses methods that can be used to build predictive models that use gene expression for diagnostic or prognostic purposes.

Chapter 10: Experimental Design, looks at a number of issues in relation to how to design a microarray experiment, including how to determine the number of replicates you would need to use.

Chapter 11: Data Standards, Storage and Sharing, describes the computer technologies needed to run a microarray laboratory, and the standards by which microarray experiments and data can be annotated and shared.

Acknowledgments

Who is wise? One who can learn from everybody.

– Shimon son of Zoma, 1st Century CE

My whole life has been extremely simple. For the first half I sat facing a blackboard and for the second half I stood back to a blackboard. With regard to a blackboard I have made only one complete turn – and with that my biography is complete.

– Nishida Kitaro, 1870–1945

This is the first book I have written, and it is the most difficult project I have taken on in my life. It would not have been possible to start the book, let alone complete it, without the inspiration, guidance, help, advice and support of my family, friends, colleagues and students.

Ed Southern has supported this book in every possible way: personally, with friendship and encouragement; scientifically, by giving me access to his slides that formed the basis of Chapter 1, and by providing feedback on other chapters; and financially, via access to an MRC research fellowship. It is a rare privilege to work with such an outstanding individual, and one that I deeply treasure.

The inspiration for this book has been the Microarray Bioinformatics course that is run at the Department of Continuing Education at Oxford University, and at the Roslin Institute in Edinburgh. Every chapter of this book contains knowledge and understanding that I have taken from the course and its lecturers; thank you to Andy Greenfield, Clare Pritchard, David Wild, Ed Southern, Francesco Falciani, Kate Rice, Lorenz Wernisch, Louise Pedlar, Pete Underhill, Rob Andrews and Sarah Webb. The course would not be possible without the administrative support of Roni McGowan and Rachel Cox, who have also been deeply helpful and supportive of this book. To this date, more than 100 delegates have come through the course; I have learned from each and every one of you, though you are too numerous to list in this short section.

Many figures in this book have been reproduced from lectures given on the microarray course. To have had access to such excellent lecture slides and figures has been a tremendous bonus for which I am deeply appreciative. I thank Oxford University for permission to use these figures, and specifically Clare Pritchard (Figures 1.1a and 1.10), Ed Southern (Figures 1.1b, 1.1c, 1.3, 1.4, 1.5, 1.6, 1.7 and 1.9), Francesco Falciani (Figures 1.1d and 1.11), Rob Andrews (Figure 11.3), Roni McGowan (Figures 1.1a and 1.10) and Sarah Webb (Figures 1.12, 1.13, 1.14, 4.2, 4.4 and 4.5). Pete Underhill helped produce Figures 4.1a, 4.3 and 4.5 at the MRC Mouse Genetics Unit in Harwell.

Figure 1.4 is reproduced with permission from Affymetrix, and Figure 11.3 with permission from the Gene Ontology (GO) Consortium.

I have also been very fortunate to have been able to call on so many friends and colleagues to read chapters and provide feedback on them. Thank you to Ann Git, Chris Littler, Claire Parker, Clare Pritchard, David Wild, Ed Southern, Pete Underhill, Richard Talbot, Rob Andrews, Sarah Webb and Yoav Git, without whom this book would contain many more errors and inconsistencies. Any remaining mistakes are mine alone.

Anyone who has written a book knows that it is a long and lonely process. Nicky Press was a constant source of strength, support and companionship, not to mention hot lunches and cups of tea when I have most needed them. My writer friends: Daniel Litvin, Margaret Myers, Samantha Ellis and Sandra Shulman have a particular understanding of the exercise, and I thank them for their support, encouragement and advice.

Finally, and most importantly, I would like to thank my parents and sister for their unconditional love and support.

MICROARRAY BIOINFORMATICS

CHAPTER ONE

Microarrays: Making Them and Using Them

SECTION 1.1 INTRODUCTION

A DNA microarray consists of a solid surface, usually a microscope slide,[1] onto which DNA molecules have been chemically bonded. The purpose of a microarray is to detect the presence and abundance of labelled nucleic acids in a biological sample, which will hybridise to the DNA on the array via Watson–Crick duplex formation, and which can be detected via the label. In the majority of microarray experiments, the labelled nucleic acids are derived from the mRNA of a sample or tissue, and so the microarray measures gene expression. The power of a microarray is that there may be many thousands of different DNA molecules bonded to an array, and so it is possible to measure the expression of many thousands of genes simultaneously.

This book is about the bioinformatics of DNA microarrays: the mathematics, statistics and computing you will need to design microarray experiments; to acquire, analyse and store your data; and to share your results with other scientists. One of the features of microarray technology is the level of bioinformatics required: it is not possible to perform a meaningful microarray experiment without bioinformatics involvement at every stage.

However, this chapter is different from the remainder of the book. While the other chapters discuss bioinformatics, the aim of this chapter is to set out the basics of the chemistry and biology of microarray technology. It is hoped that someone new to the technology will be able to read this chapter and gain an understanding of the laboratory process and how it impacts the quality of the data. The chapter is arranged into two further sections:

Section 1.2: Making Microarrays, describes the main technologies by which microarrays are manufactured.
Section 1.3: Using Microarrays, describes what happens in a microarray laboratory when a microarray experiment is performed.

SECTION 1.2 MAKING MICROARRAYS

There are two main technologies for making microarrays: **robotic spotting** and **in-situ synthesis**.

[1] Historically, microarrays have also been produced using nylon filters and larger glass slides.

(a) 

(b)

(c) **(d)**

Figure 1.1: Spotting robot. (a) An example of a spotting robot. There are many different robots on the market; this is a Genetix spotting robot located at the Mouse Genetics Unit in Harwell, Oxfordshire. **(b)** The pins are held in a cassette in a rectangular grid, which in turn is held on a robot arm that can be moved between the microtiter well plates and the glass arrays to deposit liquid. **(c)** The number of pins in the cassette can vary. The more pins, the greater the throughput of the robot, but the greater the propensity for pin-to-pin variability. Each pin will spot a different grid on the array (Chapter 4). **(d)** Most pins in modern use have a reservoir that holds sample and so can print multiple features – usually on different arrays – from a single visit to the well containing probe. Earlier robots use solid pins, which can only print one feature before needing to collect more DNA from the well.

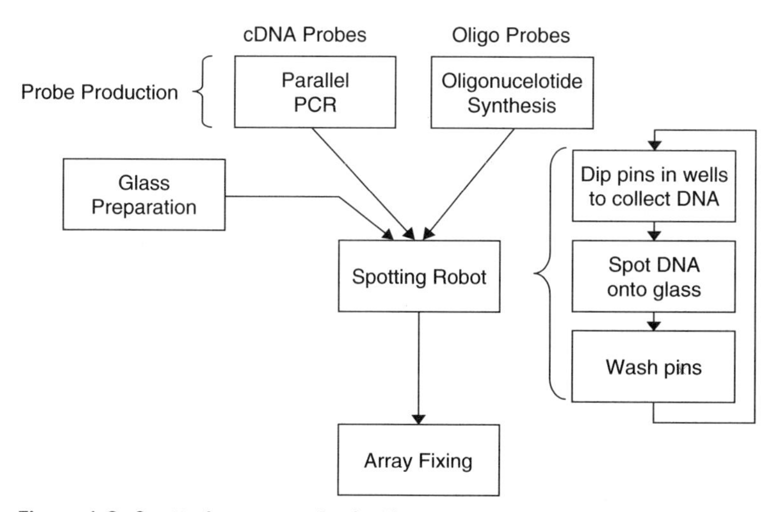

Figure 1.2: Spotted array synthesis. There are several steps involved in the synthesis of spotted arrays. First is the production of the probes. cDNA probes are made via highly parallel PCR; oligonucleotide probes have to be presynthesised. The next step is the spotting step. The robot arm moves the cassette containing the pins over one of the microtiter plates containing probe and dips the pins into the wells to collect DNA. The arm is then moved over the first array and the cassette is moved down so that the pins touch the glass and DNA is deposited on the surface. If more than one array is being synthesised, the cassette is moved to the subsequent arrays. Before collecting the next DNA to be spotted, the pins are washed to ensure no contamination. The final step of array production is fixing, in which the surface of the glass is modified so that no additional DNA can stick to it.

Spotted Microarrays

This is the technology by which the first microarrays were manufactured. The array is made using a spotting robot (Figure 1.1a) via three main steps (Figure 1.2):

1. Making the DNA probes[2] to put on the array
2. Spotting the DNA onto the glass surface of the array with the spotting robot
3. Postspotting processing of the glass slide

There are three main types of spotted array (Table 1.1), which can be subdivided in two ways: by the type of DNA probe, or by the attachment chemistry of the probe to the glass.

The DNA probes used on a spotted array can either be polymerase chain reaction (PCR) products or oligonucleotides. In the first case, highly parallel PCR is used to amplify DNA from a clone library, and the amplified DNA is purified. In the second case, DNA oligonucleotides are presynthesised for use on the array.

[2] There are now three camps in the microarray community as to what to call the DNA on the array and the DNA in solution. Throughout this book, we will use the "Southern" terminology and refer to the DNA on the array as *probes* and the labeled DNA in solution as *target*. Other researchers refer to the DNA on the array as *target* and the labeled DNA in solution as *probe*. More recently, MIAME (Minimal Information About a Microarray Experiment) introduced a new convention of referring to the DNA on the array as *reporters* and the DNA in solution as the *hybridisation extract*. MIAME conventions are described in full in Chapter 11, and MIAME terms are detailed in the Appendix.

TABLE 1.1

	Surface Chemistry	
DNA Probes	**Covalent**	**Non-covalent**
Oligonucleotides	✓	
cDNAs	✓	✓

Note: There are three types of spotted microarrays, which can be thought about in two different ways. The DNA probes can be oligonucleotides or cDNAs; the surface chemistry can be covalent or non-covalent. Oligonucleotide probes can only be attached covalently; cDNA probes can be attached either covalently or non-covalently. Covalent attachment is via an aliphatic amine (NH_2) group added to the 5′ end of the DNA probe, and consequently the probes are tethered to the glass from the 5′ end. Non-covalent attachment is via electrostatic attraction between amine groups on the glass slide and the phosphate groups on the DNA probe backbone; thus the DNA probe is attached to the glass by its backbone.

The attachment chemistry can either be covalent or non-covalent. With covalent attachment, a primary aliphatic amine (NH_2) group is added to the DNA probe and the probe is attached to the glass by making a covalent bond between this group and chemical linkers on the glass. With oligonucleotide probes, the amine group can be added to either end of the oligonucleotide during synthesis, although it is more usual to add it to the 5′ end of the oligonucleotide. With cDNA probes, the amine group is added to the 5′ end of the PCR primer from which the probes are made. Thus the cDNA probes are always attached from the 5′ end.

With non-covalent attachment, the bonding of the probe to the array is via electrostatic attraction between the phosphate backbone of the DNA probe and NH_2 groups attached to the surface of the glass. The interaction takes place at several locations along the DNA backbone, so that the probe is tethered to the glass at many points. Because most oligonucleotide probes are shorter than cDNAs, these interactions are not strong enough to anchor oligonucleotide probes to glass. Therefore, non-covalent attachment is usually only used for cDNA microarrays.

The DNA probes are organised in microtiter well plates, typically 384 well plates. Most modern spotting robots will use a number of plates to print arrays, so the plates are arranged in a "hotel," whereby the robot is able to gain successive access to each of the plates. The spotting robot itself consists of a series of pins arranged as a grid and held in a cassette (Figures 1.1b and 1.1c). The pins are used to transfer liquid from the microtiter plates to the glass array.[3]

There are a number of different designs of pins. The first spotting robots used solid pins (Figure 1.1b); these can only hold enough liquid for one spot on the array, thus requiring the pin cassette to return to the plate containing probe before printing the next spot. Most array-making robots today have pins with a reservoir that holds

[3] Not every spotting robot is a pin-based system: Perkin Elmer sell some robots which use a piezo-electric system to fire tiny drops of liquid onto the arrays. These are in the minority in the microarray field.

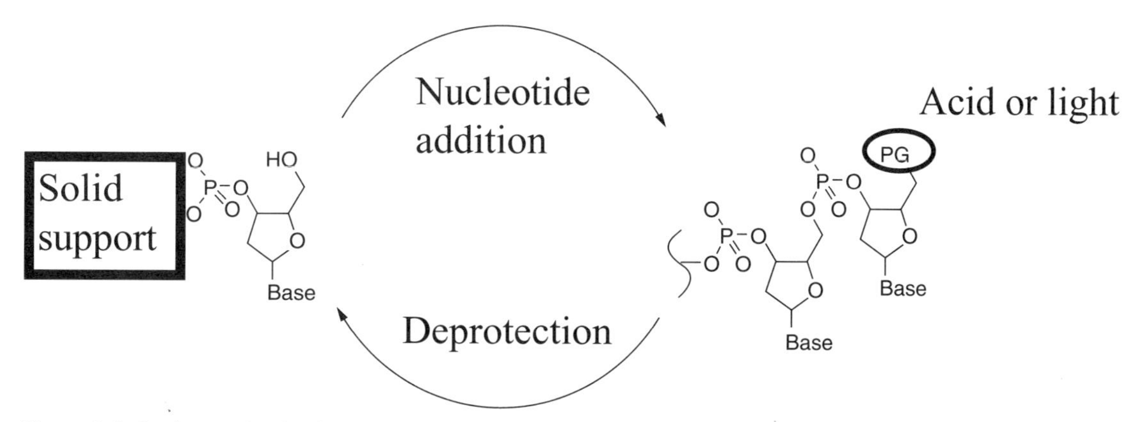

Figure 1.3: In-situ synthesis of oligonucleotides. The oligonucleotides are built on the glass array one base at a time. At each step, the base is added via the reaction between the hydroxyl group 5′ of the terminal base and the phosphate group of the next base. There is a protective group on the 5′ of the base being added, which prevents the addition of more than one base at each step. Following addition, there is a deprotection step at which the protective group is converted to a hydroxyl group to allow addition of the next base.

liquid (Figure 1.1d). This enables higher throughput production of arrays because each probe can be spotted on several arrays without the need to return the pins to the sample plates.

The typical printing process follows five steps (Figure 1.2):

1. The pins are dipped into the wells to collect the first batch of DNA.
2. This DNA is spotted onto a number of different arrays, depending on the number of arrays being made and the amount of liquid the pins can hold.
3. The pins are washed to remove any residual solution and ensure no contamination of the next sample.
4. The pins are dipped into the next set of wells.
5. Return to step 2 and repeat until the array is complete.

In the final phase of array production, the surface of the array can be *fixed* so that no further DNA can attach to it. There are many fixing processes that depend on the precise chemistry on the surface of the glass. The desired outcome is always the same: we do not want DNA target from the sample to stick to the glass of the array during hybridisation, so the surface must be modified so this does not happen. It is also common to modify the surface so that the glass becomes more hydrophilic because this aids mixing of the target solution during the hybridisation stage. Some microarray production facilities do not fix their arrays.

In-Situ Synthesised Oligonucleotide Arrays

These arrays are fundamentally different from spotted arrays: instead of presynthesising oligonucleotides, oligos are built up base-by-base on the surface of the array (Figure 1.3). This takes place by covalent reaction between the 5′ hydroxyl group of

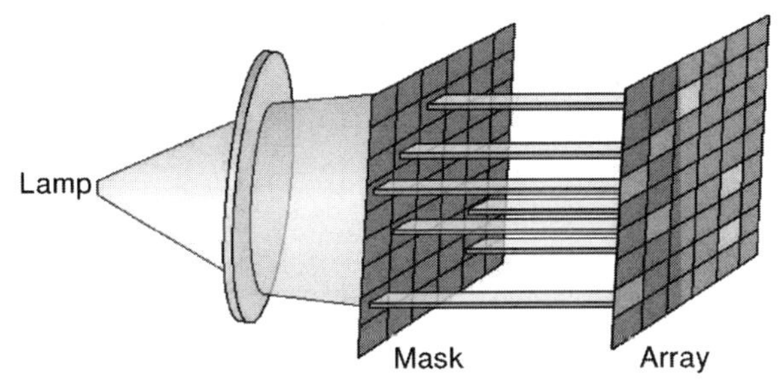

Figure 1.4: Affymetrix technology. Affymetrix arrays are manufactured using in-situ synthesis with a light-mediated deprotection step. During each round of synthesis, a single base is added to appropriate parts of the array. A mask is used to direct light to the appropriate regions of the array so that the base is added to the correct features. Each step requires a different mask. The masks are expensive to produce, but once made, it is straightforward to use them to manufacture a large number of identical arrays. (Reproduced with Permission from Affymetrix Inc.)

the sugar of the last nucleotide to be attached and the phosphate group of the next nucleotide. Each nucleotide added to the oligonucleotide on the glass has a protective group on its 5′ position to prevent the addition of more than one base during each round of synthesis. The protective group is then converted to a hydroxyl group either with acid or with light before the next round of synthesis. The different methods for deprotection lead to the three main technologies for making in-situ synthesised arrays:

1. Photodeprotection using masks: this is the basis of the Affymetrix® technology.
2. Photodeprotection without masks: this is the method used by Nimblegen and Febit.
3. Chemical deprotection with synthesis via inkjet technology: this is the method used by Rosetta, Agilent and Oxford Gene Technology.

Affymetrix Technology

Affymetrix arrays use light to convert the protective group on the terminal nucleotide into a hydroxyl group to which further bases can be added. The light is directed to appropriate features using *masks* that allow light to pass to some areas of the array but not to others (Figure 1.4). This technique is known as photolithography and was first applied to the manufacture of silicon chips. Each step of synthesis requires a different mask, and each mask is expensive to produce. However, once a mask set has been designed and made, it is straightforward to produce a large number of identical arrays. Thus Affymetrix technology is well suited for making large numbers of "standard" arrays that can be widely used throughout the community.

Maskless Photodeprotection Technology

This technology is similar to Affymetrix technology in that light is used to convert the protective group at each step of synthesis. However, instead of using masks, the light is directed via micromirror arrays, such as those made by Texas Instruments. These are solid-state silicon devices that are at the core of some data projectors: an array of mirrors is computer controlled and can be used to direct light to appropriate parts

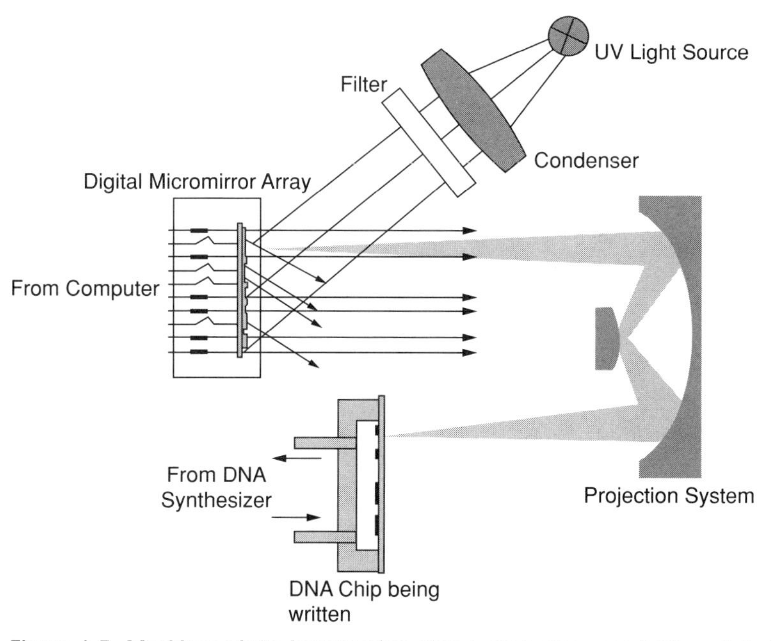

Figure 1.5: Maskless photodeprotection. This system also uses light-mediated deprotection. However, instead of using a physical mask, the array is synthesised using a computer-controlled micromirror array. This consists of a large number of mirrors embedded on a silicon chip, each of which can move between two positions: one position to reflect light, and the other to block light. At each step, the mirrors direct light to the appropriate parts of the array. This technology is used by Nimblegen and Febit.

of the glass slide at each step of oligonucelotide synthesis (Figure 1.5). This is the technology used by Nimblegen and Febit.

Inkjet Array Synthesis

Instead of using light to convert the protective group, deprotection takes place chemically, using the same chemistry as a standard DNA synthesiser. At each step of synthesis, droplets of the appropriate base are fired onto the desired spot on the glass slide via the same nozzles that are used for inkjet printers; but instead of firing cyan, magenta, yellow and black ink, the nozzles fire A, C, G and T nucleotides (Figure 1.6).

One of the main advantages of micromirror and inkjet technologies over both Affymetrix technology and spotted arrays is that the oligonucleotide being synthesised on each feature is entirely controlled by the computer input given to the array-maker at the time of array production. Therefore, these technologies are highly flexible, with each array able to contain any oligonucleotide the operator wishes. However, these technologies are also less efficient for making large numbers of identical arrays.

Synthesis Yields

The different methods of oligonucleotide synthesis have different coupling efficiencies: this is the proportion of nucleotides that are successfully added at each step of synthesis. Photodeprotection has a coupling efficiency of approximately 95%, whereas

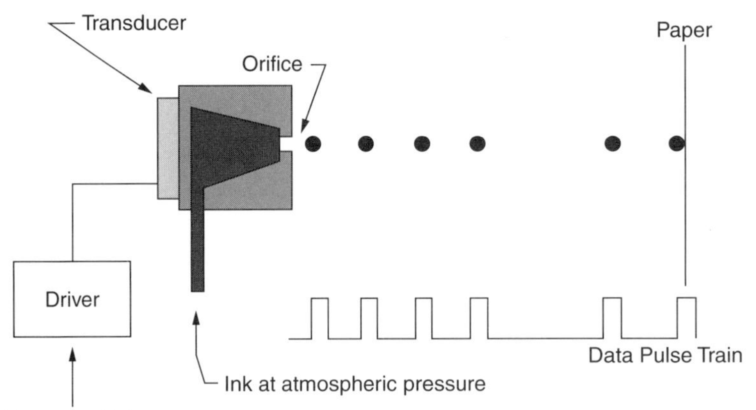

Figure 1.6: Inkjet array synthesis. This technology uses chemical deprotection to synthesise the oligonucleotides. The bases are fired onto the array using modified inkjet nozzles, which, instead of firing different coloured ink, fire different nucleotides. At each step, the appropriate nucleotide is fired onto each spot on the array. The nozzles are computer controlled, so any oligonucleotides can be synthesised on the array simply by specifying the sequences in a computer file. This is the technology used by Rosetta, Agilent and Oxford Gene Technology.

acid-mediated deprotection of dimethoxytrityl protecting groups has a coupling efficiency of approximately 98%. The effect on the yield of full-length oligos is dependent on the length of the oligonucleotide being synthesised: the longer the oligonucleotide, the worse the yield. This dependence is multiplicative, so that even a small difference in coupling efficiency can make a large difference in the yield of long oligonucleotides (Table 1.2).

The composition of the final population of oligonucleotides produced depends on whether or not a capping reaction is included during synthesis. Capping is used by Affymetrix and prevents further synthesis on a failed oligonucleotide. As a result,

TABLE 1.2

Oligonucleotide Length (s)	Coupling Efficiency (p)	Oligonucleotide Yield (p^s)
25	95%	28%
25	98%	60%
60	95%	5%
60	98%	30%

Note: The yield of in-situ synthesised oligonucleotides of desired length s depends on the coupling efficiency p according to the formula yield $= p^s$. So the longer the oligonucleotide, the worse the yield. Photodeprotection has a coupling efficiency of approximately 95%, while chemical deprotection has a coupling efficiency of approximately 98%. For a 25-base oligonucleotide, the yields are 28 and 60%, respectively. For a 60-base oligonucleotide, the yields are 5 and 30%, respectively. This is why Affymetrix is restricted to making 25-base oligonucleotides: the coupling efficiency is too low to produce longer oligos. Companies using chemical deprotection are able to synthesise 60-base oligos with similar yield to Affymetrix's 25-base oligos.

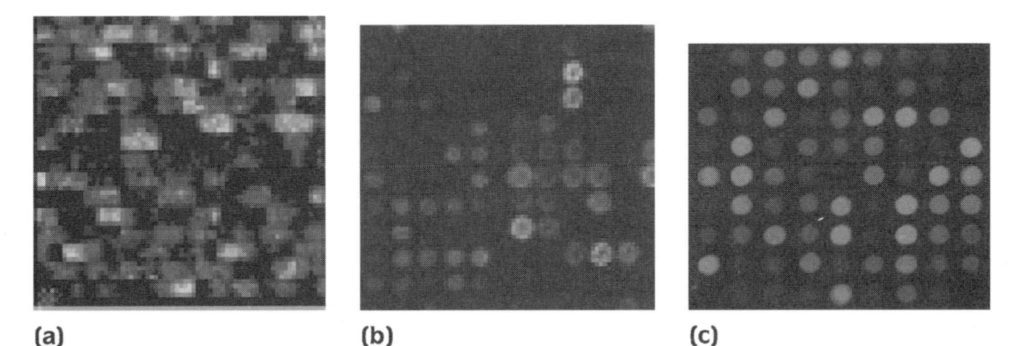

Figure 1.7: Array quality. (a) On Affymetrix arrays the features are rectangular regions. The masks refract light, so there is leakage of signal from one feature to the next. The Affymetrix image-processing software compensates for this by using only the interior portions of the features. **(b)** Spotted arrays produce spots of variable size and quality. This image shows some of this variation; we cover image processing of spotted arrays in detail in Chapter 4. **(c)** Inkjet arrays tend to be of the highest quality, with regular, even spots. (*Please see also the color section at the middle of the book.*)

all oligonucleotides on a feature will have the same start, but will be of different lengths (e.g., with a coupling efficiency of 95%, each feature will be 4.8% monomers, 4.5% dimers, 4.3% trimers, etc.). In contrast, uncapped oligonucleotides allow further synthesis to take place. Therefore, all the oligonucleotides on a feature will be of similar length but may contain random deletions (e.g., with a coupling efficiency of 95% and synthesis of 20 mers, the average probe length would be 19 bases, with such probes containing one deletion).

Spot Quality

The quality of the features depends on the method of array production (Figure 1.7). Spotted array images can be of variable quality, and Chapter 4 is dedicated to the bioinformatics of image processing associated with these arrays. Affymetrix arrays have the problem that the masks refract light, so light leaks into overlapping features; Affymetrix compensates for this with their image-processing software, so the user need not worry about this problem. Inkjet arrays tend to produce the highest quality features.

SECTION 1.3 USING MICROARRAYS

There are four laboratory steps in using a microarray to measure gene expression in a sample (Figure 1.8):

1. Sample preparation and labelling
2. Hybridisation
3. Washing
4. Image acquisition

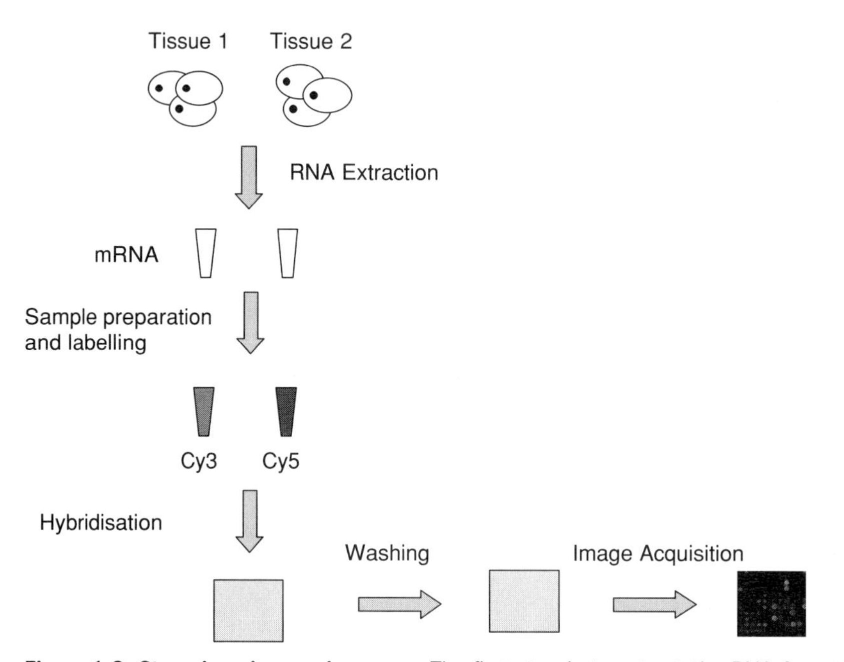

Figure 1.8: Steps in using a microarray. The first step is to extract the RNA from the tissue(s) of interest. With most technologies, it is common to prepare two samples and label them with two different dyes, usually Cy3 (green) and Cy5 (red). The samples are hybridised to the array simultaneously and incubated for between 12 and 24 hours at between 45 and 65°C. The array is then washed to remove sample that is not hybridised to the features.

Sample Preparation and Labelling

There are a number of different ways in which a sample can be prepared and labelled for microarray experiments. In all cases, the first step is to extract the RNA from the tissue of interest. This procedure can be difficult to reproduce, and there is much variability among the individual scientists performing the extraction.

The labelling step depends on the technology used. For the Affymetrix platform, one constructs a biotin-labelled complementary RNA for hybridising to the GeneChip®. The protocols are very carefully defined by Affymetrix,[4] so every Affymetrix laboratory should be performing identical steps. This has the advantage that it is easier to compare the results of experiments performed in different Affymetrix laboratories, because the procedures they will have followed should be the same.

Although it is possible to hybridise complementary RNA to other types of microarrays, it is much more common to hybridise a complementary DNA to these arrays. In the past, the DNA has been radioactively labelled, but now most laboratories use fluorescent labelling, with the two dyes Cy3 (excited by a green laser) and Cy5 (excited by a red laser). In the most common experiments, two samples are hybridised to the arrays, one labelled with each dye; this allows the simultaneous measurement of both

[4] See the reference to the Affymetrix manual at the end of the chapter.

samples. In the future, it is possible that more than two labelled samples could be used.

There are three common ways to make labelled cDNA. The most common method is direct incorporation by reverse transcriptase. The mRNA is primed with a poly-T primer: this starts the reverse transcription from the polyadenylation signal at the 3′ untranslated region (UTR) of the mRNA. In addition to the nucleic acids added for the transcription reaction (dA, dC, dG and dT), a proportion of dCTP (or sometimes dUTP used in place of dT) to which the fluorescent Cy dye has been covalently attached is added to the solution. This means that a proportion of the "C"s in the cDNA product have Cy fluorophors attached to them.

The transcripts produced by this method can be between 0 and 3,000 bases long, and are typically a few hundred bases. They will always be complementary to the 3′ end of the mRNA, so oligonucleotide or cDNA probes on the array must be in the 3′ region of the mRNA, otherwise they will not detect the labelled target produced. Fortunately, the 3′ UTR also tends to be the most variable region of genes, so this is an aid in designing specific probes (Chapter 3).

The next most common method is indirect labelling. This method also uses a reverse transcription reaction, primed from the 3′ end of the mRNA with a poly-T primer. However, instead of using fluorescently labelled dC, the reaction takes place with an amino-allyl-modified dC. This is a much smaller molecule than the Cy-modified dCTP, so the reverse transcription is more efficient. Following reverse transcription, the cDNA is reacted with an active ester of the dye, so that the dye becomes attached to the modified dCs in the cDNA. This method also has the advantage that each target has the same "foreign" base incorporated at the same rate. This contrasts with direct incorporation, where Cy5 is incorporated less well by reverse transcriptase than Cy3.

As with direct incorporation, indirect labelling produces transcripts of a few hundred bases, complementary to the 3′ end of the mRNA. This has similar implications for the design of probes for the array.

The third and least common method for labelling is by random primed labelling using the Klenow fragment of DNA polymerase I. The first step is a reverse transcription reaction, which generates a single-stranded cDNA. The cDNA is then primed with random primers and extended using the Klenow fragment of DNA polymerase I in the presence of labelled dC. The product is a mixture of shorter labelled transcripts, complementary to both strands of the gene.

Because the labelled fragments are on both strands, there is greater potential for cross-hybridisation, and so it is important to check for cross-hybridisation on both strands when designing probes for the arrays (Section 3.3).

Hybridisation

Hybridisation is the step in which the DNA probes on the glass and the labelled DNA (or RNA) target form heteroduplexes via Watson–Crick base-pairing (Figure 1.9). Hybridisation is a complex process that is not fully understood. It is affected by many

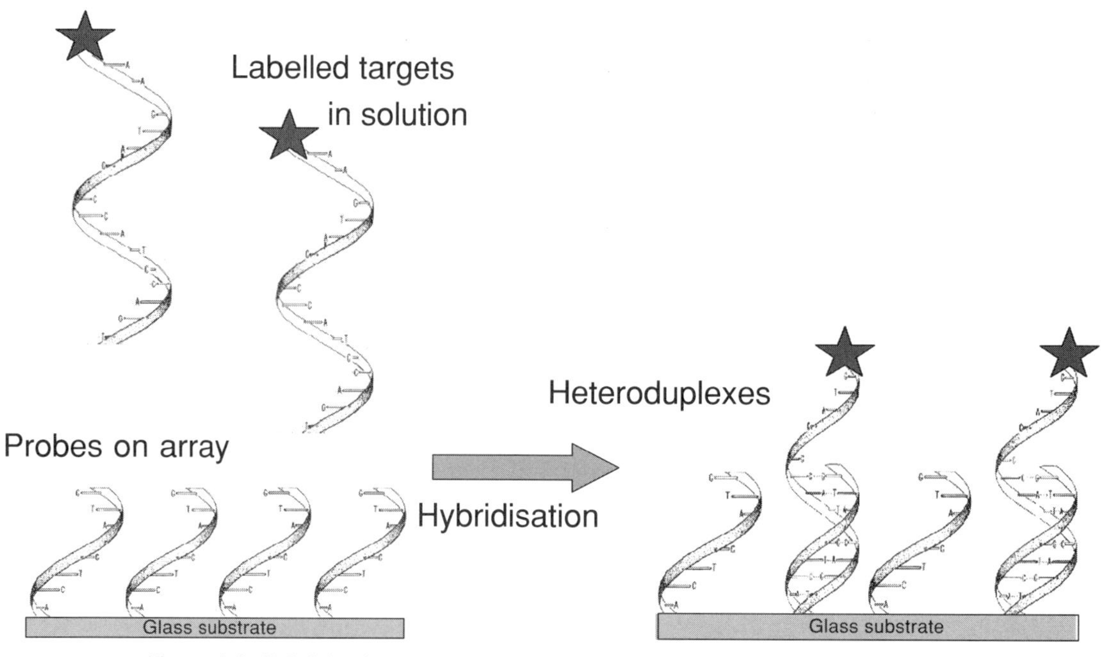

Figure 1.9: Hybridisation. In the hybridisation process, labelled target in solution forms heteroduplexes with probes on the array via Watson–Crick base-pairing between the probes and the target. Unbound target is then washed off the array, so that the only fluorescent signal on the array is in the heteroduplexes. The microarray measures the level of fluorescence on each of the features, and from this we infer the absolute or relative amount of DNA bound to each feature on the array.

conditions, including temperature, humidity, salt concentrations, formamide concentration, volume of target solution and operator.

There are two main methods for hybridisation: manual and robotic. In a manual hybridisation, the array is placed in a hybridisation chamber (Figure 1.10a). The scientist

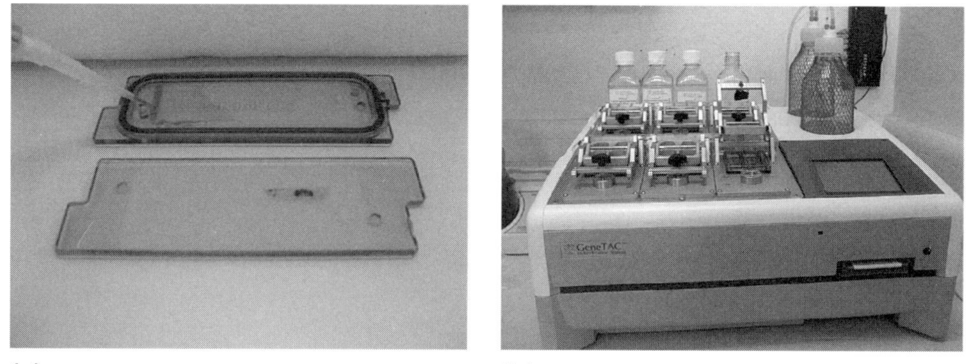

(a) **(b)**

Figure 1.10: Hybridisation systems. (a) Manual hybridisation takes place in chamber; the target is injected onto the slide; the chamber is sealed with a cover slip and placed in an incubator. **(b)** Hybridisation station; this station has six chambers in which all hybridisation, incubation and washing takes place. Robotic hybridisation reduces the variability of microarray experiments, but care must be taken not to break the arrays in the robot. This GeneTac station is located at the Mouse Genetics Unit in Harwell, Oxfordshire.

injects the hybridisation solution containing the target onto the array under a cover slip before sealing the chamber. The chamber is placed in an incubator which keeps the array at the correct temperature; some incubators also agitate the array to ensure mixing of the hybridisation solution over the surface of the array. Most hybridisations take place over a period of 12 to 24 hours.

Alternatively, hybridisation can be performed robotically by a hybridisation station (Figure 1.10b). Robotic hybridisation has the advantage over manual hybridisation in that it provides much better control of the temperature of the target and slide. The consistent use of a hybridisation station also reduces the variability between hybridisations and operators.

Most hybridisations are performed at between 45 and 65°C, depending on the type of array used. With oligonucleotide arrays, arrays with different-length oligonucleotides may require different temperatures. The addition of formamide enables mixing of the hybridisation solution of the target over the array but has the effect of decreasing the apparent melting temperature of duplexes. This has the positive benefit of reducing spatial hybridisation irregularities on the array – a matter that is discussed in Chapters 4 and 5. Different laboratories have used a wide range of formamide concentration in the target solution, between 0 and 50%. The use of no formamide at 65°C gives approximately equivalent thermodynamic conditions as the use of 50% formamide at 45°C.

It is also usual to include Na^+ in the hybridisation solution.[5] The less Na^+ present, the greater the stringency of the hybridisation; the thermodynamic effect of Na^+ concentration is well characterised and described in Chapter 3. Most hybridisations take place in approximately 1M Na^+ [between 3 and 5 standard saline citrate (SSC)].

It is also common to add DNA to the hybridisation solution that blocks unwanted cross-hybridisation. The two most common additions are some type of repetitive DNA that masks genomic repeat sequence, such as COT-1, and either poly-A or poly-T to mask the polyadenylation sites on the cDNA.

Washing

After hybridization, the slides are washed. There are two reasons for this. The first is to remove excess hybridisation solution from the array. This ensures that the only labelled target on the array is the target that has specifically bound to the features on the array and thus represents the DNA that we are trying to measure.

The second reason is to increase the stringency of the experiment by reducing cross-hybridisation. This can be achieved either by washing in a low-salt wash (e.g., 0.1 SSC and 0.1 SDS, a detergent that removes grease) or with a high-temperature wash. In either case, the aim is that only the DNA complementary to each of the features will remain bound to the features on the array. Most automatic hybridisation stations include a washing cycle as part of the automated process.

[5] It is also possible, but much less common, to use K^+.

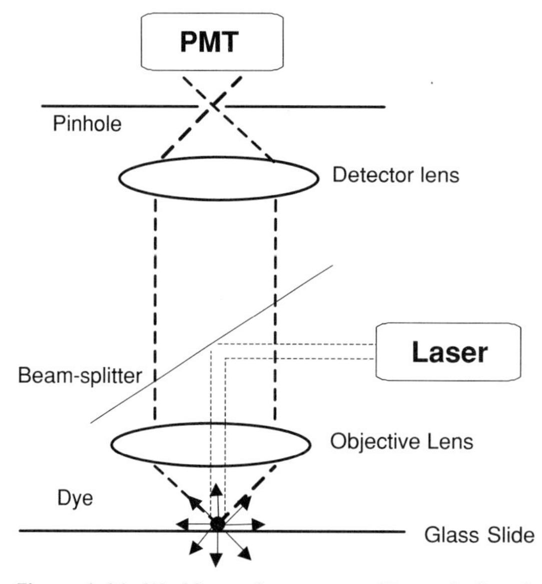

Figure 1.11: Workings of a scanner. The majority of microarray scanners work in a similar way. A laser is used to excite the dyes incorporated into the heteroduplexes on the surface of the array; the fluorescence of the dye is then measured by a PMT and converted to digital signal. Each pixel on the scanned image represents a single point of measurement of fluorescence from the laser. The slide (or in the case of some scanners, the optics) is moved so that the laser excites the whole of the slide. Two-colour arrays are scanned twice: once with a green laser (for Cy3; excitation wavelength is 550 nm and emission wavelength is 581 nm) and once with a red laser (for Cy5; excitation wavelength is 649 nm and emission wavelength is 670 nm).

Image Acquisition

The final step of the laboratory process is to produce an image of the surface of the hybridised array. The heteroduplexes on the array, where the target has bound to the probe, contain dye that fluoresces when excited by light of an appropriate wavelength. The slide is placed in a **scanner**, which is a device that reads the surface of the slide. Most scanners have a similar design (Figure 1.11). The scanner contains one or more lasers that are focussed onto the array: most scanners for two-colour arrays use two lasers.

Each pixel on the digital image represents the intensity of fluorescence induced by focussing the laser at that point on the arrray (Figure 1.12a). The dye at that point will be excited by the laser and will fluoresce; this fluorescence is detected by a photomultiplier tube (PMT) in the scanner. In order to scan the whole array, the laser must be focussed on every point on the array. This is achieved either by moving the slide so that the laser can focus on different points, or by shifting the optics to achieve the same result.

It is usual for the size of the physical space represented by the pixel to be the same as the spot size of the laser. When this is the case, the majority of the light measured at that pixel comes from that point on the array (Figure 1.12b). It is also possible to set the pixel size to be smaller than the laser spot. In that case, much of the light at each pixel comes from neighbouring areas on the array (Figure 1.12c). This has the effect of blurring the image and can mask minor irregularities in the feature (Figures 1.12d and 1.12e).

With two-colour arrays, the output of the scanner is usually two monochrome images: one for each of the two lasers in the scanner (Figure 1.13a). These are combined to create the familiar red–green false colour images of microarrays. Both the monochrome and two-colour images are usually stored as tagged image file format (TIFF). The array data is stored in 16 bits. This means that the intensity of each pixel in each channel is quantified as a 16-bit number, which takes values between 0 and $2^{16} - 1$, which is equal to 65,535. Since background is approximately 100, and saturation can occur when the average pixel intensity is larger than 50,000, the microarray can detect intensities over an approximately 500-fold dynamic range. With 10-μm pixel sizes, a typical microarray image will be 7,500 × 2,200 pixels. This means that each of the two

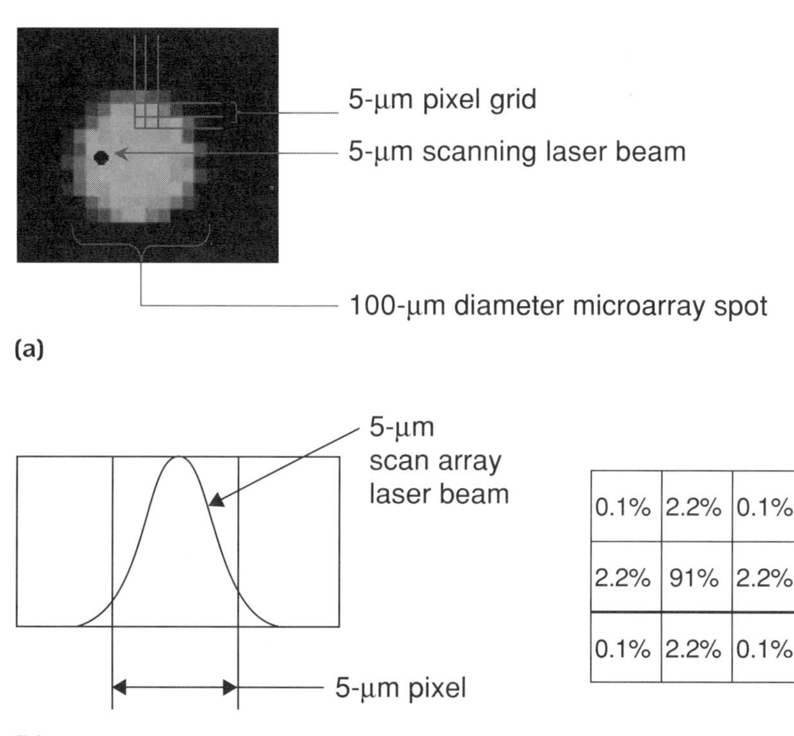

Figure 1.12: The pixels comprising a feature. **(a)** A false-colour image of the pixels from a single scan of a 100-μm microarray feature. The size of the laser spot is 5 μm. The pixel size has been set to 5 μm so that each pixel represents the area from the size of the laser spot. **(b)** The intensity of light from a laser is normally distributed. With a 5-μm laser size and 5-μm pixel size, 91% of the emission from the array resulting from the laser is measured at that pixel. **(c)** With a 10-μm laser size and 5-μm pixel size, the majority of light from the laser is measured in neighbouring pixels. This has the effect of blurring the image. **(d)** Two neighbouring features on an array with a streak through them, measured with a laser spot size of 5 μm and a pixel size of 5 μm. The streak is clear on both spots and so the spot can be identified as problematic. **(e)** The same features scanned with laser spot size of 10 μm and a pixel size of 5 μm. The streak has become blurred. (*Please see also the color section at the middle of the book.*)

(*continued*)

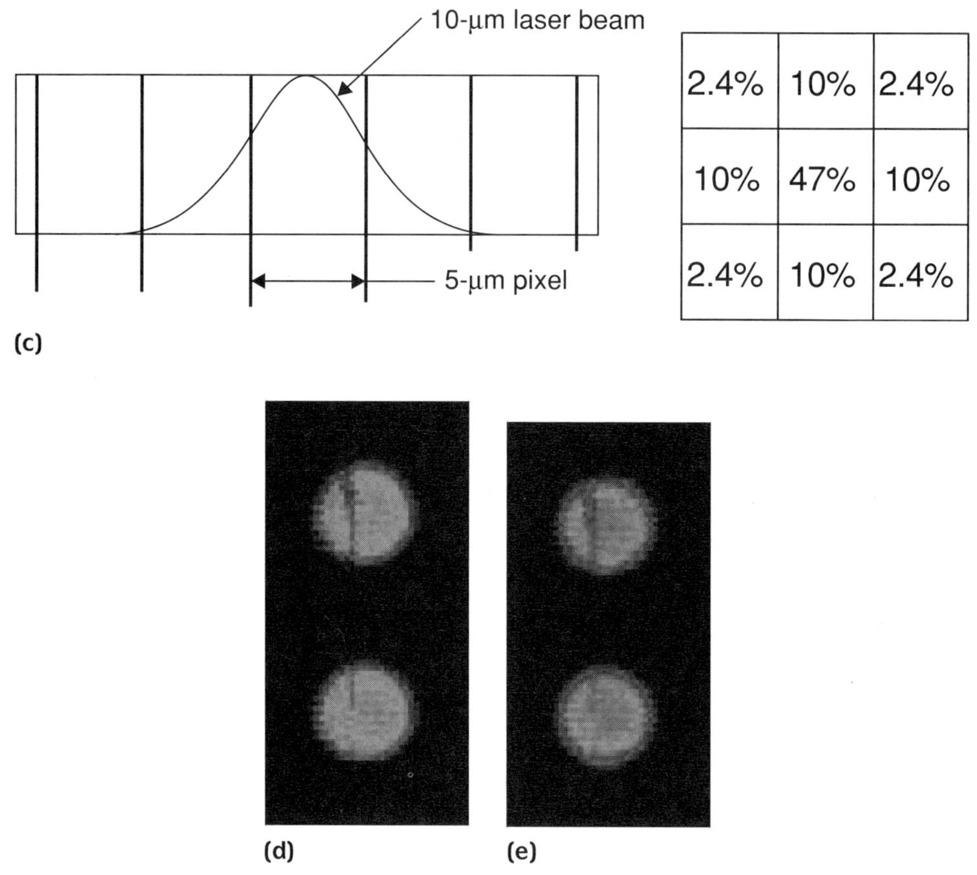

Figure 1.12: (*continued*)

TIFF images is 32 Mb; these are large files so if you are producing a large number of microarray images, data storage becomes an important consideration (Chapter 11).

The pixel resolution of the image should be chosen so that each feature has sufficient pixels to make the measurement of the intensity of the feature robust from pixel-to-pixel noise. It is normally recommended that there should be at least 50 pixels per feature on the array.

KEY POINTS SUMMARY

- Making and using microarrays is a complex laboratory process.
- There are many sources of variability in microarray experiments.
- The main microarray technologies are
  - Spotted cDNAs, the most common type of microarray;
  - Spotted oligonucleotides: increasingly common and better quality than spotted cDNAs;

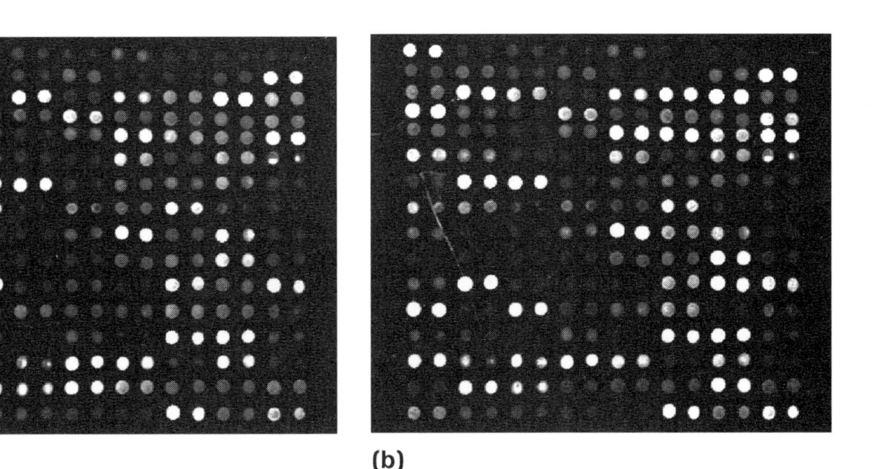

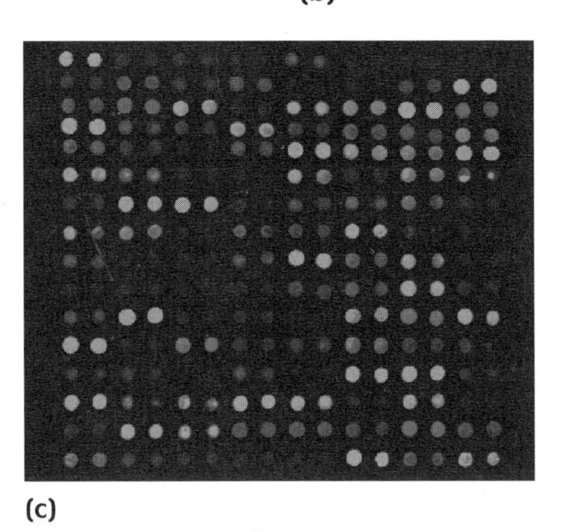

Figure 1.13: Output of scanners. (a) This is the scanner output for a part of a microarray – in this case one of twelve 16 × 16 blocks of features. This is the monochrome image of the Cy3 (green) channel. **(b)** The scanner output for the same part of the array but using the Cy5 (red) channel. **(c)** It is usual to combine the two monochrome images into a composite false-colour image of the array. Green features correspond to features that are expressed more in the sample labelled with Cy3 than the sample labelled with Cy5, and so will be bright in (a) and dark in (b). Similarly, red spots will be bright in (b) and dark in (a). Yellow features have a similar level of expression in both samples. Dark features are low expressed in both samples. (*Please see also the color section at the middle of the book.*)

  - Light-directed in-situ synthesised arrays: e.g., Affymetrix, the most common commercial platform; and
  - Inkjet in-situ synthesised arrays, the highest quality arrays but not widely available.
- The steps in using a microarray are
  - Target preparation, which introduces most variability;
  - Hybridisation, also a source of variability, which can be reduced by robotics;
  - Washing, which can increase stringency – variability can be reduced by robotics; and

- image acquisition: using a scanner to produce a digital image of the array that is stored on computer.

RECOMMENDED READING

Schena, M., Shalon, D., Davis, R.W., and Brown, P.O. 1995. Quantitative monitoring of gene expression patterns with a complementary DNA microarray. *Science* 270(5235): 467–70.

The paper that started the microarray revolution.

Genechip Expression Analysis Technical Manual

Describes the protocols for an Affymetrix experiment. Available from http://www.affymetrix.com/technology/ge_analysis/index.affx

Hughes, T., Mao, M., Jones, A.R., Burchard, J., Marton, M.J., Shannon, K.W., Lefkowitz, S.M., Ziman, M., Schelter, J.M., Meyer, M.R., Kobayashi, S., Davis, C., Dai, H., He, Y.D., Stephaniants, S.B., Cavet, G., Walker, W.L., West, A., Coffet, E., Shoemaker, D., Stoughton, R., Blanchard, A., Friend, S., and Linsley, P.S. 2001. Expression profiling using microarrays fabricated by an ink-jet oligonucleotide synthesiser. *Nature Biotechnology* 19: 342–47.

Paper describing the use of inkjet technology containing technical information about hybridisation conditions and stringency.

INTERNET RESOURCES

There are several excellent web sites that give a general overview of microarray technology, the laboratory process, and often links to useful software. Of particular interest are

- The Institute for Genomic Research microarray pages: http://www.tigr.org/tdb/microarray/
- The microarray project at the National Human Genome Research Institute: http://research.nhgri.nih.gov/microarray/main.html
- Microarrays.org is a public source for microarray protocols and software: http://www.microarrays.org/

SCANNERS

A comprehensive review of microarray scanners, prepared by Y.F. Leung at the Chinese University, Hong Kong, is available at

http://www.lab-on-a-chip.com/files/mascanner.pdf

CHAPTER TWO

Sequence Databases for Microarrays

SECTION 2.1 INTRODUCTION

Chapter 1 introduced microarray technologies and discussed the use of microarrays in the laboratory. The remainder of the book is dedicated to microarray bioinformatics. This chapter, together with the next chapter, discusses the bioinformatics required to design a DNA microarray. In this chapter, we look at the sequence databases that are used to select and annotate the genes that the microarray detects and, thus, the sequences that will appear on the array. Chapter 3 looks at the computer design of oligonucleotide probes for oligonucleotide arrays.

There are two broad questions and one more specific consideration that this chapter seeks to address:

1. What resources could I use to design my own custom array?
If you are designing a custom microarray to study a particular disease, tissue or organism, you will need to identify the genes that might be expressed in your samples and identify the sequences of those genes. One of the aims of this chapter is to give an understanding of which databases you could use to select such genes.

2. How can I find more information about the sequences of the genes on my array?
DNA microarrays contain sequences that will have derived from DNA sequence databases. The output file containing the numerical results of the microarray experiment that you will analyse also contains a number of fields that relate these sequences to the databases from which they derive. This chapter describes the meanings of these fields and the nature of the databases.

EXAMPLE 2.1 DATA FILES

Table 2.1 shows the gene identification fields from the files containing data from four published experiments whose data are available on the World Wide Web.[1] There are three two-colour arrays from human, arabidopsis and yeast, and an example of Affymetrix data. In the example you can see a number of field names:

- Clone IDs
- Gene names

[1] References to the work from which the data derives are included at the end of the chapter.

TABLE 2.1: Examples from Files Containing Gene Information from Four Different Microarray Experiments

(a)

NAME	Clone ID	Gene Symbol	Gene Name	Cluster ID	Accession Number	Preferred Name
59970	IMAGE:1352658	SRPK1	SFRS protein kinase 1	Hs.75761	AA830216	SRPK1=serine kinase

(b)

NAME	Clone ID	Description	Accession Number	Gene Info
34358	4D3T7P		T04790	(AC006931) putative MAP kinase [*Arabidopsis thaliana*]

(c)

NAME	Gene Name	Chromosome	Strand	Beginning Coordinate	Ending Coordinate	Process	Function	DS_GENE_1	DS_GENE_2
YAL010C	MDM10	1	C	135663	134182	mitochondrion inheritance*	molecular_function unknown	YAL009W (SPO7)	

(d)

Gene Description	Gene Accession Number
Osteomodulin	AB000114_at

(a) Data from Alizadeh and co-workers from a study of human diffuse large B-cell lymphomas. *Name* is a unique identifier for the array and does not reflect any public database. *Clone ID, Cluster ID* and *Accession Number* are identifiers for IMAGE, Unigene and EMBL/GenBank, respectively. *Gene Symbol, Gene Name* and *Preferred Name* are all common names for the gene. (b) Data from Schaffer and co-workers from a study of differentially regulated genes in *Arabidopsis*. The *Name* field is a unique identifier for the gene on the array. The *Clone ID* and *Accession Number* are database IDs. The *Gene Info* field is free text describing the gene. (c) Data from Chu and co-workers from a study of yeast. Because the genome is small, there are start and end coordinates for this gene in the yeast genome. (d) Data from Golub and co-workers from a study of human leukaemia using Affymetrix arrays. The *Gene Description* is free text, and the *Gene Accession Number* contains an EMBL/GenBank database ID with the suffix "_at".

Sources: (a) Alizadeh et al., 2000; (b) Schaffer et al., 2001; (c) Chu et al., 1998; (d) Golub et al., 1999.

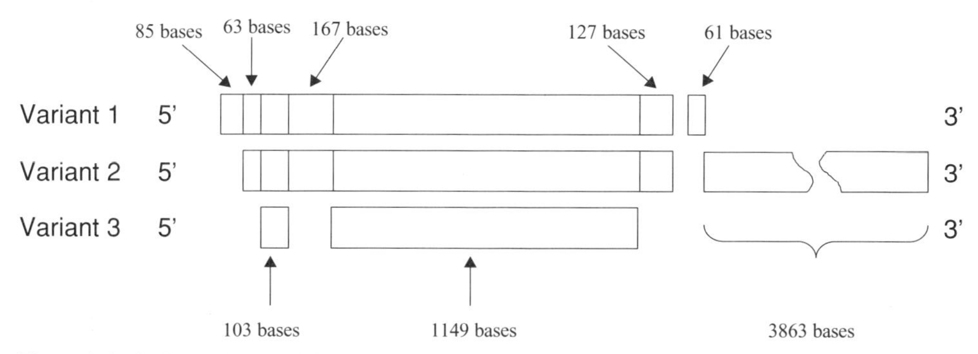

Figure 2.1: Splice variants of CNR1. *Homo sapiens* Cannabinoid Receptor 1 is a 7-*trans*-membrane protein that responds to THC, the psychoactive ingredient in cannabis. It has three splice variants: variants 1 and 2 encode the same protein, and variant 3 encodes a shorter protein. Variant 1 (1,755 bases) has a unique 85-base sequence in the 5′ UTR, and a unique 61-base sequence at the 3′ UTR. Variants 1 and 2 share a 63-base sequence that is at the 5′ end of variant 2, and a 167-base sequence that is spliced at a site between bases 103 and 104 of variant 3. Variant 2 (5,472 bases) has a unique 3,863-base sequence at its 3′ UTR. All of the bases in variant 3 (1,252 bases) are found in variants 1 and 2. The 3′ UTRs for variants 1 and 2 are unique, so it would be possible to design probes that would be specific for these variants. Variant 3, however, has a 3′ UTR that is part of the sequence of variants 1 and 2. Furthermore, there are only 188 bases from the 3′ end of variant 1 to the 3′ end of variant 3. If the target is labelled via poly-T primed reverse transcription, then target of variants 1 and 2 will both hybridise to a probe for variant 3. The only sequence in variant 3 that is unique to variant 3 is the sequence across the splice site into which the 167 bases that feature in variants 1 and 2 are spliced. In order for such a probe to detect variant 3 target, it would be necessary to make labelled target using random priming of the sample mRNA.

- GenBank accession numbers
- UniGene cluster IDs
- Gene information and descriptions
- Genomic information

3. What information is available on splice variant genes and how does this impact array design?

A gene has **splice variants** if the organism can make different transcripts of the gene by using different exons. It is thought that many genes from eukaryotic organisms have splice variants. The different splice variants of a gene have different sequences; therefore, when designing and using microarrays, it is important to know what sequences are on the arrays. Is there a single probe on the array for the gene of interest that will measure all variants, or would you want specific probes for each of the variants?

EXAMPLE 2.2 SPLICE VARIANTS OF CNR1

The gene Cannabinoid Receptor 1 has three splice variants in humans (Figure 2.1):

- Splice Variant 1, accession NM_001840 (1,755 bases)
- Splice Variant 2, accession NM_016083 (5,472 bases)
- Splice Variant 3, accession NM_033181 (1,252 bases)

In this example, all three splice variants have different 3′ UTRs and any microarray that will measure this gene will need to be carefully designed.

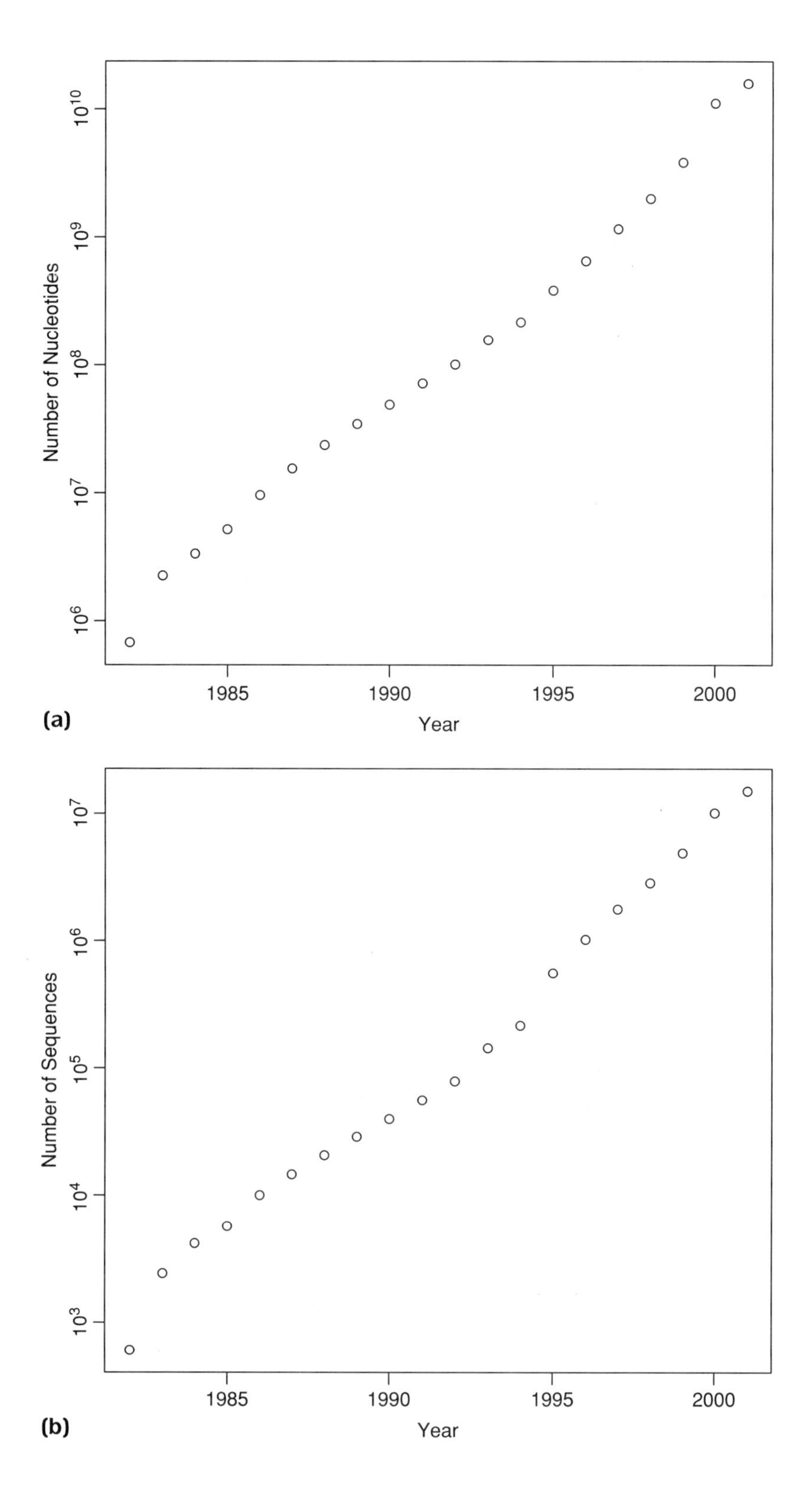

Number of Nucleotides
10^6
10^7
10^8
10^9
10^{10}
1985
1990
1995
2000
Year
(a)
Number of Sequences
10^3
10^4
10^5
10^6
10^7
1985
1990
1995
2000
Year
(b)

The remainder of this chapter looks at the databases available that will allow you to answer these or similar questions. There are three further sections:

Section 2.2: Primary Sequence Databases, discusses the sequence databases that hold all published sequence data – EMBL, GenBank and the DNA Data Bank of Japan (DDBJ).

Section 2.3: Secondary Sequence Databases, discusses two Expressed Sequence Tag (EST) resources that are most commonly used for microarray design – UniGene and the TIGR Gene Indices – and also looks at the Reference Sequence project for high-quality mRNA sequences.

Section 2.4: Genomic Database Resources, looks at genomic databases – Ensembl for complex organisms such as humans, and databases for organisms with small genomes, namely, microbes and yeast.

SECTION 2.2 PRIMARY SEQUENCE DATABASES

These are the international sequence databases that contain all published sequences. They date back to 1982, when it became clear that there was a need to publish and share DNA sequences. The American initiative, GenBank, and the European initiative, EMBL (European Molecular Biology Laboratory), were launched simultaneously in June 1982, each with approximately 600 sequences. Since that time, the sizes of the databases have grown exponentially, doubling approximately every 17 months (Figure 2.2). In December 2002, GenBank had more than 22 million sequences and more than 28 billion nucleotides (release 133).

In 1987, the DDBJ was started as a Japanese equivalent of GenBank and EMBL. In 1992, the three databases entered into a collaboration to share all sequences. Since that date, the three databases contain almost identical sequence information. Any sequence submitted to one of the databases will automatically be added to the other two.

The success of these sequence databases has resulted not only from the advances in sequencing technology, but also from the advances in computer technology. The databases require significant computing power and storage to operate, but most important is the role of the Internet. Everyone with access to the Internet has the ability to submit sequences to these databases, and so they are truly representative of global research. Similarly, everyone with access to the Internet can query or download these databases, so the entire world has unrestricted access to any sequence submitted by any laboratory.

Figure 2.2: Growth of public sequence databases. (a) Log plot of the number of nucleotides in the GenBank database between 1982 and 2001. The straight line is indicative of exponential growth in the number of nucleotides, with a doubling time of approximately 17 months. The most recent release of GenBank (release 133) has approximately 2.8×10^{10} (28 billion) nucleotides. **(b)** Log plot of the number of sequences in the GenBank database between 1982 and 2001. There is similar exponential growth. The most recent release of GenBank (release 133) has approximately 2.2×10^{7} (22 million) sequences.

BOX 2.1 GenBank File for AA830216

Example entry of a GenBank format file for the sequence AA830216; this sequence is an EST. The entry gives a wide range of information about the sequence, including the type of sequence, species, date submitted, the publication associated with the sequence and the sequence itself. The file format has a text identifier at the beginning of each section to state what type of information it contains.

```
LOCUS       AA830216                 407 bp    mRNA    linear   EST 18-MAR-1998
DEFINITION  oc45c10.s1 NCI_CGAP_GCB1 Homo sapiens cDNA clone IMAGE:1352658 3',
            mRNA sequence.
ACCESSION   AA830216
VERSION     AA830216.1 GI:2903315
KEYWORDS    EST.
SOURCE      Homo sapiens (human)
ORGANISM    Homo sapiens
            Eukaryota; Metazoa; Chordata; Craniata; Vertebrata; Euteleostomi;
            Mammalia; Eutheria; Primates; Catarrhini; Hominidae; Homo.
REFERENCE   1  (bases 1 to 407)
  AUTHORS   NCI-CGAP http://www.ncbi.nlm.nih.gov/ncicgap.
  TITLE     National Cancer Institute, Cancer Genome Anatomy Project (CGAP),
            Tumor Gene Index
  JOURNAL   Unpublished (1997)
COMMENT     Contact: Robert Strausberg, Ph.D.
            Email: cgapbs-r@mail.nih.gov
            Tissue Procurement: Louis M. Staudt, M.D., Ph.D., David Allman,
            Ph.D., Gerald Marti, M.D.
             cDNA Library Preparation: M. Bento Soares, Ph.D., M. Fatima
            Bonaldo, Ph.D.
             cDNA Library Arrayed by: Greg Lennon, Ph.D.
             DNA Sequencing by: Washington University Genome Sequencing Center
             Clone distribution: NCI-CGAP clone distribution information can be
            found through the I.M.A.G.E. Consortium/LLNL at:
            www-bio.llnl.gov/bbrp/image/image.html
            Insert Length: 1670 Std Error: 0.00
            Seq primer: -40m13 fwd. ET from Amersham
            High quality sequence stop: 352.
FEATURES             Location/Qualifiers
     source          1..407
                     /organism="Homo sapiens"
                     /db_xref="taxon:9606"
                     /clone="IMAGE:1352658"
                     /clone_lib="NCI_CGAP_GCB1"
                     /tissue_type="germinal center B cell"
                     /lab_host="DH10B"
                     /note="Vector: pT7T3D-Pac (Pharmacia) with a modified
                     polylinker; Site_1: Not I; Site_2: Eco RI; 1st strand cDNA
                     was prepared from human tonsillar cells enriched for
                     germinal center B cells by flow sorting (CD20+, IgD-),
                     provided by Dr. Louis M. Staudt (NCI), Dr. David Allman
                     (NCI) and Dr. Gerald Marti (CBER). cDNA synthesis was
                     primed with a Not I - oligo(dT) primer
                     [5'-TGTTACCAATCTGAAGTGGGAGCGGCCGCCTCATTTTTTTTTTTTTTTTTT-3'
                     ]. Double-stranded cDNA was ligated to Eco RI adaptors
```

```
                    (Pharmacia), digested with Not I and cloned into the Not I
                    and Eco RI sites of the modified pT7T3 vector. Library
                    went through one round of normalization, and was
                    constructed by Bento Soares and M. Fatima Bonaldo."
BASE COUNT      131 a      67 c      67 g     142 t
ORIGIN
        1 tttttttttt tttttggtta ttaaataatt tgtttattgt acggcattta caaagaaaac
       61 agacaatgcc ctcagtagaa agaataaaaa tgtatttagg gctttatttt taactgacag
      121 caaatagaaa tcctttagtg agatcgtggc aatttgacag tattataatt aagctcaata
      181 aaggtacatg gggtacctgg aagatcaaga tctacagctg cctatttcca catctttcaa
      241 tccatctggc tccttaaata ggggaaaaag cccttatttg gtggagaagc atttccaaaa
      301 tgaagttaca ggttctatta aaacttactg tcacatcaac tgttaaaata gggccttttg
      361 tgttttgtta tttcacctta atatcaccag aattcctgta attccac
//
```

EXAMPLE 2.3 GENBANK ENTRY FOR AA830216

The sequence with accession number AA830216 that appears in Example 2.1 can be found in the GenBank and EMBL databases. Box 2.1 contains the GenBank entry for that sequence; this is an EST sequence that was submitted to the database on 18th March 1998. The entry contains a wealth of information about the sequence and its authorship. However, it does not include information about the gene from which the EST derives – information that is included in Table 2.1. This information is not in the primary database and is only available through secondary databases.

EXAMPLE 2.4 EMBL ENTRY FOR AJ313384

The EMBL entry with accession number AJ313384 for the aspartic proteinase gene from *Theobroma cacao* (the chocolate tree) is shown in Box 2.2. This entry has similar information to the GenBank entry but with a different file format.

Primary sequence databases are the first port of call when querying a sequence in order to obtain information about it. However, there are two reasons why primary sequence databases are not sufficient for microarray design and annotation.

First, they do not contain *meta-information*. In Example 2.3, although we could identify the sequence on the array, the primary sequence database does not include information about the gene from which this EST derived.

Second, the primary sequence databases contain too many sequences for array design. When designing an array, we would want the database to be able to provide a list of genes in which each gene that will appear on the array will appear once in the list. There are two reasons why primary sequence databases cannot provide this:

- Redundancy. Each gene can be represented several times in the database (e.g., if it were submitted by different research groups who have sequenced it).
- Replication. Each gene sequence may be in the database in several forms (e.g., as a gene sequence, as genomic sequence and as ESTs).

BOX 2.2 EMBL Entry for AJ313384

The EMBL entry for sequence with accession number AJ313384, *Theobroma cacao* mRNA for aspartic proteinase. The EMBL database contains the same type of information as GenBank. However, the file format has a two-letter identifier at the beginning of each line to state what type of information is in that line.

```
ID   TCA313384 standard; RNA; PLN; 1784 BP.
XX
AC   AJ313384;
XX
SV   AJ313384.1
XX
DT   25-JUN-2002 (Rel. 72, Created)
DT   25-JUN-2002 (Rel. 72, Last updated, Version 1)
XX
DE   Theobroma cacao mRNA for aspartic proteinase (apl gene)
XX
KW   apl gene; aspartic proteinase.
XX
OS   Theobroma cacao (cacao)
OC   Eukaryota; Viridiplantae; Streptophyta; Embryophyta; Tracheophyta;
OC   Spermatophyta; Magnoliophyta; eudicotyledons; core eudicots; Rosidae;
OC   eurosids II; Malvales; Malvaceae; Byttnerioideae; Theobroma.
XX
RN   [1]
RP   1-1784
RA   Laloi M.;
RT   ;
RL   Submitted (25-JUN-2001) to the EMBL/GenBank/DDBJ databases.
RL   Laloi M., Plant Science, Centre de Recherche Nestle, 101, avenue Gustave
RL   Eiffel, Notre Dame d'Oe, BP 9716,, 37097 Tours cedex 2, FRANCE.
XX
RN   [2]
RA   Laloi M., McCarthy J., Morandi O., Gysler C., Bucheli P.;
RT   "Molecular characterisation of aspartic endoproteinases TcAP1 and TcAP2
RT   from Theobroma cacao seeds";
RL   Unpublished.
XX
DR   GOA; Q8L6A9; Q8L6A9.
DR   SPTREMBL; Q8L6A9; Q8L6A9.
XX
FH   Key             Location/Qualifiers
FH
FT   source          1..1784
FT                   /db_xref="taxon:3641"
FT                   /organism="Theobroma cacao"
FT   CDS             63..1607
FT                   /db_xref="GOA:Q8L6A9"
FT                   /db_xref="SPTREMBL:Q8L6A9"
FT                   /gene="apl"
FT                   /product="aspartic proteinase"
FT                   /function="endoprotease"
FT                   /protein_id="CAC86003.1"
```

```
FT                   /translation="MGRIVKTTTVTLFLCLLLFPIVFSISNERLVRIGLKKRKFDQNYR
FT                   LAAHLDSKEREAFRASLKKYRLQGNLQESEDIDIVALKNYLDAQYFGEIGIGTPPQNFT
FT                   VIFDTGSSNLWVPSSKCYFSIACYLHSRYKSSRSSTYKANGKPADIQYGTGAISGFFSE
FT                   DNVQVGDLVVKNQEFIEATREPSITFLVAKFDGILGLGFQEISVGNAVPVWYNMVNQGL
FT                   VKEPVFSFWFNRDPEDDIGGEVVFGGMDPKHFKGDHTYVPITRKGYWQFDMGDVLIGNQ
FT                   TTGLCAGGCSAIADSGTSLITGPTAIIAQVNHAIGASGVVSQECKTVVSQYGETIIDML
FT                   LSKDQPLKICSQIGLCTFDGTRGVSTGIESVVHENVGKATGDLHDAMCSTCEMTVIWMQ
FT                   NQLKQNQTQERILEYINELCDRLPSPMGESAVDCSSLSTMPNVSFTIGGKIFELSPEQY
FT                   VLKVGEGDVAQCLSGFTALDVPPPRGPLWILGDVFMGQFHTVFDYGNLQVGFAEAA"
XX
SQ   Sequence 1784 BP; 482 A; 326 C; 428 G; 548 T; 0 other;
     tctgctcagc ttttcttgtc gaaatcatca ctaaaaccat ttgcggactt gcagttatca        60
     gaatggggag aatagtcaaa actactacag tcactctttt tctttgtctt cttctgtttc       120
     ctatcgtatt ttccatatcc aatgagagat tggtcagaat tggactgaaa aagagaaagt       180
     tcgatcaaaa ctatcggttg gctgcccacc ttgattccaa ggagagagag gcatttagag       240
     cttctcttaa aaagtatcgt cttcaaggga acttacaaga gtctgaggac attgatattg       300
     tggcactaaa gaactacttg gatgctcagt actttggtga gattggtatt ggcacacctc       360
     cacagaactt cactgtgatt tttgacactg gtagttctaa tttgtgggtc ccttcatcta       420
     agtgctattt ctcgatagct tgctatctcc attcaagata taaatcaagc cgttcaagca       480
     cctacaaggc taatggtaaa ccagccgata tccaatacgg gactggagct atttctggat       540
     tctttagtga ggacaatgta caagttggtg atcttgtagt taaaaatcag gaatttatcg       600
     aggcaacaag ggagcccagc ataacatttt tggtggccaa gtttgatggg atacttggac       660
     ttggatttca agagatttcg gttggaaatg ctgtgcctgt gtggtacaat atggtcaatc       720
     aaggtcttgt taaggaacct gttttctcat tttggtttaa ccgcgatcct gaggatgata       780
     taggtgggga agttgttttt ggtggaatgg atccaaaaca tttcaagggg gatcacactt       840
     acgttcctat aacgcggaaa ggatactggc agtttgatat gggtgatgtc ctgattggta       900
     accaaacaac tggactttgt gctggtggct gcagtgcaat tgctgattct gggacttcct       960
     tgataaccgg tcctacggct attattgctc aagtcaatca tgctattgga gcatcagggg      1020
     ttgtaagtca agaatgcaag actgtagttt cacagtatgg agagacaata attgatatgc      1080
     ttttatctaa ggaccaacca ctgaaaattt gctcacaaat aggtttgtgc acatttgatg      1140
     gaactcgagg tgtaagtacg gggattgaaa gtgttgtgca tgagaatgtt gggaaagcca      1200
     ctggtgattt gcatgatgca atgtgttcta cttgtgagat gacagttata tggatgcaaa      1260
     accagcttaa gcagaaccag acacaggagc gtatacttga gtacatcaat gagctctgtg      1320
     atcggttgcc tagtccaatg ggagaatcag ctgttgattg tagcagtcta tctaccatgc      1380
     ctaatgtctc gttcacaatt ggtggaaaga tatttgagct cagccccgag cagtatgtcc      1440
     tgaaagtggg tgagggagat gtagctcaat gcctcagtgg attcactgct ctggatgtgc      1500
     cacctcctcg tggacctctc tggatcttgg gcgacgtctt tatgggccag ttccatacag      1560
     tatttgacta tggcaacctg caagttggat ttgccgaggc tgcataagtg aaactttctg      1620
     cttttataaa caacttcatg ttatgcagtg ctagtagtac ccttagaact gtggggatta      1680
     agtatcaaat gataattgca tgtaaatatc tatgcaaaca tgatctgtga tcttcactgg      1740
     atcgttgagt gtgatgcact ttgtttaaga atttcatgtg atcc                        1784
//
```

So although primary sequence databases are used for annotating sequences that appear on arrays, they are not used for array design. This is the domain of secondary sequence databases.

SECTION 2.3 SECONDARY SEQUENCE DATABASES

In this section we describe the three secondary sequence databases that are commonly used for microarray design: UniGene, the TIGR Gene Indices and RefSeq.

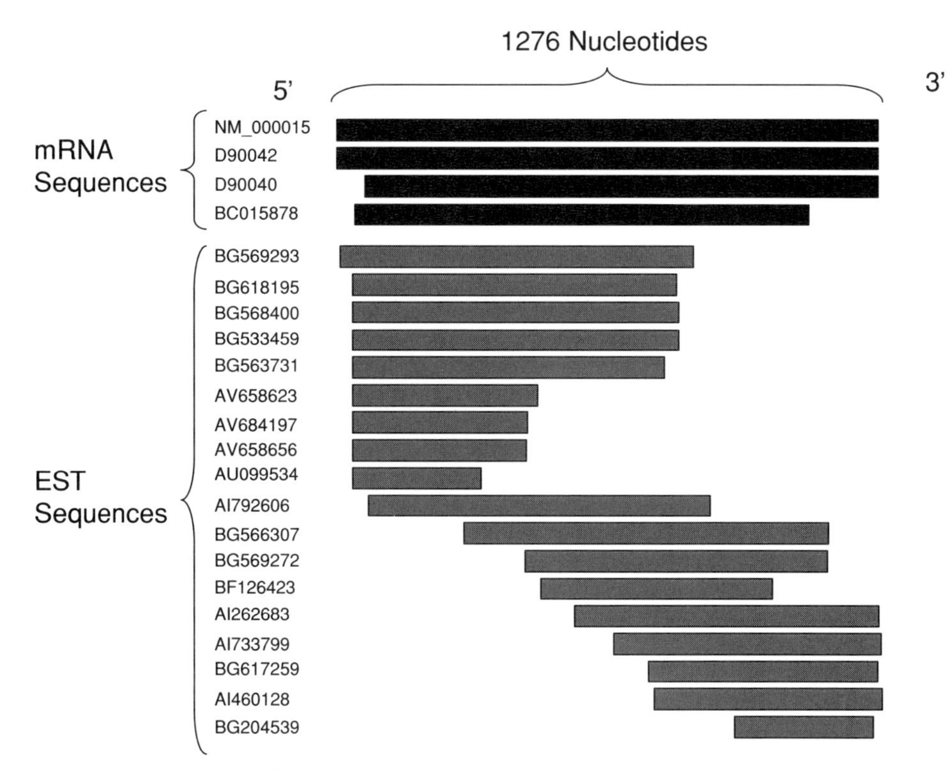

Figure 2.3: Example UniGene cluster. The first cluster in the UniGene database, Hs.2, is the gene *Homo sapiens* NAT2. The cluster has 4 mRNA sequences (shown in black) and 18 EST sequences (shown in grey). The gene itself is 1,276 nucleotides long; two of the mRNAs are that length, and two are slightly shorter. The 18 ESTs cover the full length of the gene, but tend to be found at the 5′ and 3′ ends of the gene. This example is a known gene and so it is straightforward to assign the ESTs to the cluster. Many clusters do not contain mRNAs and so are assembled as collections of overlapping ESTs without a reference mRNA.

UniGene

UniGene is the database with the greatest historical use for selecting sequences for microarrays. It is an attempt to partition GenBank sequences into **clusters**, each of which is intended to represent a unique gene. The clusters themselves may contain both mRNA sequences and ESTs, so that they represent both known genes and putative genes based on expressed material that has been sequenced.

The clusters are built by comparing all mRNA and EST sequences in GenBank and assigning overlapping sequences to the same cluster (Figure 2.3). In clusters that contain full-length mRNAs, the task is straightforward, because all ESTs deriving from the gene will align with the mRNAs. However, many clusters in UniGene contain only ESTs; the algorithms by which UniGene is built assemble the clusters out of overlapping ESTs in order to produce a picture of the gene from which the ESTs have putatively derived.

UniGene is available for a range of species (Table 2.2). Although (in December 2002) there were 20 species in the database, there is only broad coverage of the main research species. The human database has approximately 110,000 clusters

TABLE 2.2: Unigene Species

Scientific Name	Common Name	Number of Clusters
Animals		
Anopholes gambiae	Mosquito	2,584
Bos taurus	Cow	10,765
Caenorhabditis elegans	Worm	20,401
Ciona intestinalus	Sea Squirt	14,262
Danio rerio	Zebrafish	15,968
Drosophila melanogaster	Fruit Fly	14,660
Homo sapiens	Human	115,523
Mus musculus	Mouse	87,543
Rattus norvegicus	Rat	63,430
Sus scrofa	Pig	14,321
Xenopus laevis	Frog	19,441
Plants		
Arabidopsis thanalia	Thale Cress	27,159
Chlamydomonas reinhardtii	Green Algae	6,582
Glycine max	Soya	8,772
Hordeum vulgare	Barley	7,933
Lycopersicon esculentum	Tomato	3,737
Medicago truncatula	Barrel Medic	5,828
Oryza sativa	Rice	16,914
Triticum aestivum	Wheat	22,188
Zea mays	Maize	12,624

Note: The species represented in UniGene with number of clusters as of 31st December 2002. Although there are 13 species in the database, UniGene is centered on the 4 main research species: human, mouse, rat and cress. The other species in the database have patchy coverage and are better represented in other databases, such as the TIGR Gene Indices.

(Table 2.3): each of these clusters is supposed to represent a potentially different gene. Since current thinking is that there are approximately 30,000 genes in the human genome, it is likely that many of these clusters belong together. Of the 110,000 clusters, approximately 32,000 contain at least one mRNA and so represent known genes.

EXAMPLE 2.5 UNIGENE ENTRY FOR CLUSTER Hs.2

The UniGene entry for the first cluster in the database, Hs.2, is shown in Box 2.3. The entry contains a mine of information about the cluster, including species, gene name, chromosome location, tissue distribution and sequence information. This cluster is the human gene **N-acetyltransferase 2** (NAT2) and has been sequenced in liver, colon, adenocarcinoma and hepatocellular carcinoma.

The information about tissue distribution in the database allows the user to query UniGene to find clusters that contain sequences that have been found in a particular tissue. Thus UniGene is an excellent resource for designing tissue-specific microarrays for focussed research projects.

The sequence information for each cluster gives details about every sequence in the cluster. The first mRNA sequence in Example 2.5 is a RefSeq sequence (see the

TABLE 2.3: Human UniGene Database Statistics

(a)

Sequences		Clusters	
mRNA sequences	112,518	Clusters containing at least one mRNA	32,090
EST sequences	3,911,348	Clusters containing at least one EST	110,743
		Clusters containing mRNAs and ESTs	27,310
Total	4,023,866	Total	115,523

(b)

Number of Sequences	Number of Clusters
1	50,476
2	15,509
3–4	14,695
5–8	9,543
9–16	5,567
17–32	4,073
33–64	3,694
65–128	4,034
129–256	4,190
257–512	2,527
513–1,024	843
1,025–2,048	248
2,049–4,096	78
4,097–8,192	35
8,193–16,384	10
16,385–32,768	1

Note: Statistics for *Homo sapiens* UniGene build number 157, issued in October 2002. (a) The number of sequences and clusters in the database. The majority of the sequences in the database are EST sequences. Approximately 32,000 clusters contain at least one mRNA and thus represent known genes. The other 83,000 clusters contain only ESTs and thus contain less reliable but potentially novel information. The fact that there are many more EST clusters than the number of genes thought to be in the human genome implies that probably many of these clusters are unreliable and belong together. (b) The distribution of the number of sequences in clusters. Approximately 50,000 clusters have only one sequence; these are typically EST sequences that have no overlap with other EST sequences and are likely to be the least reliable information in the database.

following discussion) and, as such, will be of high quality. The EST sequences have Integrated Molecular Analysis of Gene Expression (IMAGE) clone IDs where they are available; this enables the user to purchase these clones from the IMAGE Consortium. Thus UniGene allows the user to go from a tissue-specific query to being able to purchase clones for spotting onto a microarray.

In UniGene, different splice variants of a gene are grouped together in the same cluster. For example, all three splice variants of the gene CNR1 (Example 2.2) are clustered together in the cluster Hs.75110. This is a significant disadvantage with using UniGene as a resource for microarray design.

BOX 2.3 UniGene Cluster Hs.2

The cluster Hs.2 as it appears on the NCBI web site. There is a wealth of information about this cluster: species, gene names, closely related genes in other organisms, chromosome location and STS sites, tissue distribution and sequence information. The tissue distribution gives the types of libraries from which the EST and mRNA sequences were derived. This is very powerful information, as it allows the user to select genes that have been seen to be expressed in particular tissues and, thus, construct tissue-specific microarrays. There are 4 mRNA sequences and 18 EST sequences for this cluster (the relationship between them is shown in Figure 2.3). The first mRNA sequence is a RefSeq sequence and will thus be the highest quality. There is substantial information about the EST sequences. The clone IDs that refer to IMAGE clones allow the clones to be purchased from the IMAGE consortium, which can then be used for spotting the sequences onto arrays.

```
NCBI UniGene UniGene Cluster Hs.2 Homo sapiens

NCBI UniGene UniGene Cluster Hs.2 Homo sapiens

NAT2 N-acetyltransferase 2 (arylamine N-acetyltransferase)

SEE ALSO
 LocusLink:  10
 OMIM:       243400
 HomoloGene: Hs.2

SELECTED MODEL ORGANISM PROTEIN SIMILARITIES
organism, protein and percent identity and length of aligned region
 H.sapiens:     pir:B34585 - B34585 arylamine N-acetyltransferase (EC  100 % / 289 aa
                2.3.1.5) 2 - human                                     (see ProtEST )
 M.musulus:     sp:P50295 - ARY2_MOUSE Arylamine N-acetyltransferase 2 74 % / 289 aa
                (Arylamide acetylase 2) (N-acetyltransferase type 2)   (see ProtEST )
                (NAT
 R.norvegicus:  ref:NP_446306.1 - N-acetyltransferase 2 (arylamine N-   73 % / 289 aa
                acetyltransferase) [Rattus norvegicus]                 (see ProtEST )
 E.coli:        ref:NP_415980.1 - putative N-hydroxyarylamine O-       24 % / 254 aa
                acetyltransferase [Escherichia coli K12]               (see ProtEST )

MAPPING INFORMATION
 Chromosome:     8
 OMIM Gene Map:  8p22
 Whitehead map:  WI-7224
 UniSTS entries: GDB:386004  Genomic Context:  Map View
 UniSTS entries: WI-7224  Genomic Context:  Map View
 UniSTS entries: stSG40  Genomic Context:  Map View
 UniSTS entries: SHGC-130680  Genomic Context:  Map View

EXPRESSION INFORMATION
 cDNA sources: liver ;corresponding non cancerous liver tissue ;colon ;colon,
               2 pooled adenocarcinomas ;hepatocellular carcinoma
 SAGE :        Gene to Tag mapping
mRNA SEQUENCES (4)
 NM 000015 Homo sapiens N-acetyltransferase 2 (arylamine N-acetyltransferase)    P
           (NAT2), mRNA
```

BOX 2.3 (*continued*)

D90042	Human liver arylamine N-acetyltransferase (EC 2.3.1.5) gene	P
D90040	Human mRNA for arylamine N-acetyltransferase (EC 2.3.1.5)	P
BC015878	Homo sapiens, N-acetyltransferase 2 (arylamine N-acetyltransferase), clone MGC:27492 IMAGE:4716636, mRNA, complete cds	P A

EST SEQUENCES (18)

BG618195	cDNA clone IMAGE:4767316	liver	5′ read	P M
BG569293	cDNA clone IMAGE:4722596	liver	5′ read	P M
BG568400	cDNA clone IMAGE:4716802	liver	5′ read	P M
BG563731	cDNA clone IMAGE:4712210	liver	5′ read	P M
BG533459	cDNA clone IMAGE:4072143	liver	5′ read	P M
AI792606	cDNA clone IMAGE:1870937	colon, 2 pooled adenocarcinomas	5′ read 1.2 kb	P
AI733799	cDNA clone IMAGE:1870937	colon, 2 pooled adenocarcinomas	3′ read 1.2 kb	P A
AI262683	cDNA clone IMAGE:1870937	colon, 2 pooled adenocarcinomas	3′ read 1.2 kb	P A
BG204539	cDNA clone (no-name)			
BG617259	cDNA clone IMAGE:4734378	liver	5′ read	P
BG569272	cDNA clone IMAGE:4722638	liver	5′ read	P
AU099534	cDNA clone HSI08034			P
BF126423	cDNA clone IMAGE:4071536	liver	5′ read	P
AV658656	cDNA clone GLCFOG07	corresponding non cancerous liver tissue	3′ read	P
AV658623	cDNA clone GLCFOD10	corresponding non cancerous liver tissue	3′ read	P
AI460128	cDNA clone IMAGE:2151449	colon	3′ read	P A
AV684197	cDNA clone GKCFZH06	hepatocellular carcinoma	5′ read	P
BG566307	cDNA clone IMAGE:4712733	liver	5′ read	P

Key to Symbols

P Has similarity to known Proteins (after translation)
A Contains a poly-Adenylation signal
M Clone is putatively CDS-complete by MGC criteria

TABLE 2.4: Advantages and Disadvantages of UniGene

Advantages	Disadvantages
✓ Freely available on the WWW	× Restricted range of species
✓ Good information for human, mouse, rat and cress	× No consensus sequences
✓ Information on both known genes and ESTs	× No splice variant information
✓ Tissue distribution information	× Clusters can be unstable and change from build to build
✓ Real sequences: can get clones from IMAGE	× Probably many clusters belong together
✓ Links well with other databases	

A further consideration of UniGene is that the cluster numbers and assignments change with each database update. Therefore, arrays made using UniGene must have a record of the version of UniGene used, and in addition to the cluster ID, the accession number of the sequence put on the array should also always be recorded. If only the cluster ID is recorded, it becomes difficult to trace the sequences in new versions of UniGene when the data are being analysed.[2] The advantages and disadvantages of UniGene are summarised in Table 2.4.

The TIGR Gene Indices

The Gene Indices (GI) at the Institute for Genetics Research (TIGR) are a resource that is similar in scope to UniGene. As with UniGene, the TIGR GI are arranged according to species (Table 2.5). The TIGR GI covers more species than UniGene, with 19 animal species, 18 plant species, 13 protist species and 7 fungal species. Also, the TIGR GI includes a greater number of sequences for most of the species that are also represented in UniGene. For example, UniGene's Bovine index (Build 38) contains 130,311 sequences and 10,787 clusters, while TIGR's Bovine index (Version 7.0) contains 231,316 sequences and 26,931 clusters.

The TIGR human gene index contains a similar number of sequences to the UniGene human database (Table 2.6). However, it is arranged into approximately 180,000 clusters – substantially more than UniGene. As with UniGene, this is much greater than the number of predicted genes in the human genome, so it is likely that this database will change over the next few years.

Unlike UniGene, TIGR contains consensus sequences for each of the clusters. From the perspective of designing microarrays, this has both advantages and disadvantages. On the positive side, a consensus sequence is a higher quality sequence and is therefore a better starting point for oligonucleotide design. On the negative side, the UniGene sequences are all real clones and can be purchased from the IMAGE Consortium for use with a spotted array.

TIGR also intends to include full information about splice variants in their database. In December 2002, there was very limited splice variant information in the TIGR GI, and no information on human splice variants. This will probably change in the

[2] I've done this myself and it was painful.

TABLE 2.5: TIGR Gene Indices Species

Animals		Plants	
Amblyomma variegatum	478	*Arabidopsis*	22,485
Brugia malayi	1,846	Barley	18,552
Catfish	1,190	*Chlamydomonas reinhardtii*	9,785
Cattle	26,931	Cotton	6,441
C. elegans	14,262	Ice plant	1,977
Chicken	9,425	Lettuce	7,977
C. intenstinalis	14,057	*Ljaponicus*	3,790
Drosphila	16,926	Maize	18,715
Honey bee	2,893	*Medicago truncatula*	16,086
Human	181,539	*Pinus*	7,732
Mosquito	10,865	Potato	11,388
Mouse	105,520	Rice	14,026
O. laptipes	5,386	Rye	1,294
Oncochocerca volvulis	915	Sorghum bicolor	11,336
Pig	17,354	Soybean	24,750
Rat	41,023	Sunflower	4,374
Schistosoma mansoni	1,920	Tomato	15,211
Xenopus laevis	24,246	Wheat	24,124
Zebrafish	18,811		
Fungi		**Protists**	
Aspergillus nidulans	1,750	*Cryptosporidium parvum*	123
Coccidioides immitis	366	*Dictyostelium discoideum*	6,180
Cryptococcus	2,168	*Eimeria tenella*	1,181
Magnaporthe grisea	1,407	*Leishmania*	330
Neurospora crassa	2,406	*Neospora caninum*	338
Saccharomyces cerevisiae	4,050	*Plasmodium berghei*	678
Schizosaccharomyces pombe	2,499	*Plasmodium falciparum*	1,917
		Plasmodium yoelii	2,144
		Sarcocystis neurona	663
		Trypanosoma brucei	577
		Trypanosoma cruzi	1,672
		Toxoplasma gondii	2,481
		Tetrahymena thermophila	1,166

Note: Species and number of clusters in TIGR Gene Indices as of 31st December 2002. There are 19 animal species, 18 plant species, 13 protist species and 7 fungal species, representing much broader coverage than UniGene. The species that appear in both UniGene and TIGR have better coverage in TIGR than UniGene.

next couple of years and, if implemented, will make the GI a powerful resource for microarray design. The advantages and disadvantages of TIGR are summarised in Table 2.7.

RefSeq

The third secondary database resource we describe for the construction of microarrays is the NCBI's reference sequence project, or RefSeq. The reference sequence project aims to collect high-quality, well-annotated sequences of many types,

TABLE 2.6: Statistics for the TIGR Human Gene Index Version 10.0

(a)

	In Clusters	Singletons	Total
ESTs	3,708,778	467,295	4,176,073
mRNAs	95,129	5,176	100,305
Total	3,803,907	472,471	4,276,378

(b)

Number of clusters	181,539
Singleton mRNAs	5,176
Singleton ESTs	467,295
Total	654,010

(a) As with UniGene, the database contains both EST and mRNA sequences. Unlike UniGene, singleton sequences that do not overlap with any other sequences in the database are kept separately. With the human gene index, the number of sequences in TIGR is comparable with the number of sequences in UniGene. (b) However, there are many more clusters in TIGR than UniGene, resulting from differences in the algorithm. Because it is thought that there are only about 30,000 genes in the human genome, it is possible that the TIGR clusters will change over the next few years.

including complete genomes, complete chromosomes, genomic regions, mRNAs, other types of RNA, genome contigs and proteins.

The mRNA section of RefSeq is of particular interest for microarray design and is available for human, mouse, fruit fly, rat and zebrafish (Table 2.8). RefSeq does not provide a complete picture of expressed material for any of these species. For example, as of December 2002, there are only about 19,000 RefSeq entries for humans, compared with about 32,000 UniGene clusters containing at least one mRNA. However, the sequences in RefSeq represent the highest possible quality mRNA sequences in the database, and so they are used where possible as the basis for microarray and other work.

RefSeq mRNAs are recognisable by accession numbers starting with the prefix NM_. For example, in Example 2.2, the three cannabinoid receptor splice variants are all RefSeq genes. In Example 2.5, the first sequence in the UniGene cluster Hs.2 is also a RefSeq sequence and would represent the best sequence to use to represent the gene in this cluster (Box 2.3).

TABLE 2.7: Advantages and Disadvantages of TIGR

Advantages	Disadvantages
✓ Freely available on the WWW for noncommercial users	× Licenses required for commercial users
✓ Wide range of species	× Probably many clusters belong together
✓ Information on both known genes and ESTs	× Splice variants not yet included
✓ Consensus sequences	
✓ Future inclusion of splice variants	

TABLE 2.8: RefSeq Species and Statistics

Species	Provisional	Predicted	Reviewed	Total
Human	7,186	4,085	7,159	19,043
Mouse	9,144	4,221	83	14,099
Drosophila	7,580	6,930	1,221	16,450
Rat	4,089	106	14	4,210
Zebrafish	846		1	852

Note: Statistics for the five species currently covered by RefSeq, as of 31st December 2002. Provisional sequences have not yet been subject to individual review. Predicted means that while there is evidence that this is a valid locus, such as a cDNA sequence, some aspect of the record is predicted, including possibly the protein sequence. Reviewed sequences have been individually reviewed and annotated by staff at the NCBI and represent very high-quality sequence records.

TABLE 2.9: Advantages and Disadvantages of RefSeq

Advantages	Disadvantages
✓ Freely available on the WWW	× Limited number of species
✓ High-quality sequence information for known genes	× Limited coverage of genes
✓ Inclusion of splice variants	× Does not include ESTs or predicted genes

TABLE 2.10: Ensembl Species and Statistics

	Human	Mouse	Zebrafish	Fugu	Mosquito
Ensembl Gene Predictions	22,980	22,444	1,511	31,059	15,088
Genscan Gene Predictions	73,218	109,654	2,406	32,615	
Gene Exons	204,094	191,290	7,524	181,098	54,788
GeneTranscripts	27,628	28,097	1,828	33,609	15,101
Base Pairs	3,342,501,203	2,726,795,854	36,330,021	332,504,233	278,253,050

Note: The Ensembl database is available for five species. Ensembl gene predictions represent known and predicted genes for which there is experimental evidence. Genscan gene predictions include genes which have been predicted in silico but which may not necessarily have supporting experimental evidence. Gene exons are the number of exons, and gene transcripts represent the number of transcripts which can include multiple splice variants for the same gene.

TABLE 2.11: Advantages and Disadvantages of Ensembl

Advantages	Disadvantages
✓ Freely available on the WWW	× Limited number of species
✓ Contains all sequence information from genomic perspective	× Requires more advanced bioinformatics programming skills to access at low level
✓ Contains known and computer-predicted genes	
✓ Includes information on exons and splice variants	

Splice variants of genes are fully represented in RefSeq, making it a very powerful resource for the design of arrays for known splice variants of known genes. For example, the three splice variants of Cannabinoid Receptor 1 in Example 2.2 are all separate RefSeq entries linked to the same locus on the human genome. The advantages and disadvantages of RefSeq are summarised in Table 2.9.

SECTION 2.4 GENOMIC DATABASE RESOURCES

Sections 2.2 and 2.3 looked at sequences that could be used for microarray experiments from a gene-centric perspective. The databases in those sections start at the low level of mRNA and EST sequences, which are kept in primary databases and are assembled and annotated into higher level secondary databases. This section looks at resources that can be used for choosing sequences for microarrays from a genomic perspective: to start with the whole genome and then choose gene sequences for the array based on the annotation of that genome. For small organisms, such as bacteria and yeast, this is the most natural approach. But even for complex organisms such as humans, there are resources that allow this approach to microarray design and annotation.

Ensembl

Ensembl is a joint project between the European Bioinformatics Institute (EBI) and the Wellcome Trust Sanger Institute to provide complete annotation of eukaryotic genomes. Originally established to cover the human genome, at the time of this writing it also included coverage of mouse, rat, zebrafish, fugu and mosquito (Table 2.10).

The reason for setting up Ensembl is to provide a single, seamless resource for querying and mining completed genomes, such as the human genome. When a genome is sequenced, it is sequenced in small chunks. Ensembl assembles these chunks into chromosome sequences so that each chromosome appears as a single "virtual" sequence, also known as the **Golden Path**.

The real power of Ensembl as a resource for microarray design is in its annotation. The Ensembl project links all available data about human sequences, so that information on known genes, known proteins and ESTs are included as part of the genome annotation. It also provides annotation on the results of gene prediction algorithms. This is important for microarray design because it allows oligonucleotide probes to be designed for predicted genes and exons in addition to known expressed sequences. The advantages and disadvantages of Ensembl are summarised in Table 2.11.

EXAMPLE 2.6 ENSEMBL ENTRY FOR ENSG00000118432

Cannabinoid Receptor 1 has Ensembl accession number ENSG00000118432. The Ensembl entry has two predicted transcripts (Box 2.4) and references to the three RefSeq entries for this gene (Example 2.3).

BOX 2.4 Ensembl Entry for Cannabinoid Receptor 1

The text entry for the gene with ENSG00000118432. In the Ensembl database, there are two predicted transcripts, which do not correspond exactly to the three splice variants in RefSeq. However, lower in the entry there are references to all three RefSeq transcripts, as well as sequences in the EMBL database, GO (Gene Ontologies: see Chapter 10), HUGO, LocusLink, MIM and protein identifiers. The HTML link takes the user to a page showing graphical information about the gene, including the exon structure of the gene.

```
EnsEMBL gene ENSG00000118432 has 2 transcripts: ENST00000303726,
ENST00000237199
CANNABINOID RECEPTOR 1 (CB1) (CB-R) (CANN6). [Source:SWISSPROT;Acc:P21554]
The gene has the following external identifiers mapped to it:
EMBL: U73304, X54937, AF107262, X81120, X81121
GO: GO:0005887, GO:0004949, GO:0007187, GO:0007610
HUGO: CNR1, 2159
LocusLink: 1268
MIM: 114610
protein_id: AAD34320, CAA57018, AAB18200, CAA57019, CAA38699
RefSeq: NM_016083, NM_001840, NM_033181
SWISSPROT: CB1R_HUMAN, P21554
http://www.ensembl.org:80/Homo_sapiens/geneview?gene=ENSG00000118432
```

EXAMPLE 2.7 ENSEMBL REPORT FOR ENSG00000146263

This is an Ensembl gene prediction located in a 100-Mb region on chromosome 6. A portion of the Ensembl report that would appear on the Ensembl web site for this predicted gene is shown in Box 2.5. The predicted genes in Ensembl could be included in an exploratory microarray experiment or as part of an attempt to confirm these sequences as genes and ascertain some understanding of their possible function.

Microbial Genomes

Microbial genomes are small – typically with genomes between 2 and 5 megabases, and between 2,000 and 5,000 genes. This makes microbes very attractive organisms for microarray analysis: it is possible to place probes for every gene in the organism on a single array and perform powerful and exciting experiments.

Microbial genomes are readily accessible from two databases: GenBank and the TIGR Comprehensive Microbial Resource (CMR). In December 2002, there were 102 genomes in GenBank and 96 genomes in TIGR. Data are exchanged between the two databases: most genomes are in both databases, but the genomes that are sequenced at TIGR are published in the TIGR database before they reach GenBank, and genomes sequenced elsewhere are published in GenBank before they reach TIGR. Of the 102 genomes in GenBank, there are 85 different organisms, with 12 organisms having multiple strains in the database. For example, there are two strains of *Yersinia pestis* (the bacterium responsible for the Black Death in the Middle Ages): CO92 and KIM.

BOX 2.5 Ensembl Report for Gene ENSG00000146263

Part of the Ensembl report for a predicted gene. The gene is predicted using the exon prediction programs Genscan and Genewise on the human chromosome sequence. The predicted exons are then compared with EST, cDNA and protein databases and exons with similar sequences in these databases are confirmed predictions. This particular gene consists of 11 exons and is 1,692 bases long. An exploratory microarray experiment could include probes specific to this gene.

```
Ensembl gene ID      ENSG00000146263
Genomic Location     View gene in genomic location: 97462884 - 97502767 bp (97.5 Mb)
                   on chromosome 6
                     This gene is located in sequence: AL023656.8.1.112361
Description          No description
Prediction Method    This gene was predicted by the Ensembl analysis pipeline from
                   either a GeneWise or Genscan prediction followed by confirmation
                   of the exons by comparisons to protein, cDNA and EST databases.
Predicted            1:    ENST00000275053 [View supporting evidence]    [View protein
Transcripts        information]
InterPro           IPR001092 Basic helix-loop-helix dimerization domain bHLH [View
                   other Ensembl genes with this domain]
Protein Family       ENSF00000015640 : UNKNOWN
                     This cluster contains 1 Ensembl gene member(s)
Export Data          Export gene data in EMBL, GenBank or FASTA
Homology Matches    No homologues identified for this gene.

                             Transcript 1: ENST00000275053
Transcript cDNA Sequence                 Total length: 1692 bp    No. Exons: 11
>ENST00000275053
AGTATGCATCAACAATTGTGTCAGGAACTTCAAAGGGACAATGTGGACCT
ATTTGTACAGTCTTCATTATCGGCTAAAGAGCGCCACCTTGCTGCAGTTG
CCAGTGCACTGTGGAGACATTTCTTTTCATTTTTGAAGAGTCAGAGAATG
TCACAGGTAGTGCCTTTCTCACAACTTGCGGATGCAGCTGCAGACTTTAC
TTTGCTAGCAATGGACATGCCAAGCACAGCTCCATCAGATTTTCAGCCTC
AGCCAGTTATATCAATTATTCAACTTTTTGGTTGGGATGATATCATCTGC
CCTCAAGTTGTAGCAAGATATTTAAGTCATGTCCTACAAAATAGCACATT
ATGTGAAGCACTTTCTCATTCAGGCTATGTATCTTTTCAAGCCTTAACCG
TAAGATCATGGATTCGTTGTGTTTTGCAAATGTATATTAAAAACCTCTCT
GGGCCTGATGATTTGCTCATAGATAAAAATCTGGAAGAGGCAGTTGAAAA
AGAGTACATGAAACAGTTGGTCAAACTGACAAGATTACTATTTAATCTCT
CAGAAGTAAAGAGTATTTTCTCAAAGGCCCAAGTTGAATATTTATCCATC
TCAGAAGACCCTAAAAAAGCACTTGTTCGATTCTTTGAGGCTGTTGGTGT
AACTTACGGGAACGTCCAGACACTTTCTGATAAATCTGCCATGGTCACAA
AGTCCTTGGAATACCTTGGTGAAGTATTAAAATATATTAAGCCTTATTTG
GGAAAAAAAGTTTTCAGTGCAGGGCTGCAGCTGACTTATGGAATGATGGG
AATTCTTGTGAAATCATGGGCACAAATCTTTGCCACTTCTAAAGCCCAAA
AATTACTATTCCGGATCATAGATTGTTTACTGCTGCCACATGCAGTATTA
CAGCAAGAGAAGGAACTGCCTGCACCTATGTTGTCAGCAATTCAGAAAAG
TCTTCCTTTGTATCTCCAGGGCATGTGTATCGTGTGTTGTCAATCTCAAA
ATCCGAATGCCTATTTGAATCAATTGCTAGGGAATGTTATTGAGCAGTAT
ATTGGGCGATTTCTTCCAGCTTCACCATATGTTTCAGATCTTGGACAACA
TCCTGTTTTGCTGGCATTGAGAAACACAGCCACTATTCCACCAATATCAT
CTCTAAAGAAATGCATTGTGCAAGTCATAAGGAAATCCTACCTTGAGTAT
```

BOX 2.5 (continued)

```
AAGGGGTCCTCACCTCCTCCTCGCTTAGCATCCATTCTGGCCTTCATCCT
CCAACTCTTCAAGGAAACTAACACAGACATTTATGAAGTTGAACTACTCC
TCCCTGGCATTTTAAAATGCTTGGTGTTAGTCAGTGAACCACAAGTTAAA
AGGCTGGCCACAGAGAACCTGCAATACATGGTAAAAGCCTGCCAAGTGGG
GTCAGAAGAAGAACCTTCCTCCCAGCTGACTTCTGTGTTTAGGCAGTTTA
TCCAGGATTATGGTATGAGGTACTATTACCAGGTTTACAGCATTTTAGAA
ACAGTAGCAACATTGGACCAGCAGGTTGTCATCCACTTGATTTCTACCCT
TACTCAGTCTCTGAAGGATTCAGAGCAGAAATGGGGCCTTGGCAGGAATA
TAGCACAAAGGGAAGCCTATAGCAAACTTTTGTCTCACCTTGGACAGATG
GGACAAGATGAGATGCAGAGACTGGAAAATGATAATACTTAA
//
```

The two databases have different annotation for the same genomes. As a result, an array built from the sequences downloaded from each of these data resources may have slightly different genes.

EXAMPLE 2.8 GENBANK AND TIGR ANNOTATIONS OF *E. COLI* K12

The GenBank annotation of *E. coli* strain K12 has accession number U00096. This is a GenBank format file (see Box 2.2) for the whole genome of this organism. It has 4,639,221 bases and 4,403 genes [including predicted Open Reading Frames (ORFs)]. The TIGR annotation of this genome has 5,295 genes (having more predicted ORFs). The details of the first few genes of the GenBank annotation of this genome are shown in Table 2.12. Because of the small number of genes in the genome, it is straightforward to produce an array with approximately 10,000 features containing every gene or ORF in duplicate.

Yeast

Baker's yeast, *Saccharomyces cerevisiae*, is a eukaryotic organism with a short genome making it highly tractable for microarray experiments. Because of this, yeast was used

TABLE 2.12: *E. Coli* K12 Genes

Start	Finish	Strand	Identifier	Name	Gene Product
190	255	+	b0001	thrL	thr operon leader peptide
337	2799	+	b0002	thrA	aspartokinase I, homoserine dehydrogenase I
2801	3733	+	b0003	thrB	homoserine kinase
3734	5020	+	b0004	thrC	threonine synthase
5234	5530	+	b0005		ORF, hypothetical protein
5683	6459	–	b0006	yaaA	ORF, hypothetical protein

Note: The first six genes in the GenBank annotation of *E. coli* strain K12. The GenBank file gives the start and finish positions in the genome, the strand on which the gene lies, a standard gene identifier and a gene name if it exists. The annotation includes both known and hypothetical genes. Bacterial genomes are very efficient and almost all codons are used. For example, the first codon for the gene thrC is immediately after the last codon for thrB.

TABLE 2.13: Yeast Genes from the Saccharomyces Genome Database

ORF	Gene	SGDID	GO_Aspect	GO_Term	GOID
YAL001C	TFC3	3573	transcription initiation from Pol III promoter	RNA polymerase III transcription factor	YAL001C
YAL002W	VPS8	3531	not yet annotated	molecular_function unknown	YAL002W
YAL003W	EFB1	987	translational elongation	translation elongation factor	YAL003W
YAL004W		648	biological_process unknown	molecular_function unknown	YAL004W
YAL005C	SSA1	1929	protein folding	chaperone	YAL005C

Note: Five genes with their annotation from the Saccharomyces Genome Database. The first column is the yeast ORF name; the second column the common gene name; the third column is a unique ID in the SGD. The GO_Aspect and GO_Term columns are gene ontology terms describing the process and function of the gene, respectively (see Section 11.4 for more information on ontologies). The GOID in the final column is a gene ontology ID for the gene.

in the first major microarray paper of DeRisi, Iyer and Brown (1997); on their arrays, they had 6,102 yeast ORFs representing all known and predicted ORFs. Yeast remains a commonly used organism for microarray experiments, so arrays are readily available from a number of academic and commercial sources. The Saccharomyces genome database (SGD) at Stanford University is an excellent resource for navigating the yeast genome.

The yeast genome is arranged into 16 chromosomes. The strain whose sequence appears in the SGD is the s288c strain and has 12,057,495 bases. There are 4,988 registered genes in yeast, and 6,267 ORFs, including predicted ORFs, in the SGD. Both the registered genes and the ORFs can be viewed and downloaded from the SGD web site. Table 2.13 gives information for five of the genes in the database.

KEY POINTS SUMMARY

- Primary gene sequence databases (GenBank, EMBL, DDBJ) hold all published sequences and are the basis of all other databases.
- Secondary gene sequence databases (UniGene, TIGR GI, RefSeq) are excellent resources for designing DNA microarrays.
- Genomic databases (Ensembl, TIGR CMR, SGD) are excellent resources for designing arrays for small organisms and can also be used for more complex organisms.

INTERNET RESOURCES

GenBank:
http://www.ncbi.nlm.nih.gov/
EMBL:
http://www.ebi.ac.uk/embl/index.html
DDBJ:
http://www.ddbj.nig.ac.jp/
UniGene:
http://www.ncbi.nlm.nih.gov/UniGene/

TIGR Gene Indices:
http://www.tigr.org/tdb/tgi/
RefSeq:
http://www.ncbi.nlm.nih.gov/LocusLink/refseq.html
Ensembl:
http://www.ensembl.org
TIGR Comprehensive Microbial Resource:
http://www.tigr.org/tigr-scripts/CMR2/CMRHomePage.spl
Saccharomyces Genome Database:
http://genome-www.stanford.edu/Saccharomyces/

RESEARCH PAPERS

The papers from which the data in Table 2.1 were derived:

Alizadeh, A.A., Eisen, M.B., Davis, R.E., Ma, C., Lossos, I.S., Rosenwald, A., Boldrick, J.C., Sabet, H., Tran, C., Powell, J.I., Yang, L., Marti, G.E., Moore, T., Hudson, J. Jr., Lu, L., Lewis, D.B., Tibshirani, R., Sherlock, G., Chan, W.C., Greiner, T.C., Weisenberger, D.D., Armitage, J.O., Warnke, R., Levy, R., Wilson, W., Grever, M.R., Byrd, J.C., Botstein, D., Brown, P.O., and Staudt, L.M. 2000. Distinct types of diffuse large B-cell lymphoma identified by gene expression profiling. *Nature* 403: 503–11.

Schaffer, R., Landgraf, J., Accerbi, M., Simon, V.V., Larson, M., and Wisman, E. 2001. Microarray analysis of diurnal and circadian-regulated genes in *Arabidopsis*. *Plant Cell* 13: 113–23.

Chu, S., DeRisi, J., Eisen, M., Mulholland, J., Botstein, D., Brown, P.O., and Herskowitz, I. 1998. The transcriptional program of sporulation in budding yeast. *Science* 282: 699–705.

Golub, T.R., Slonim, D.K., Tamayo, P., Huard, C., Gaasenbek, M., Mesirov, J.P., Coller, H., Loh, M.L., Downing, J.R., Caligiuri, M.A., Bloomfield, C.D., and Lander, E.S. 1999. Molecular classification and class prediction by gene expression monitoring. *Science* 286: 531–37.

The research paper that started use of microarrays and which used yeast as a model organism:

DeRisi, J.L., Iyer, V.R., and Brown, P.O. 1997. Exploring the metabolic and genetic control of gene expression on a genomic scale. *Science* 278: 680–86.

CHAPTER THREE

Computer Design of Oligonucleotide Probes

SECTION 3.1 INTRODUCTION

An oligonucleotide probe is a short piece of single-stranded DNA complementary to the target gene whose expression is measured on the microarray by that probe. In most microarray applications, oligonucleotide probes are between 20 and 60 bases long. The probes are either spotted onto the array or synthesised in situ, depending on the microarray platform (Chapter 1).

Usually, oligonucleotide probes for microarrays are designed within several hundred bases of the 3′ end of the target gene sequence. So for a fixed oligonucleotide length, there are several hundred potential oligonucleotides, one for each possible starting base. Some of these oligonucleotides work better than others as probes on a microarray. This chapter describes methods for the computer selection of good oligonucleotide probes.

What Makes a Good Oligonucleotide Probe?

Good oligonucleotide probes have three properties: they are **sensitive**, **specific** and **isothermal**.

A sensitive probe is one that returns a strong signal when the complementary target is present in the sample. There are two factors that determine the sensitivity of a probe:

- The probe does not have internal secondary structure or bind to other identical probes on the array.
- The probe is able to access its complementary sequence in the target, which could potentially be unavailable as a result of secondary structure in the target.

A specific probe is one that returns a weak signal when the complementary target is absent from the sample; i.e., it does not cross-hybridise. There are two factors that determine the specificity of a probe:

- Cross-hybridization to other targets as a result of Watson–Crick base-pairing
- Non-specific binding to the probe; e.g., as a result of G-quartets

Isothermal probes behave similarly under the hybridization conditions of the microarray experiment: temperature, salt concentrations and formamide concentration. Usually, we demand that all the probe–target duplexes on the array have similar melting temperatures.

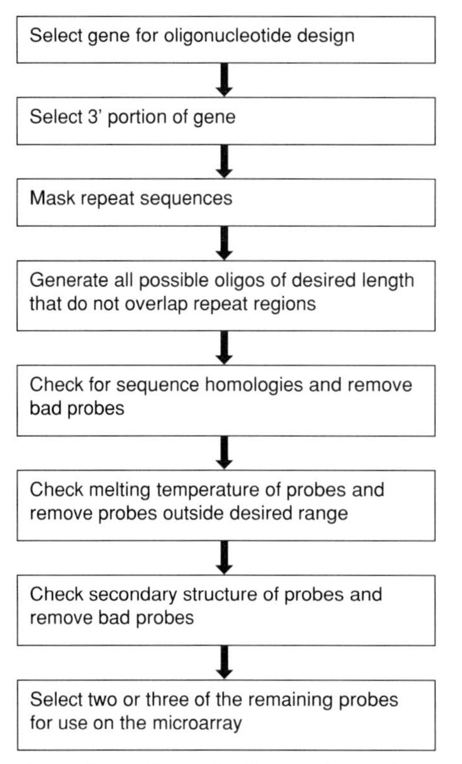

Figure 3.1: Oligonucleotide probe design methodology. The design methodology that we follow in this chapter. At each step, we filter more of the probes until we are left with a subset of probes that we can use on the array. It is not necessary to perform the steps in this order.

Of the five factors listed that determine these three properties, it is straightforward to use computer algorithms to predict the melting temperature, any internal secondary structure of a probe and the similarity of a probe to other targets. The other factors – secondary structure in the target, and non-specific interactions – are more difficult to predict *in silico* and will not be discussed in this chapter.

The remainder of this chapter will follow the structure of most probe design algorithms and focus on the prediction of these properties of oligonucleotide probes. It is arranged into the following four sections:

Section 3.2: The Filtering of Low-Complexity Sequence, discusses the identification of repeat regions in targets; probes designed against repeat regions would be likely to cross-hybridise and should be avoided.

Section 3.3: Prediction of Cross-Hybridisation to Related Genes, looks at the use of homology search algorithms to rule out probes that might also hybridise to genes with similar sequence.

Section 3.4: The Thermodynamics of Nucleic Acid Duplexes and the Prediction of Melting Temperature, describes the base-stacking model for the prediction of thermodynamic properties of nucleic acid duplexes and the use of this model for determining duplex free energies and melting temperatures.

Section 3.5: Probe Secondary Structure, describes software that can be used to rule out probes that will self-hybridise and thus not form duplexes with the target.

In these sections, we follow the probe design process for a single gene, as a concrete example of how to select oligonucleotides for a DNA microarray (Figure 3.1).

SECTION 3.2 THE FILTERING OF LOW-COMPLEXITY SEQUENCE

EXAMPLE 3.1 LOW-COMPLEXITY PROBES

Throughout this chapter, we shall use the gene *Homo sapiens* alcohol dehydrogenase beta2 subunit (ADH2) as an example gene for probe design. The sequence we choose has accession number AF153821 and, at the time of writing this book, is the unique UniGene sequence for cluster Hs.4 (Box 3.1a). Probes are usually designed to be complementary to the 3′ end of genes because of the widespread use of poly-T primers to generate the labelled cDNA target.

The following are two 30mer probes that are complementary to the ADH2 sequence. Are they good probes?

(a) TTTTTAATTTTTTTTTTTTTAAGCAGTAAT
(b) TATATATATATATATATATATATACAATCA

It should be clear that these are not good probes: (a) is complementary to the polyadenylation site at the 3′ end of the mRNA and so contains a large run of Ts that would potentially cross-hybridise to a wide range of other targets; (b) contains a TATA repeat that will also cross-hybridise to a wide range of targets.

Both these probes contain examples of what we call **low-complexity** sequence, which is a term used to describe repetitive sequence. We do not want probes that contain low-complexity sequence because these are likely to cross-hybridise to other targets.

RepeatMasker

It is straightforward to write computer programs to detect this type of repetitive sequence. We will describe software called RepeatMasker that detects all types of repeat sequences: low-complexity regions such as in Example 3.1, and longer genomic repeats such as ALU and LINE sequences.

There are two ways in which RepeatMasker can be accessed. It is available over the Internet via a web interface. This allows the user to submit sequences that will be filtered for low-complexity regions and other repeats. It is also available as a command-line Unix program that can be used in a high-throughput method using simple scripts. In Table 3.1, we show the advantages and disdvantages of using software such as

BOX 3.1 RepeatMasker on the ADH2 Sequence

(a) ADH2 Gene Sequence: 600 bases from the 3′ end of *Homo sapiens* alcohol dehydrogenase beta2 subunit (ADH2). Probes are usually designed at the 3′ end of sequences because of the use of poly-T priming to generate the cDNA target.

```
>AF153821 600 bases 3′ Homo sapiens alcohol dehydrogenase beta2 subunit (ADH2)
ATATAGTTAAGTTGATTGTATATATATATATATATATATATGTATTCCAG
TGTAGAAAGTATTGGCTGGGTCTGTAATCAAGTATTCCACAGAAGTGCCT
TCATAGGTGCTGAAACAGGATTATGAGAGTGACAGACACAGGACTGAAGA
CAGGGAGAAAATGTGCATGGCTGTGAGAATATATTGCTTTGATTCTTTGA
CTTCTCATGATATATGAAAACAAATAAAGTGGGGGGGGGTGACTTCAATA
ATATCACGAACATATTTAAAATGAAATTCTGAGCATGGAGAAAAATTTCA
AATTCTGGTATAAACAAATAATCTAAAAATATAGATAAGGAAAAAATGTT
TTCATTTTTTTGGTCATATTGATTTGGGTCATATATGATTGACCAAAAAA
AAAGATATTTCTAAAATCCTGGTGCTTTTTCATTTAAAAATAGACAACCA
AAGAGGAAAATGAACAATGATGGAAATTCGTGGAAGAGGAAACTAAAATG
GCAAAATAGCATATTAAATGGCACTGAGCCTTGGTGTGATCAAAAAAAAA
AAAATAAAGCAAAATGAGATATTACTGCTTAAAAAAAAAAAAATTAAAAA
```

(b) ADH2 Masked Sequence: Results of running RepeatMasker on the 600-base sequence for ADH2 shown in Figure 3.1. Two repetitive regions have been replaced with Ns: the TATA repeat on the first line and the polyadenylation site on the last line. RepeatMasker has not masked shorter runs of repeats, such as the poly-G on the fifth line, or the poly-A at the end of the penultimate line and beginning of the bottom line.

```
>AF153821 600 bases 3′ Homo sapiens alcohol dehydrogenase beta2 subunit
(ADH2) Masked Sequence
ATATAGTTAAGTTGATTGNNNNNNNNNNNNNNNNNNNNNNNNNNNNTCCAG
TGTAGAAAGTATTGGCTGGGTCTGTAATCAAGTATTCCACAGAAGTGCCT
TCATAGGTGCTGAAACAGGATTATGAGAGTGACAGACACAGGACTGAAGA
CAGGGAGAAAATGTGCATGGCTGTGAGAATATATTGCTTTGATTCTTTGA
CTTCTCATGATATATGAAAACAAATAAAGTGGGGGGGGGTGACTTCAATA
ATATCACGAACATATTTAAAATGAAATTCTGAGCATGGAGAAAAATTTCA
AATTCTGGTATAAACAAATAATCTAAAAATATAGATAAGGAAAAAATGTT
TTCATTTTTTTGGTCATATTGATTTGGGTCATATATGATTGACCAAAAAA
AAAGATATTTCTAAAATCCTGGTGCTTTTTCATTTAAAAATAGACAACCA
AAGAGGAAAATGAACAATGATGGAAATTCGTGGAAGAGGAAACTAAAATG
GCAAAATAGCATATTAAATGGCACTGAGCCTTGGTGTGATCAAAAAAAAA
AAAATAAAGCAAAATGAGATATTACTGCNNNNNNNNNNNNNNNNNNNNNN
```

RepeatMasker, BLAST (Section 3.3) and mfold (Section 3.5) over the web relative to using a local command-line program.

RepeatMasker takes a DNA (or RNA) sequence as an input and returns the same sequence, in which the different repeats have been replaced by runs of "N"s. When designing oligonucleotide probes for a microarray, we would avoid selecting probes in any of the regions masked with the Ns. RepeatMasker includes a number of options:

- The sequence to be masked can be supplied either via a file name or pasted into a pane (if using the web interface).

TABLE 3.1

Advantages and disadvantages of using software such as RepeatMasker, BLAST and mfold via a user interface on the Internet as opposed to obtaining a local copy and running it as a command-line program.

User Interface on the Internet	Local Copy of Software
✓ Free access over the Internet ✓ Easy to use user interfaces × Can only be applied to one sequence at a time and results need to be parsed visually × Dependent on speed and stability of the Internet connection and remote server	✓ Good for high-throughput analysis as can be applied to many sequences with a simple script ✓ Can run in background on local server while you do other work × Need to obtain license agreements from software suppliers × Need Unix skills to install and run the programs, and scripting skills to parse the results

- When using the web interface, the results can be returned either via the web or via email. Usually, with short sequences it is possible to return the results via the web, but it is worth using email when submitting longer (e.g., genomic) sequences.
- There are three sensitivities that can be applied. We recommend using the slowest setting, which gives the most robust results.
- Different species have different families of repeat sequences. RepeatMasker uses different databases of repeat sequences for different species. At the time of this writing, the user can select databases for primates, rodents, other mammals, other vertebrates, *Arabidopsis*, grasses or *Drosophila*.
- There are several options to switch off masking of different types of repeats. For the design of DNA probes, we mask all types of repeats.

EXAMPLE 3.2 REPEATMASKER ON ADH2

The 600-base-pair sequence (Box 3.1a) is submitted to the RepeatMasker server. RepeatMasker has identified two repeat regions and replaced them with Ns (Box 3.1b): these are both regions from which we selected the bad probes in Example 3.1. Note that RepeatMasker has not masked shorter runs of repeats, such as the poly-G on the fifth line, or the poly-A at the end of the penultimate line and beginning of the bottom line; more stringent probe design algorithms might also exclude these regions. In this example, we will not select probes for ADH2 in the regions masked by RepeatMasker.

SECTION 3.3 PREDICTION OF CROSS-HYBRIDISATION TO RELATED GENES

Genes are related to each other. They have evolved from common ancestors, and the majority of genes form part of closely related gene families, with sequence similarity between family members. For an oligonucleotide probe to be specific, it must not hybridise to other targets, whether gene family members or unrelated genes sharing similar sequence.

As with low-complexity sequence, it is possible to use computer algorithms to identify parts of the target sequence that show similarity to other genes. These algorithms are called **homology searches**. A homology search usually takes two inputs:

1. A DNA (or protein) query sequence which, in the case of oligonucleotide probe design, will be the potential oligonucleotide sequences.
2. A database of DNA (or protein) sequences, which would contain other target genes that we do not want to bind to the probe we select for the array.

A homology search return sequences in the database that match either part or the whole of the query sequence. Typically, the alignments between the query sequence and the database hits are returned along with numerical information describing the significance of each hit.

There are many different homology searching algorithms, each of which performs different tasks in different ways. The most commonly used are called BLAST, FASTA and Smith–Waterman alignments, but there are others also available. These are all quite different from each other: in this section, we shall show how BLAST can be used to identify probes that show minimum cross-hybridisation to related probes.

BLAST is a program that is freely available to all, both via web interfaces and as an executable program that can be downloaded for local use. It is hosted by the NCBI. The query sequences to the BLAST program will be the potential probes for the gene target sequence. We could also use the target sequence itself, because BLAST automatically looks for homologies on both strands.

Great care needs to be taken when choosing the database sequences against which to check the query sequences. Many sequence databases contain multiple entries for the same gene. For example, a gene in GenBank may have multiple submissions to a database from different laboratories. Also, the same sequence may be present in a database in different forms, e.g., as an mRNA, a chromosome sequence and as a BAC clone (see Example 3.3).

Therefore, we need to select a database in which the query sequence appears as itself exactly once. The simplest way to ensure this is to use the same database for checking for cross-hybridising homologies as the one used for selecting genes for the array. For example, if we use UniGene (Section 2.3) as the database for selecting genes to put on the array, we would also use UniGene as the database against which we check for cross-hybridising probes.

Because we will be using a specialised BLAST database, it is impractical to perform these searches without using a locally installed version of BLAST at the command line. At the end of the chapter, we give the FTP site where you can download the

BOX 3.2 BLAST on the Web

The sequence TGATTACAGACCCAGCCAATACTTTCTACA, a potential 30mer probe for ADH2, is BLASTed against the non-redundant database on the NCBI server. We show the summary information for the top four hits, and the first alignment.

The summary information has three columns. The first gives an accession number and description of the gene in the database to which a homology has been found. The second is a bit score; with nucleotide sequences, this is equal to twice the number of matched bases, so a bit score of 60 represents a 100% identity with a 30-base-pair sequence. Finally, the E value is a measure of likelihood of seeing this alignment with random sequences.

The alignment gives further information, including the number of identities, the orientation of the alignment, and the positions of the matching sequences in both the query and database sequence.

Of these hits, the first three are the same gene in different forms: as a BAC clone, as the gene itself, and as part of genomic DNA. The fourth hit is the closest homology to a different gene, and spans 18 bases. We show this 18-base alignment.

```
                                                              Score
                                                              (bits)  E Value
Sequences producing significant alignments:
gi|18072230|gb|AC097530.3| Homo sapiens BAC clone RP11-696N...  60      5e-08
gi|5002378|gb|AF153821.1|AF153821 Homo sapiens alcohol dehy...  60      5e-08
gi|9293862|dbj|AP002027.1|AP002027 Homo sapiens genomic DNA...  60      5e-08
gi|18370038|gb|AC074378.4| Homo sapiens chromosome 4 clone...   36      0.79
>gi|18370038|gb|AC074378.4| Homo sapiens chromosome 4 clone RP11-4I17,
complete sequence
          Length = 154112
Score = 36.2 bits (18), Expect = 0.79
Identities = 18/18 (100%)
Strand = Plus / Plus
Query: 11   cccagccaatactttcta 28
            ||||||||||||||||||
Sbjct: 4953 cccagccaatactttcta 4970
```

BLAST executable. In Chapter 2, we showed where you can download commonly used sequence databases, such as UniGene, TIGR GI, RefSeq and Ensembl.

EXAMPLE 3.3 USING BLAST ON THE WEB

BLAST is available on a number of web sites. One good site is at the NCBI:

`http://www.ncbi.nlm.nih.gov/BLAST/`

There are a number of options for blasting nucleotides or proteins. We use BLASTN (standard nucleotide BLAST) to check the probe sequence

`TGATTACAGACCCAGCCAATACTTTCTACA`

against the non-redundant database provided by the NCBI. In this example, the top three sequences found are all the same ADH2 gene from which the 30mer probe derives (Box 3.2). The first sequence is the gene on a BAC clone, the second sequence

TABLE 3.2

Results of a BLAST search of 30mer oligonucleotides complementary to the 600 bases at the 3′ end of ADH2 that do not contain repeat sequences detected by RepeatMasker. In the first column is the position of the 5′ base of the probe sequence from the 3′ end of the target sequence. In the second column is the oligo sequence. In the third column is the identifier of the sequence in the database with the closest hit. In the fourth column is the bit score of that hit. In the majority of these cases, the bit score will be twice the length of the maximum alignment, so a bit score of 30 means that there is 100% identity with a 15-base sequence. The E value is a measure of likelihood of seeing that bit score by random chance.

Observe that there are no BLAST results for the sequences at positions 41, 191 and 351. This is because BLAST also performs a low-complexity filter and has not processed those sequences because of the poly-T and poly-C present in the central portions of them, respectively. We will therefore not select these probes for use on a microarray.

In the final column, we have selected those probes with minimum sequence homology for the next phase of the probe selection. We have used a threshold of E value greater than 1.0, or bit score less than or equal to 30. In other applications, for example, when using longer probes, different thresholds would be appropriate.

3′	Oligo		Bit Score	E Val	Use?
31	ATCTCATTTTGCTTTATTTTTTTTTTTTTG	gnl|UG|Hs#S1607371	28	4.3	✓
41	GCTTTATTTTTTTTTTTTTGATCACACCAA				
51	TTTTTTTTTGATCACACCAAGGCTCAGTGC	gnl|UG|Hs#S2139687	30	1.1	✓
61	ATCACACCAAGGCTCAGTGCCATTTAATAT	gnl|UG|Hs#S848092	34	.07	
71	GGCTCAGTGCCATTTAATATGCTATTTTGC	gnl|UG|Hs#S1056596	36	.018	
81	CATTTAATATGCTATTTTGCCATTTTAGTT	gnl|UG|Hs#S1056596	40	.001	
91	GCTATTTTGCCATTTTAGTTTCCTCTTCCA	gnl|UG|Hs#S2826142	32	.28	
101	CATTTTAGTTTCCTCTTCCACGAATTTCCA	gnl|UG|Hs#S2826142	32	.28	
111	TCCTCTTCCACGAATTTCCATCATTGTTCA	gnl|UG|Hs#S3439044	34	.07	
121	CGAATTTCCATCATTGTTCATTTTCCTCTT	gnl|UG|Hs#S3439044	34	.07	
131	TCATTGTTCATTTTCCTCTTTGGTTGTCTA	gnl|UG|Hs#S2512174	32	.28	
141	TTTTCCTCTTTGGTTGTCTATTTTTAAATG	gnl|UG|Hs#S2512174	32	.28	
151	TGGTTGTCTATTTTTAAATGAAAAAGCACC	gnl|UG|Hs#S1646953	36	.018	
161	TTTTTAAATGAAAAAGCACCAGGATTTTAG	gnl|UG|Hs#S1646953	36	.018	
171	AAAAAGCACCAGGATTTTAGAAATATCTTT	gnl|UG|Hs#S1692901	34	.07	
181	AGGATTTTAGAAATATCTTTTTTTTTGGTC	gnl|UG|Hs#S1459454	30	1.1	✓
191	AAATATCTTTTTTTTTGGTCAATCATATAT				
201	TTTTTTGGTCAATCATATATGACCCAAATC	gnl|UG|Hs#S2511077	30	1.1	✓
211	AATCATATATGACCCAAATCAATATGACCA	gnl|UG|Hs#S2264745	30	1.1	✓
221	GACCCAAATCAATATGACCAAAAAAATGAA	gnl|UG|Hs#S2874360	28	4.3	✓
231	AATATGACCAAAAAAATGAAAACATTTTTT	gnl|UG|Hs#S3507254	28	4.3	✓
241	AAAAAATGAAAACATTTTTTCCTTATCTAT	gnl|UG|Hs#S2649446	36	.018	
251	AACATTTTTTCCTTATCTATATTTTTAGAT	gnl|UG|Hs#S3507655	38	.004	
261	CCTTATCTATATTTTTAGATTATTTGTTTA	gnl|UG|Hs#S3507655	38	.004	
271	ATTTTTAGATTATTTGTTTATACCAGAATT	gnl|UG|Hs#S1731899	36	.018	
281	TATTTGTTTATACCAGAATTTGAAATTTTT	gnl|UG|Hs#S1162424	32	.28	
291	TACCAGAATTTGAAATTTTTCTCCATGCTC	gnl|UG|Hs#S4015371	36	.018	
301	TGAAATTTTTCTCCATGCTCAGAATTTCAT	gnl|UG|Hs#S4015371	32	.28	
311	CTCCATGCTCAGAATTTCATTTTAAATATG	gnl|UG|Hs#S712102	32	.28	
321	AGAATTTCATTTTAAATATGTTCGTGATAT	gnl|UG|Hs#S1040378	32	.28	
331	TTTAAATATGTTCGTGATATTATTGAAGTC	gnl|UG|Hs#S472703	34	.07	
341	TTCGTGATATTATTGAAGTCACCCCCCCCC	gnl|UG|Hs#S222457	34	.07	

3′	Oligo		Bit Score	E Val	Use?
351	TATTGAAGTCACCCCCCCCCACTTTATTTG				
361	ACCCCCCCCCACTTTATTTGTTTTCATATA	gnl|UG|Hs#S3988102	36	.018	
371	ACTTTATTTGTTTTCATATATCATGAGAAG	gnl|UG|Hs#S3988102	36	.018	
381	TTTTCATATATCATGAGAAGTCAAAGAATC	gnl|UG|Hs#S1693871	32	.28	
391	TCATGAGAAGTCAAAGAATCAAAGCAATAT	gnl|UG|Hs#S1726320	32	.28	
401	TCAAAGAATCAAAGCAATATATTCTCACAG	gnl|UG|Hs#S3854638	30	1.1	✓
411	AAAGCAATATATTCTCACAGCCATGCACAT	gnl|UG|Hs#S938742	30	1.1	✓
421	ATTCTCACAGCCATGCACATTTTCTCCCTG	gnl|UG|Hs#S3422346	32	.28	
431	CCATGCACATTTTCTCCCTGTCTTCAGTCC	gnl|UG|Hs#S1388024	34	.07	
441	TTTCTCCCTGTCTTCAGTCCTGTGTCTGTC	gnl|UG|Hs#S2139611	36	.018	
451	TCTTCAGTCCTGTGTCTGTCACTCTCATAA	gnl|UG|Hs#S2294332	32	.28	
461	TGTGTCTGTCACTCTCATAATCCTGTTTCA	gnl|UG|Hs#S1732357	32	.28	
471	ACTCTCATAATCCTGTTTCAGCACCTATGA	gnl|UG|Hs#S1732357	32	.28	
481	TCCTGTTTCAGCACCTATGAAGGCACTTCT	gnl|UG|Hs#S1730614	30	1.1	✓
491	GCACCTATGAAGGCACTTCTGTGGAATACT	gnl|UG|Hs#S3988737	32	.28	
501	AGGCACTTCTGTGGAATACTTGATTACAGA	gnl|UG|Hs#S3542723	30	1.1	✓
511	GTGGAATACTTGATTACAGACCCAGCCAAT	gnl|UG|Hs#S226217	30	1.1	✓
521	TGATTACAGACCCAGCCAATACTTTCTACA	gnl|UG|Hs#S4027871	30	1.1	✓

is the gene sequence itself and the third sequence is the chromosome sequence from the genomic DNA containing the gene. This demonstrates the care that needs to be taken when selecting a BLAST database to ensure that the gene is represented only once. The longest alignment to a different gene is 18 bases.

EXAMPLE 3.4 HIGH-THROUGHPUT BLAST FOR ADH2 PROBE SELECTION

We continue our probe selection methodology for ADH2 that we started in Example 3.2 by choosing 30mer oligonucleotides from the 600 bases at the 3′ end of ADH2. In this example we examine 30mers with 10-base intervals between probes in order to fit the probes into the tables of the chapter. In a real probe design application, it would be better to look at every possible probe.

In Table 3.2 we show all the probes that do not contain repeat-masked sequence, together with the results of the BLAST search against the human UniGene database Build 148. BLAST has a built-in low-complexity filter and was not able to perform a homology search on the probes numbered 41, 191 and 351 because of the poly-T and poly-C bases towards the middle of the sequence. We therefore rule out these sequences as potential probes.

The bit score is in most cases twice the length of the longest perfect alignment. In this example, we have selected those probes with an e-value of at least 1.0, equivalent to a bit score of less than or equal to 30. This corresponds to a perfect alignment over 15 or fewer bases. We have filtered out those probes with higher bit scores. This leaves us with 13 possible probes for the next step of the oligonucleotide design procedure. When designing longer oligonucleotides, or oligonucleotides for genes with close family members, you will need to use different thresholds.

SECTION 3.4 THE THERMODYNAMICS OF NUCLEIC ACID DUPLEXES AND THE PREDICTION OF MELTING TEMPERATURE

In order to predict the melting temperature and the stability of secondary structure in probes (Section 3.5), we need an understanding of the thermodynamics of nucleic acid duplexes. There are three key concepts that we use: the changes in **enthalpy**, **entropy** and **Gibbs free energy** of chemical reactions. With nucleic acid duplexes or structures, the chemical reactions in which we are interested are the formation or disassociation of hydrogen bonds and stacking interactions between the base pairs.

The change in enthalpy of a chemical reaction, known as ΔH, is equal to the heat absorbed by the reaction at constant pressure.

The change in entropy of a chemical reaction, known as ΔS, is a measure of the loss of capacity of a system to do work. It can be thought of as a measure of the loss of degrees of freedom in the system.

The change in Gibbs free energy of a chemical reaction, known as ΔG, is defined by the following equation:

$$\Delta G = \Delta H - T\Delta S \qquad \text{(Eq. 3.1)}$$

This measures the stability of the chemical reaction at the temperature T, assuming constant pressure and temperature. Equation 3.1 summarises the balance between the tendency of a system to minimise its enthalpy (releasing heat in accordance with the first law of thermodynamics) and to maximise its entropy (in accordance with the second law of thermodynamics).

The **melting temperature** of a duplex depends on the concentration of the reactants and is given by the following equation:

$$T_{\mathrm{m}} = \Delta H/(\Delta S - R\ln(C/4)) \qquad \text{(Eq. 3.2)}$$

In this equation, R is the molar gas constant, equal to 0.001987 kcal mol^{-1}, C is the molar concentration of the target[1] and ln denotes a natural logarithm.

We calculate the thermodynamics of nucleic acid duplexes by calculating their enthalpy and entropy. From these quantities, we derive the free energy at the desired hybridisation temperature, or the melting temperature at an appropriate concentration.

The calculation of both the enthalpy and entropy of a nucleic acid duplex is performed using the base-stacking model. This is a linear model that calculates these thermodynamic properties as a sum of contributions from the individual base pairs in the duplex. The sum is based on parameters for each base pair that depend not only

[1] It is important to remember that this equation is derived for chemical reactions that occur in solution. With DNA microarrays, one of the DNA molecules is attached to solid support and it is not clear to what extent this equation may be valid. At the time of this writing, there have been no published comprehensive studies of the thermodynamics of nucleic acid hybridisation on solid support, so most people continue to use this equation in the absence of a better alternative.

```
5'-AA-3'  5'-AC-3'  5'-AG-3'  5'-AT-3'
   ||        ||        ||        ||
3'-TT-5'  3'-TG-3'  3'-TC-5'  3'-TA-5'

5'-CA-3'  5'-CC-3'  5'-CG-3'
   ||        ||        ||
3'-GT-5'  3'-GG-5'  3'-GC-5'

5'-GA-3'  5'-GC-3'
   ||        ||
3'-CT-5'  3'-CG-5'

5'-TA-3'
   ||
3'-AT-5'
```

(a)

```
5'-AA-3'  5'-AC-3'  5'-AG-3'  5'-AT-3'
   ||        ||        ||        ||
3'-UU-5'  3'-UG-3'  3'-UC-5'  3'-UA-5'

5'-CA-3'  5'-CC-3'  5'-CG-3'  5'-CT-3'
   ||        ||        ||        ||
3'-GU-5'  3'-GG-5'  3'-GC-5'  3'-GU-5'

5'-GA-3'  5'-GC-3'  5'-GG-3'  5'-GT-3'
   ||        ||        ||        ||
3'-CU-5'  3'-CG-5'  3'-CC-5'  3'-CU-5'

5'-TA-3'  5'-TC-3'  5'-TG-3'  5'-TT-3'
   ||        ||        ||        ||
3'-AU-5'  3'-UG-5'  3'-UC-5'  3'-UU-5'
```

(b)

Figure 3.2: (a) DNA–DNA base-stacking interactions. There are 10 different interactions because the parameters are the same when reading either strand from 5′ to 3′. For example, the interaction with 5′–CT–3′ on the top strand binding with 3′–GA–5′ on the bottom strand is identical to the interaction 5′–AG–3′ on the top strand binding with 3′–TC–5′ on the bottom strand. Each base-stacking interaction has entropy and enthalpy parameters associated with it. **(b)** DNA–RNA base-stacking interactions. All 16 base-stacking interactions are possible. The top strand is the DNA (probe) and the bottom strand is the RNA (target). Each of these base-stacking interactions has enthalpy and entropy parameters associated with it.

on the base pair formed (A–T or G–C) but also on the base pair of the previous base on the 5′ strand. Added to the model are parameters for the initiation and termination of the helix, which depend on the base pairs at either end of the duplex.

EXAMPLE 3.5 BASE-STACKING FOR A 10-BASE DUPLEX

The following is a 10-base DNA–DNA duplex:

```
5'-TAACCACGAT-3'
   ||||||||||
3'-ATTGGTGCTA-5'
```

There are nine base-stacking interactions and two initiation terms. To calculate the ΔH and ΔS of this duplex, we add base-stacking parameters for TA, AA, AC, CC, CA, AC, CG, GA and AT interactions, and two parameters for A–T initiation.

Base-Stacking and Initiation Parameters

The number of possible base-stacking interactions depends on whether we are considering DNA–DNA, DNA–RNA or RNA–RNA duplexes. Since microarrays use DNA probes, we will not consider the case of RNA–RNA duplexes. Spotted microarrays are usually hybridised with cDNA targets, and Affymetrix GeneChips are usually hybridised with cRNA targets. Therefore we consider both DNA–DNA and DNA–RNA duplexes.

With DNA–DNA hybridisation, there are 10 possible base-stacking interactions (Figure 3.2a). The reason why there are only 10 interactions is that a DNA–DNA

TABLE 3.3

Parameters for all base-stacking interactions for DNA–DNA and DNA–RNA duplexes. The parameters are taken from the papers SantaLucia (1998) and Sugimoto et al. (1995) – both references are given in full at the end of the chapter. Note that the DNA parameters are symmetric, e.g., the parameters for the probe sequence AA binding to target sequence TT are the same as for probe TT binding to target AA.

These parameters have been calculated by looking at melting curves for oligonucleotide duplexes in solution. On a microarray, the probe is anchored to glass support. It is not clear to what extent these parameters truly reflect the chemistry of hybridisation on microarrays: it is probable that at least the entropy parameters will differ, and possibly even the enthalpy parameters. However, there have been no published studies to date of comprehensive measurement of these parameters on solid support. Therefore, people generally continue to use these parameters as they represent the best available data.

Probe Sequence (5′ 3′)	DNA Target Sequence (5′ 3′)	DNA ΔH (kcal mol^{-1})	DNA ΔS (cal mol^{-1} K^{-1})	RNA Target Sequence (5′ 3′)	RNA ΔH (kcal mol^{-1})	RNA ΔS (cal mol^{-1} K^{-1})
AA	TT	−7.9	−22.2	UU	−11.5	−36.4
AC	GT	−8.4	−22.4	GU	−7.8	−21.6
AG	CT	−7.8	−21.0	CU	−7.0	−19.7
AT	AT	−7.2	−20.4	AU	−8.3	−23.9
CA	TG	−8.5	−22.7	UG	−10.4	−28.4
CC	GG	−8.0	−19.9	GG	−12.8	−31.9
CG	CG	−10.6	−27.2	CG	−16.3	−47.1
CT	AG	−7.8	−21.0	AG	−9.1	−23.6
GA	TC	−8.2	−22.2	UC	−8.6	−22.9
GC	GC	−9.8	−24.4	GC	−8.0	−17.1
GG	CC	−8.0	−19.9	CC	−9.3	−23.2
GT	AC	−8.4	−22.4	AC	−5.9	−12.3
TA	TA	−7.2	−21.3	UA	−7.8	−23.2
TC	GA	−8.2	−22.2	GA	−5.5	−13.5
TG	CA	−8.5	−22.7	CA	−9.0	−26.1
TT	AA	−7.9	−22.2	AA	−7.8	−21.9
Initiation (G · C)		0.1	−2.8		1.9	−3.9
Initiation (A · T)		2.3	4.1		1.9	−3.9

duplex has a level of symmetry; we can think of the duplex forming 5′–3′ relative to either strand. With DNA–RNA hybridisation, all base-stacking interactions are possible (Figure 3.2b). The most up-to-date thermodynamic parameters for the different base-stacking interactions and initiation parameters are given in Table 3.3.

It is important to notice that the base-stacking model supersedes the use of base composition (e.g., specifying % GC content) in determining the melting temperature and stability of a nucleic acid duplex. The base-stacking model implicitly includes base composition but is more sophisticated because it also takes into account the order of the bases in the sequences.

EXAMPLE 3.6 THERMODYNAMIC CALCULATIONS FOR A 10-BASE DUPLEX

We use the base-stacking parameters from Table 3.3 to calculate the thermodynamic properties of the duplex in Example 3.5. The enthalpy and entropy are the sum of the appropriate base-stacking and intiation parameters:

$$\Delta H = -74.4 \text{ kcal mol}^{-1}$$

$$\Delta S = -200.7 \text{ cal mol}^{-1}\text{K}^{-1}$$

The free energy and melting temperature are calculated using Equations 3.1 and 3.2. For example, we shall calculate the ΔG at 37°C and the melting temperature with a target concentration of 1pM.

$$\Delta G = -74.4 + (273.2 + 37) \times 200.7 \times 10^{-3} \text{ kcal mol}^{-1} = -12.1 \text{ kcal mol}^{-1}$$

$$T_m = -74.4/(-200.7 \times 10^{-3} + 0.001987 \times \ln(10^{-12}))\text{ K} = 291.1\text{ K} = 17.9°\text{C}$$

Adjustments for Salt Concentrations

The parameters in Table 3.3 are based on studies performed in a solution of 1M NaCl. It is possible to make an approximate adjustment for other salt concentrations. The entropy, ΔS, is adjusted for the molar Na^+ concentration, using the following formula:

$$\Delta S\,(\text{Na}^+\text{corrected}) = \Delta S\,(\text{uncorrected}) + 0.368 \times N \times \ln\,[\text{Na}^+] \qquad \text{(Eq. 3.3)}$$

In this equation, N is the number of phosphate groups in the duplex divided by two. When each base has a phosphate group, it is simply the length of the duplex.

EXAMPLE 3.7 SALT CORRECTION FOR OLIGONUCLEOTIDE

Consider the same oligonucleotide as in Example 3.6, but this time with duplex formation at the more stringent conditions of 0.1M NaCl. Using Equation 3.3, we get

$$\Delta S\,(\text{Na}^+\text{corrected}) = -200.7 - 0.368 * 10 * 2.3 \text{ cal mol}^{-1}\text{K}^{-1} = -209.2 \text{ cal mol}^{-1}\text{K}^{-1}$$

We can now calculate the ΔG at 37°C and the T_m at a molar concentration of 10^{-12} using the corrected ΔS and obtain

$$\Delta\text{G} = -9.5 \text{ cal mol}^{-1}\text{K}^{-1}$$

$$T_m = 8.5°\text{C}$$

So we see that increasing the stringency of the salt conditions has both increased the ΔG at 37°C and reduced the melting temperature.

EXAMPLE 3.8 PROBE SELECTION OF ADH2 CONTINUED – MELTING TEMPERATURE

We continue the probe selection for ADH2 by computing the melting temperatures of the 30mer oligonucleotides. To compute the melting temperature, we need to supply a target concentration. This is a circular problem, because we do not know the target concentration until we have performed microarray experiments. However, in most realistic concentration ranges, the effect of the concentration term is much smaller than the sequence-specific effects. In this example, we have used a concentration of 1pM.

TABLE 3.4

Thermodynamic parameters for oligonucleotides so far selected. Enthalpy and entropy have been computed using the base-stacking parameters of Table 3.3. The melting temperature has been calculated using a 1pM target concentration.

We wish to equalise the melting temperatures on the array at 60°C, so we choose the five probes with T_m within ± 2° of 60°C. The remaining probes are filtered out.

3′	Oligo	ΔH (kcal mol^{-1})	ΔS (Kcal mol^{-1} K^{-1})	T_m (°C)	Use?
31	ATCTCATTTTGCTTTATTTTTTTTTTTTTG	−228	−634.6	56.2	
51	TTTTTTTTTGATCACACCAAGGCTCAGTGC	−235.7	−637.6	65.8	
181	AGGATTTTAGAAATATCTTTTTTTTTGGTC	−225.4	−624.9	57.0	
201	TTTTTTGGTCAATCATATATGACCCAAATC	−227.3	−624.8	59.9	✓
211	AATCATATATGACCCAAATCAATATGACCA	−223.6	−618.6	57.4	
221	GACCCAAATCAATATGACCAAAAAAAATGAA	−228.7	−632.2	58.3	✓
231	AATATGACCAAAAAAAATGAAAACATTTTTT	−225.1	−629.4	54.4	
401	TCAAAGAATCAAAGCAATATATTCTCACAG	−229	−630.7	59.5	✓
411	AAAGCAATATATTCTCACAGCCATGCACAT	−230.9	−631.2	62.0	✓
481	TCCTGTTTCAGCACCTATGAAGGCACTTCT	−232.3	−628.1	65.6	
501	AGGCACTTCTGTGGAATACTTGATTACAGA	−229.5	−626.9	62.2	
511	GTGGAATACTTGATTACAGACCCAGCCAAT	−231.1	−629.8	63.0	
521	TGATTACAGACCCAGCCAATACTTTCTACA	−228.7	−624.9	61.9	✓

In Table 3.4, we show the computation of ΔH, ΔS and T_m for the 13 probes that passed the BLAST filter of Example 3.4. As an example, we select probes on the array with melting temperatures close to 60°C. We therefore filter out all probes with T_m less than 58°C or T_m greater than 62°C. Five possible probes remain for the final step of the oligonucleotide design.

SECTION 3.5 PROBE SECONDARY STRUCTURE

In the final step of oligonucleotide probe design, we select probes that do not self-hybridise. One can think of this in two ways: either the oligonucleotide could form a stem-loop structure with itself (Figure 3.3a), or an oligonucleotide could dimerise with neighbouring, identical oligonucleotides on the surface of the array (Figure 3.3b).

Although from a chemical perspective these cases are quite different, from a computational perspective they are very similar: both cases involve the identification of palindromes[2] in potential DNA probes, and the exclusion of probes that contain palindromes for use on a microarray.

[2] Bioinformaticians refer to a DNA sequence as palindromic if it is identical to its reverse complement sequence, e.g., the sequence ACGT. This is a slight difference from the linguistic use, in which a word (or sentence) is palindromic if it is identical to its reverse, e.g., ABBA.

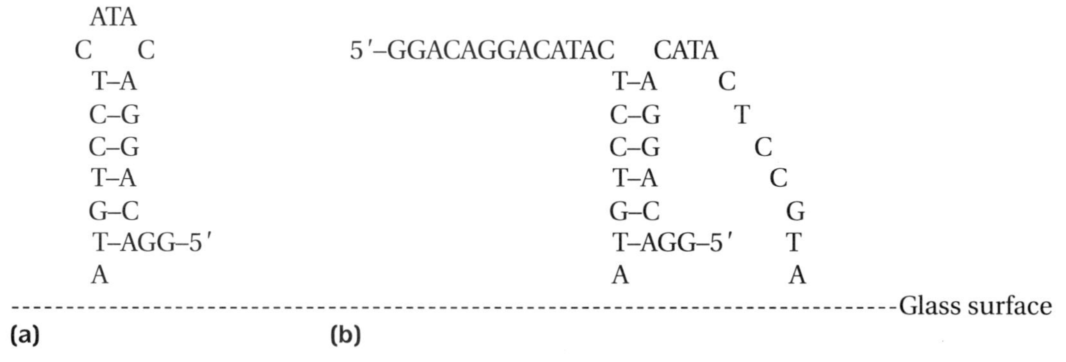

Figure 3.3: Self-hybridisation of oligonucleotide probes. A 20mer oligonucleotide probe with a self-complimentarity region. In **(a)**, the probe has formed an internal stem-loop structure that could prevent it from hybridising to target. In **(b)**, two neighbouring (and identical) probes have dimerised. In both cases, the same bases are involved in the duplex formation.

The thermodynamic stability of palindromes is calculated in exactly the same way as in Section 3.4, using the base-stacking model and any appropriate salt corrections. It is also possible to compute the stability of imperfect palindromes, using parameters for non–Watson–Crick base-pairings. We do not cover non–Watson–Crick thermodynamics in this book.[3]

Computation of Probe Secondary Structure

There are two approaches that can be employed to compute the secondary structure of potential oligonucleotide probes. The first is to write your own code to identify palindromes and compute the thermodynamics of the helices formed using the base-stacking model. This approach will provide a fast and effective search, but requires advanced programming skills. The interested reader is referred to the book by Gusfield cited at the end of the chapter, which describes algorithms to identify palindromes in sequences.

In this chapter, we shall describe the simpler but slower approach, which is to use a program called mfold that predicts secondary structure of DNA and RNA molecules from their sequence. It uses the base-stacking model to calculate thermodynamic properties of the duplexes, as well as additional thermodynamic parameters for mismatches, loops and bulges. Mfold is accessible via a web interface, and a command-line version can be obtained from Washington University, which is free for academic use.

[3] It is also possible to use thermodynamic calculations to compute the stability of the cross-hybridisation of probes to related targets. The use of base-stacking parameters for non–Watson–Crick base-pairings allows the calculation even when the homology is not exact. The implementation of such methods requires a high level of programming skill. The interested reader is referred to the book by Gusfield cited at the end of the chapter, which describes string-matching algorithms. These can be combined with thermodynamic calculations in order to compute the thermodynamic stability of imperfect cross-hybridising duplexes.

BOX 3.3 Mfold Output

Oligo 201 from Table 3.4 was submitted to the mfold DNA folding server which produced the results below. Mfold has computed ΔH and ΔS for the structure, as well as ΔG at the chosen temperature (37°C) and T_m at 1M concentration. The structure has a 5-base stem and 8-base loop that is marginally stable at 37°C (ΔG is negative).

```
Linear DNA folding at 37°C. [Na+] = 1.0 M, [Mg++] = 0.0 M.
Structure 1
Folding bases 1 to 30 of ADH2_1
dG = -3.3 dH = -43.8 dS = -130.6 Tm = 62.3
          10
TTTTTT|    ATC
      GGTCA   A
      CCAGT   T
CTAAAC^    ATA
.          20
```

Using Mfold over the Web

Mfold is available over the World Wide Web via a number of servers. The most useful interface for oligonucleotide design is the *quickfold* interface. This allows the user to fold many short sequences with a single query. The quickfold interface is available at the following URL:

http://bioinfo.math.rpi.edu/~mfold/rna/form3.cgi

Mfold takes a DNA or RNA sequence as input, and a number of parameters. It outputs the folds that minimise the free energy of the folded molecule at the supplied temperature. There are many calculations and drawings returned by mfold, but of particular importance are the free energy calculations and drawings of the folded molecules.

The inputs that you need to supply to mfold are as follows:

- A name for the group of sequences.
- The sequences themselves, separated by semicolons (";").
- Whether the nucleic acid is linear or circular. For DNA probes on a microarray, we always use linear DNA.
- The folding temperature: this may be the intended hybridisation temperature, or a slightly higher temperature.
- The ionic conditions: you can supply molar concentrations of both Na^+ and Mg^{++} ions.
- The upper bound on number of computed foldings: this should be set to 1 as we are only interested in the lowest energy fold.
- All other parameters are not relevant for microarray probe design and can be ignored.

TABLE 3.5: Results of Mfold on Oligos

Mfold was used to calculate the ΔG of the lowest energy self-complementary structure of each probe. In each probe, we have highlighted the bases that form an internal helix. Probes 201 and 401 form a 5-base helix, but the helix in probe 201 is more stable because of its base composition. Probe 201 might be less of a good probe than the others because of its stronger self-complementarity. This probe will also hybridise to neighbouring identical probes on the array via helix formation among the same bases. Probe 221 might also be ruled out because of its poly-A stretch. The remaining three probes could all be good 30mer oligos for this gene.

3′	Oligo	ΔG (kcal mol^{-1})
201	TTTTTTGGTCAATCATATATGACCCAAATC	−3.3
221	GACCCAAATCAATATGACCAAAAAAATGAA	0.3
401	TCAAAGAATCAAAGCAATATATTCTCACAG	−1.2
411	AAAGCAATATATTCTCACAGCCATGCACAT	−0.9
521	TGATTACAGACCCAGCCAATACTTTCTACA	0.9

EXAMPLE 3.9 MFOLD OF A 30MER OLIGONUCLEOTIDE

As an example, we submit the 30-base oligo for the gene ADH2, number 201 from Table 3.4, to the mfold DNA-folding server. The mfold output (Box 3.3) gives the thermodynamic parameters for the secondary structure: the enthalpy and entropy of the structure, the free energy at the chosen temperature, and the melting temperature of the structure at 1M concentration. The predicted secondary structure is shown below the thermodynamic parameters. In this example, there is a 5-base stem and 8-base loop.

EXAMPLE 3.10 FINAL PROBE SELECTION – FILTERING ON INTERNAL SECONDARY STRUCTURE

All five sequences selected from Example 3.7 are submitted to the mfold server to check for internal secondary structure. None of the probes has a particularly stable internal secondary structure (Table 3.5). Probe 201 has the most stable structure and we would not use that probe. Probe 221 might not be favoured because of its poly-A stretch. The remaining probes would all be good candidate oligonucleotide probes for use on a microarray.

KEY POINTS SUMMARY

- Oligonucleotide probes on a microarray should be
  - Sensitive
  - Specific
  - Isothermal
- We choose probes for a gene using several filters:
  - Use sequence from the 3′ end of the gene.

  - Mask repetitive sequence.
  - Remove cross-hybridising probes.
  - Remove probes with wrong melting temperatures.
  - Remove probes with internal secondary structure.
- High-throughput oligonucleotide probe design requires some programming skills.

SUPPLEMENTARY INFORMATION AND RESOURCES

Textbooks

Smith, E. B. 1999. *Basic Chemical Thermodynamics* (4th Edition). Oxford University Press.

An excellent introduction to thermodynamic principles such as enthalpy, entropy and free energy, described within the context of chemical reactions. The book presumes a basic knowledge of calculus.

Gusfield, D. 1997. *Algorithms on Strings, Trees and Sequences.* Cambridge University Press.

A beautiful book on sequence algorithms. Contains detailed algorithms with which you can compute free energies of imperfect homologies and palindromes. The book presumes a good understanding of computer algorithms, but is well worth the effort.

Papers on Nucleic Acid Thermodynamics

SantaLucia Jr., J. 1998. A unified view of polymer, dumbbell and oligonucleotide DNA nearest-neighbour parameters. *Proceedings of the National Academy of Sciences* 95: 1460–65.

A review of thermodynamic parameters for DNA–DNA duplexes. John SantaLucia also has many publications detailing thermodynamic parameters for non–Watson–Crick interactions.

Sugimoto, N. et al., 1995. Thermodynamic parameters to predict stability of RNA/DNA hybrid duplexes. *Biochemistry* 34: 11211–16.

A similar review of thermodynamic parameters for RNA–DNA duplexes.

Software Available for Download from the Internet

http://repeatmasker.genome.washington.edu/cgi-bin/RepeatMasker

There is a link from this page to an email address for the author of RepeatMasker, Arian Smit, from whom non-commercial licenses can be obtained. There is also a link to a company called GeoSpiza who distribute commercial licenses.

http://bioinfo.math.rpi.edu/~zukerm/rna/mfold-3.1.html

Information for obtaining mfold is available from this site. Again, this software is free for non-commercial use, and Washington University distributes commercial licenses for a fee.

ftp://ftp.ncbi.nih.gov/blast/

The BLAST ftp site. BLAST is available to all for free and has been compiled for many versions of Unix, including Linux, Solaris, SGI and Alpha.

CHAPTER FOUR

Image Processing

SECTION 4.1 INTRODUCTION

The image of the microarray generated by the scanner (Section 1.3) is the raw data of your experiment. Computer algorithms, known as feature extraction software, convert the image into the numerical information that quantifies gene expression; this is the first step of data analysis. The image processing involved in feature extraction has a major impact on the quality of your data and the interpretation you can place on it.

In Chapter 1 we discussed three technologies by which microarrays are manufactured: in-situ synthesis with the Affymetrix platform, inkjet in-situ synthesised arrays (Rosetta, Agilent and Oxford Gene Technology) and pin-spotted microarrays. This chapter focusses on pin-spotted arrays. Affymetrix has integrated its image-processing algorithms into the GeneChip experimental process and there are no decisions for the end-user to make. Inkjet arrays are of much higher quality than pin-spotted arrays and do not suffer from many of the image-processing difficulties of spotted arrays; also, Agilent provides image-processing software tailor-made for their platform, so there are no decisions for the end-user either.

Pin-spotted arrays, on the other hand, provide the user with a wide range of choices of how to process the image. These choices have an impact on the data, and so this chapter describes the fundamentals of these computational methods to give a better understanding as to how they impact the data.

SECTION 4.2 FEATURE EXTRACTION

The first step in the computational analysis of microarray data is to convert the digital TIFF images of hybridisation intensity generated by the scanner into numerical measures of the hybridisation intensity of each channel on each feature. This process is known as feature extraction. There are four steps:

1. Identify the positions of the features on the microarray.
2. For each feature, identify the pixels on the image that are part of the feature.
3. For each feature, identify nearby pixels that will be used for background calculation.
4. Calculate numerical information for the intensity of the feature, the intensity of the background and quality control information.

We discuss each of these steps in turn.

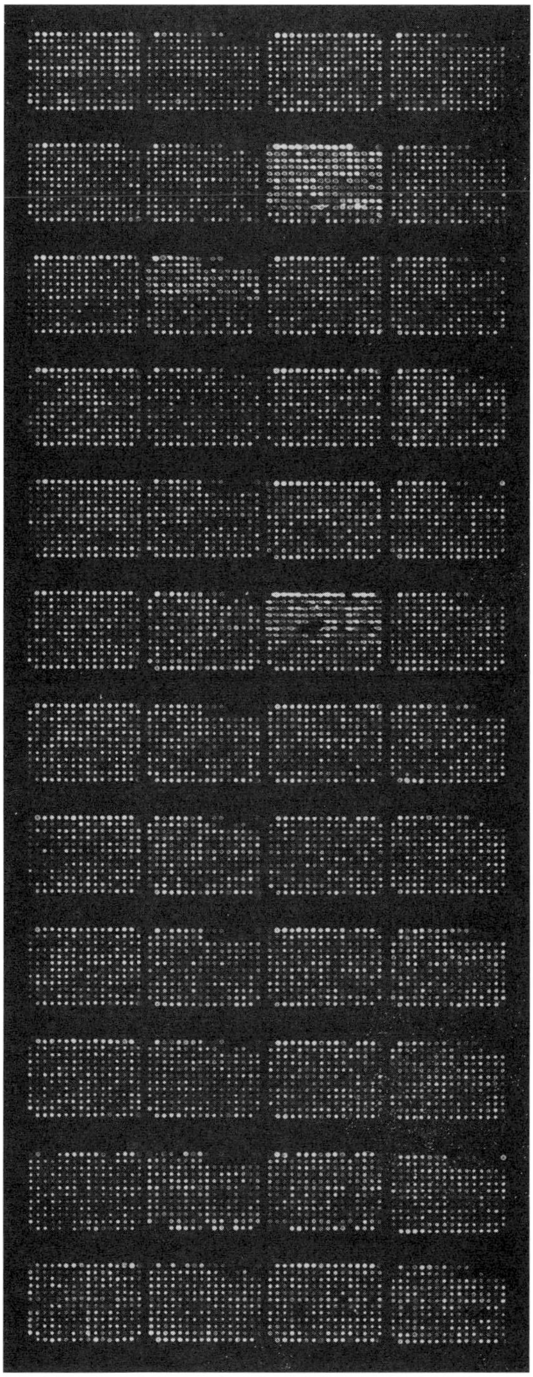

(a)

Figure 4.1: (a) An example image of a complete microarray. In this case, there are 48 grids in a 12×4 pattern, and each grid has 12×16 features. Therefore, there are a total of 9,216 features on this array. **(b)** The arrangement of grids and features according to the pins in the cassette. In this example, the spotting robot has 16 pins arranged in a 4×4 pattern. Each pin prints three grids, so that the pattern of spotting by the 16 pins is repeated three times on the array. Each grid consists of 192 features arranged in a 12×16 pattern. The total number of features on the array is 9,216. These could all be different genes (coming from 24 different 384-well plates), genes replicated in duplicate (12 different 384-well plates), or genes replicated in triplicate (8 different 384-well plates). (*Please see also the color section at the middle of the book.*)

(*continued*)

4 x 4 pattern corresponding to cassette is repeated 3 times on the array
12 x 16 pattern of features in each grid

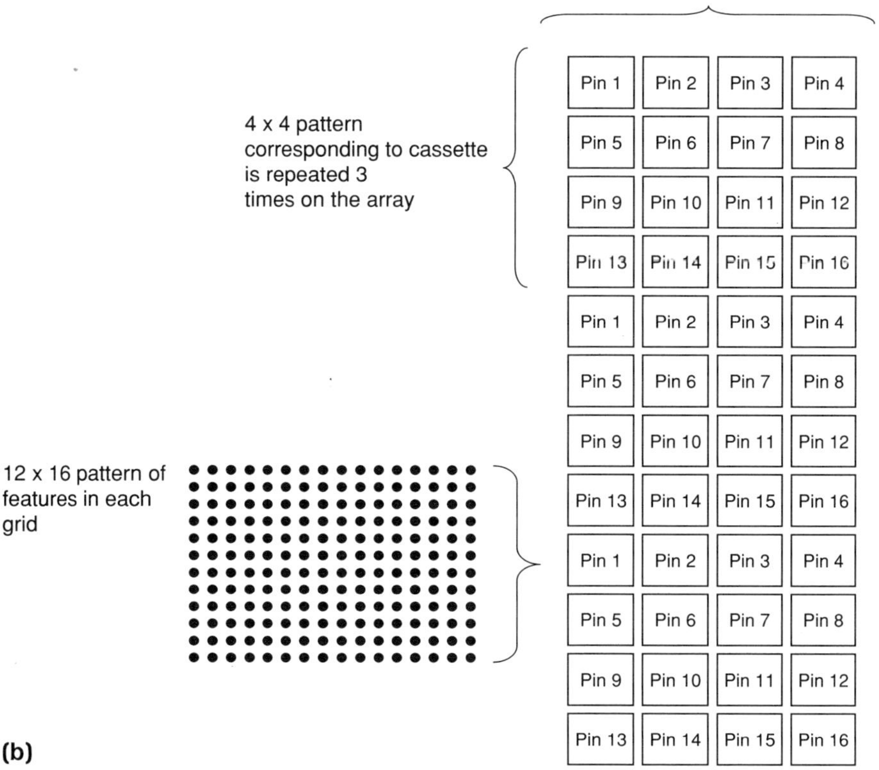

(b)

Figure 4.1: (*continued*)

Identifying the Positions of the Features

The features on most microarrays are arranged in a rectangular pattern. In general, however, the pattern is not completely regular: the features on the array are arranged in *grids*, with larger spaces between the grids than between the features within each grid (Figure 4.1). The grids come about because there are several pins on the cassette on the spotting robot (Figure 1.1); all the features in each grid have been printed by the same pin.

In order for the feature extraction software to work, it needs to be told how many grids make up the array, and the parameters associated with the grids:

- How many grids in each direction (x and y)
- How many features per grid in each direction (x and y)
- The spacing between the grids

All feature extraction software packages include the facility to provide this information.

EXAMPLE 4.1 GRIDS FROM BIOROBOTICS MICROGRID SPOTTING ROBOT

A BioRobotics Microgrid spotting robot is fitted with a pin-head holding 16 pins in a 4×4 arrangement. It is used to make an array with 9,216 features (Figure 4.1) arranged in 48 grids, each of which contains 192 features. Each pin spots three of the grids; these could be replicates of the same samples or might be different genes.

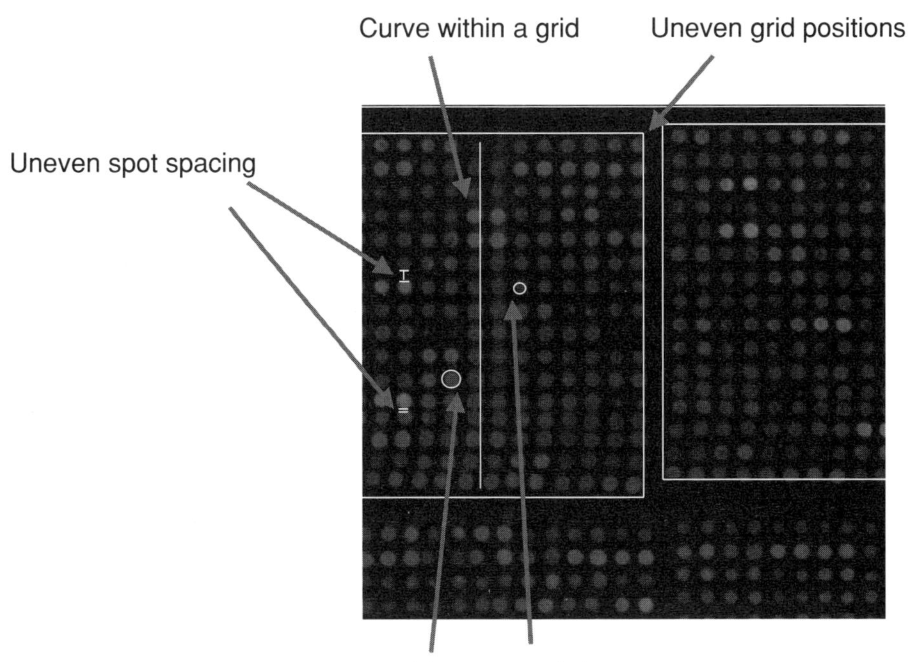

Figure 4.2: Problems with microarray images from pin-spotted arrays: (1) Uneven grid positions. The two grids are not aligned. This occurs because the pins are not perfectly aligned in the cassette. (2) Curve within the grid. Note that the centers of the features at the top of the vertical line lie on the line, but that the centers of the features at the bottom of the line are to the left of the line. This can happen if the array is not horizontal during array manufacture, or because of movement of the pins during manufacture. (3) Uneven spacing between features. This occurs because of pins moving during manufacture; this itself could result from the glass slide not being perfectly flat. (4) Uneven feature sizes. Different features can have different sizes as a result of different volumes of liquid being deposited on the array. This can also result from uneven drying of the features, so it is important to maintain constant temperature and humidity of the array during the manufacture process. (*Please see also the color section at the middle of the book.*)

A problem with identifying the positions of the features on the array is that the positions and sizes of the features within each grid may not be uniform (Figure 4.2). There are at least four difficulties that can arise:

- **Uneven grid positions.** The grids are not aligned with one another. This can happen if the pins are not perfectly aligned in the cassette.
- **Curve within a grid.** The glass slide is not completely horizontal or the pin has moved slightly in the cassette and so the features are printed in a curved pattern on the surface of the array.
- **Uneven feature spacing.** The pins have moved slightly in the cassette or the surface of the glass is not completely flat.
- **Uneven feature size.** More or less fluid has been deposited on the glass during the manufacture of the array.

All feature extraction software contains algorithms to automatically find the positions of the features. However, none of these algorithms are infallible. Current practise

TABLE 4.1: Segmentation Algorithms of Common Image-Processing Software Packages

Segmentation Method	Software Implementing Method
Fixed Circle	ScanAnalyze GenePix QuantArray
Variable Circle	QuantArray GenePix Dapple Agilent Feature Extraction
Histogram	ImaGene QuantArray
Adaptive Shape	Spot

requires manual supervision of the feature extraction process to ensure that all features are found by the software, and usually some level of manual intervention to align the software so that all features are found. The majority of feature extraction software packages have this facility.

Identifying the Pixels That Comprise the Features

The next step in the feature extraction procedure is called **segmentation**; this is the process by which the software determines which pixels in the area of a feature are part of the feature, and so their intensity will count towards a quantitative measurement of intensity at that feature. There are four commonly used methods for segmentation:

- Fixed circle
- Variable circle
- Histogram
- Adaptive shape

Different software packages implement different segmentation algorithms (Table 4.1) and some packages implement more than one algorithm, which gives the user the option to compare different algorithms on the same image.

Fixed Circle Segmentation

Fixed circle segmentation places a circle of fixed size over the region of the feature and uses all the pixels in the circle as those that form part of the feature (Figure 4.3a). The problem with fixed circle segmentation is that it gives inaccurate results if the features are of different size – which is usually the case on most microarrays. Therefore, fixed circle segmentation should be avoided if possible.

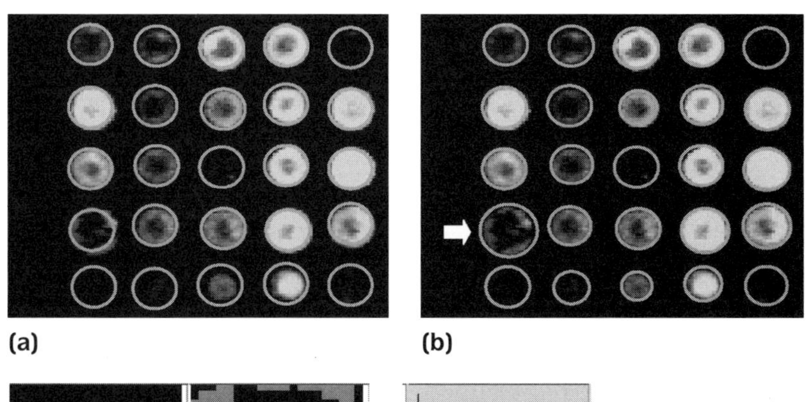

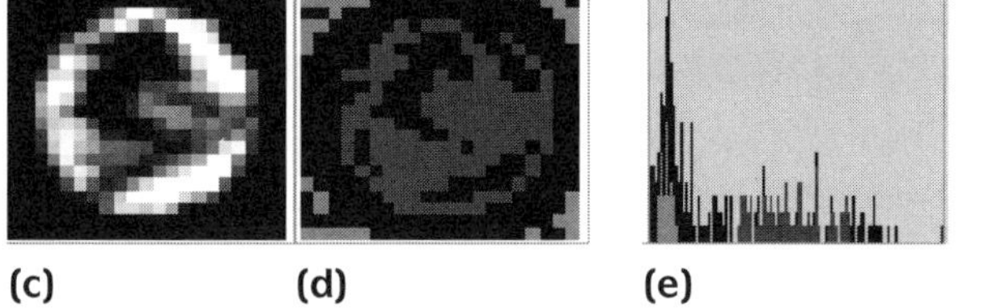

Figure 4.3: (a) Fixed circle segmentation. A circle of the same size is placed on every feature on the array and the pixels inside the circle are used to determine the intensity of the feature. This is not a good method because the circle will be too large for some features and too small for others. **(b)** Variable circle segmentation. A circle of different size is applied to each feature and the pixels inside the circle are used to determine the intensity of the feature. This performs better on different size features but does not perform so well on features with irregular shapes, for example, the irregular red feature that is marked with an arrow. **(c)** Zoom in on the red channel of the irregularly shaped feature marked with the arrow in (b). Note the black region where there is no hybridisation, probably because there is no probe attached to the glass in that area. **(d)** Histogram method applied to that feature. The red pixels are the ones that have been used to calculate the feature signal; the green pixels have been used to calculate the feature background. The black pixels are unused. The area corresponding to the black region in (c) is not used for calculating the feature intensity. The brightest features have also been excluded. The red-to-green ratio of this feature calculated by fixed circle segmentation is 1.8, variable circle segmentation is 1.9, and histogram segmentation is 2.6; so the measured differential gene expression between the samples is different with the different algorithms. Because of the irregular shape of the feature, the histogram method probably gives the most realistic measurement. **(e)** Histogram of the intensities of the pixels in the irregularly shaped feature. The red bars represent pixels used for the signal intensity; the green bars represent pixels used for the background intensity; the black bars are unused pixels. The brightest and darkest pixels are not used, thus giving a better measurement of hybridisation intensity. (*Please see also the color section at the middle of the book.*)

Variable Circle Segmentation

Variable circle segmentation fits a circle of variable size onto the region containing the feature (Figure 4.3b). This method is able to resolve features of different sizes, but performs less well on irregularly shaped features.

Histogram Segmentation

Histogram segmentation fits a circle over the region of the feature and background and then looks at a histogram of the intensities of the pixels in the feature (Figure 4.3e). The brightest and dimmest pixels are not used in the quantification of feature intensity. Histogram segmentation produces reliable results for irregularly shaped features. Histogram methods can be unstable for small features if the circular mask is too large.

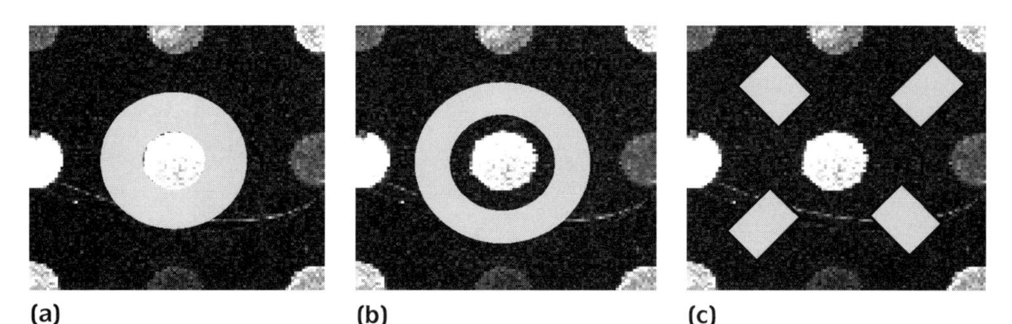

Figure 4.4: Background regions used by different software. Different software packages use different pixel regions surrounding the feature to determine the background intensity. **(a)** ScanAlyze: the region is adjacent to the feature. This will be inaccurate if the feature is larger than the fixed size of the circle used for segmentation. **(b)** ImaGene: there is a space between the feature and the background. This is a better method than (a). **(c)** Spot and GenePix: the background region is in between the features. This is also a good method.

Adaptive Shape Segmentation

This is a more sophisticated algorithm that can also resolve features with irregular shapes. The algorithm requires a smaller number of *seed pixels* in the center of each feature to start. It then extends the regions of each feature by adjoining pixels that are similar in intensity to their neighbours.

EXAMPLE 4.2 FEATURE EXTRACTION FOR AN IRREGULARLY SHAPED FEATURE

The program *QuantArray* can use fixed circle, variable circle and histogram segmentation to determine the intensity of a feature. It was used for the irregularly shaped feature shown marked with an arrow in Figure 4.3b. The red-to-green intensity ratio for the three methods is as follows:

- Fixed circle: 1.8
- Variable circle: 1.9
- Histogram: 2.6

In this case, the histogram segmentation is likely to be the most reliable result because the area of the spot that did not hybridise is not included. In this region, it is likely that there is no probe attached to the array.

Identifying the Background Pixels

The signal intensity of a feature includes contributions from non-specific hybridisation and other fluorescence from the glass. It is usual to estimate this fluorescence by calculating the **background signal** from pixels that are near each feature but are not part of any feature. Different software packages use different regions near each feature as the background pixels (Figure 4.4). The background intensity is subtracted from the feature intensity to provide a more reliable estimate of hybridisation intensity to each feature. Background subtraction is discussed in greater detail in Section 5.2.

Calculation of Numerical Information

Having determined the pixels representing each feature, the image-processing software must calculate the intensity for each feature. Image-processing software will typically provide a number of measures:

- Signal mean: the mean of the pixels comprising the feature
- Background mean: the mean of the pixels comprising the background around the feature
- Signal median: the median of the pixels comprising the feature
- Background median: the median of the pixels comprising the background
- Signal standard deviation: the standard deviation of the pixels comprising the feature
- Background standard deviation: the standard deviation of the pixels comprising the background
- Diameter: the number of pixels across the width of the feature
- Number of pixels: the number of pixels comprising the feature
- Flag: a variable that is 0 if the feature is good, and will take different values if the feature is not good

Table 4.2 shows some example output from *ImaGene*, which uses some of these fields. There are a number of ways in which this information is used.

Most important is the measure of hybridisation intensity for each feature. Here the user has a choice between the mean and the median of the pixel intensities. In general, it is preferable to use the median over the mean. The reason for this is that the median is more robust to outlier pixels than the mean: a small number of very bright pixels (arising from noise) have the potential to skew the mean, but will leave the median unchanged.

EXAMPLE 4.3 FEATURE WITH BRIGHT DUST

The feature shown in Figure 4.5 has a bright piece of dust on it. The red-to-green ratio using the mean pixel intensity is 1.9. The red-to-green ratio using the median pixel intensity is 1.3. The mean has been skewed by the bright pixels in the dust and so is very different from the median. In this case, the median is a more robust measure. This can be verified by removing the dust spot from the image and recalculating the intensities: the mean is 1.4 and the median is 1.3.

The second important numerical information is the signal standard deviation. This is used as a quality control for the array in two different ways (Figure 4.6):

- As measures of quality control of the features. If the standard deviation of a feature is greater than say 50% of the median intensity, the feature could be rejected as substandard.
- To determine whether an array is saturating. The problem with saturated features is that we do not know the true intensity of the feature, and so it is not possible to use such features as part of a quantitative analysis of gene expression.

TABLE 4.2: Example Output from ImaGene

Meta Row	Meta Column	Row	Column	Gene ID	Flag	Signal Mean	Background Mean	Signal Median	Background Median	Signal Stdev	Background Stdev	Diameter
1	1	1	1	H3126A06-3	0	29453.02	99.27228	31434	94.5	5165.786	31.88765	14
1	1	1	2	H3126A06-3	0	35591.85	134.6042	35718.5	124	3781.532	54.10425	14
1	1	1	3	H3126C06-3	0	455.3077	109.5556	436	109	112.4157	20.80846	14
1	1	1	4	H3126C06-3	0	780.3725	95.34177	786	95	136.1061	14.2317	14

Note: The first four features from an array that has been feature extracted using ImaGene. The first five columns give information about the feature itself: which grid, the position in the grid, and a Gene ID. The next column is the flag, which is zero if the feature is good, and not zero if there is a problem with the feature. The next two columns are the means of the pixels comprising the feature and the background, respectively, followed by two columns for the median of these pixels, and two columns for the standard deviation of these pixels. The final column is the number of pixels across the feature as a measure of feature size. From the diameter, one can approximate the total number of pixels in the feature – in this case about 150.

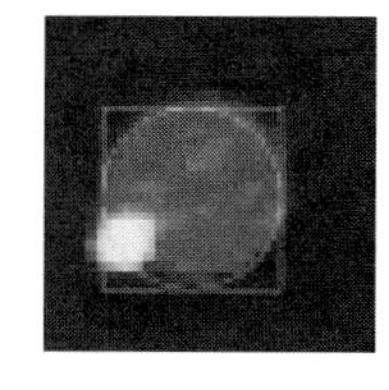

Figure 4.5: Feature with bright dust. An otherwise good feature has a bright piece of dust on it. In this case, the mean of the pixel intensities will be an unreliable measure of the intensity of the feature, because it will be skewed by the bright pixels, whereas the median will be robust to this noise.

The third piece of important information to come from the feature extraction software is the **flag** field. This is zero if the feature is good, but will be non-zero if the feature has problems. Different image-processing software use different flag values for different problems, but the typical problems are

- Bad feature. The pixel standard deviation is very high relative to the pixel mean.
- Negative feature. The signal of the feature is less than the signal of the background.

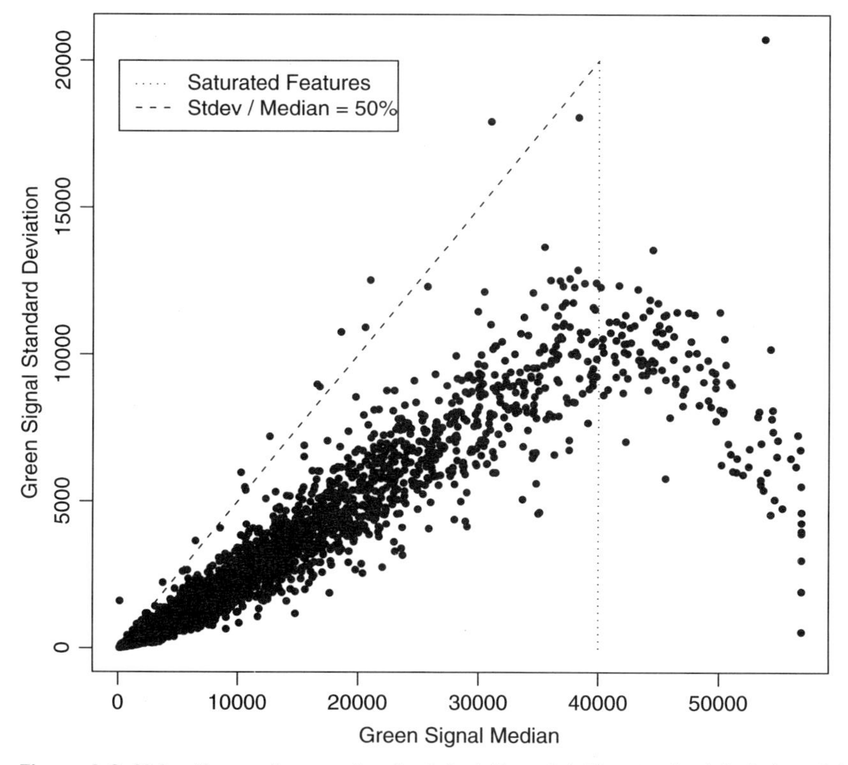

Figure 4.6: Using the median vs. standard deviation plot. The standard deviation of the pixel intensities for the Cy3 (green) signal of the features on an array is plotted against the median of the pixel intensities as a method for quality control. For features with signal intensity less than about 40,000, the standard deviation is approximately proportional to the mean, with a coefficient of variability (Chapter 6) of about 23%. However, for features with higher signal intensity, the standard deviation decreases. This is because these features have saturated pixels which all have the same intensity (with this particular scanner, the maximum pixel intensity is 56,818). The very brightest features have all pixels saturated so the standard deviation is zero. Saturated features cannot be used as part of a quantitative analysis of differential gene expression. The plot also shows a number of outlier features with very high pixel standard deviation. These features may also be unreliable and could be excluded from further analysis. The triangular region indicated represents reliable features which are not saturated and which have coefficient of variability of less than 50%. Some of the features towards the top of the triangle might also be considered outliers and excluded from analysis. (These data have been obtained privately from Ed Southern.)

- Dark feature. The signal of the feature is very low.
- Manually flagged feature. The user has flagged the feature using the image-processing software.

It is also usual to remove flagged features from further analysis (Section 5.2).

KEY POINTS SUMMARY

- The image of your array is your raw data.
- Feature extraction software calculates numerical measurements of gene expression from the image.
- The choice of feature extraction algorithms will have an impact on the data you generate.

RESOURCES AND FURTHER READING

Feature Extraction Software

- ScanAlyze:
  http://rana.lbl.gov/EisenSoftware.htm
- GenePix:
  http://www.axon.com/GN_GenePixSoftware.html
- Spot:
  http://experimental.act.cmis.csiro.au/Feature/index.php
- ImaGene:
  http://www.biodiscovery.com/imagene.asp
- QuantArray:
  http://lifesciences.dev.perkinelmer.com/areas/microarray/quantarray.asp
- Dapple:
  http://www.cs.wustl.edu/~jbuhler/research/dapple/
- Agilent:
  http://www.chem.agilent.com/Scripts/PDS.asp?lPage=2547

Papers

Adams, R. and Bischof, L. 1994. Seeded region growing. *IEEE Transactions on Pattern Analysis and Machine Intelligence* 16: 641–47.

A description of the adaptive shape algorithm used by *Spot*.

Books

Kamberova, G. and Shah, S. 2002. *DNA Array Image Analysis: Nuts & Bolts.* DNA Press.

A book that concentrates specifically on DNA microarray image analysis.

CHAPTER FIVE

Normalisation

SECTION 5.1 INTRODUCTION

Normalisation is a general term for a collection of methods that are directed at resolving the systematic errors and bias introduced by the microarray experimental platform. Normalisation methods stand in contrast with the data analysis methods described in Chapters 7, 8 and 9 that are used to answer the scientific questions for which the microarray experiment has been performed. The aim of this chapter is to give an understanding of why we need to normalise microarray data, and the methods for normalisation that are most commonly used. The chapter is arranged into three further sections:

Section 5.2: Data Cleaning and Transformation, looks at the first steps in cleaning and transforming the data generated by the feature extraction software before any further analysis can take place.

Section 5.3: Within-Array Normalisation, describes methods that allow for the comparison of the Cy3 and Cy5 channels of a two-colour microarray. This section is only relevant for two-colour arrays.

Section 5.4: Between-Array Normalisation, describes methods that allow for the comparison of measurements on different arrays. This section is applicable both to two-colour and single channel arrays, including Affymetrix arrays.

SECTION 5.2 DATA CLEANING AND TRANSFORMATION

The microarray data generated by the feature extraction software is typically in the form of one or more text files (Table 4.2). Before you use the data to answer scientific questions, there are a number of steps that are commonly taken to ensure that the data is of high quality and suitable for analysis. This section describes three stages of data cleaning and transformation:

- Removing flagged features
- Background subtraction
- Taking logarithms

Removing Flagged Features

In Chapter 4 we described four types of flagged features: bad features, negative features, dark features and manually flagged features. These are features for which the image-processing software has detected some type of problem. There are two approaches to dealing with flagged features.

The first, and most common approach, is to remove flagged features from the data set. This is a straightforward process but has the disadvantage of sometimes removing potentially valuable data.

The second method is to refer back to the original image of every flagged feature and to identify the problem that has resulted in flagging. If appropriate, the user can perform a new feature extraction on the flagged feature to obtain a more reliable measure of signal intensity. This procedure has the disadvantage of requiring time and resources, and may not always be practical.

Background Subtraction

The second step in microarray data cleaning is to subtract the background signal from the feature intensity. The background signal is thought to represent the contribution of non-specific hybridisation of labelled target to the glass, as well as the natural fluorescence of the glass slide itself. This procedure works well when the feature intensity is higher than the background intensity. However, when the background intensity is higher than the feature intensity, the result would be a negative number, which would not be meaningful. There are three approaches that are used to deal with this situation:

- Remove these features from the analysis. Since the feature intensity should be higher than the background intensity, the unusually high background is taken to represent a local problem with the array and so the intensity of the feature is regarded as unreliable. This is the most common approach.
- Use the lowest available signal-intensity measurement as the background-subtracted intensity – this will typically be the value 1.[1] The idea behind this is that if the background intensity is higher than the feature intensity, it represents a gene with no or very low expression, and so the lowest value available is used.
- Use more sophisticated (Bayesian) algorithms to estimate the true feature intensity, based on the assumption that the true feature intensity is higher than the background intensity, and so the high background represents some type of experimental error.[2]

Affymetrix Data

Data from Affymetrix GeneChips can suffer from a very similar problem, requiring similar approaches to be resolved. Gene expression is determined by comparing the signal intensity from hybridisation to probes complementary to the gene being measured

[1] The readings from the scanner are 16-bit digital signals and so are integers; 0 cannot be used because it is common to take the log of the signal at the next step.

[2] A reference detailing such a method is given at the end of the chapter.

with the signal intensity from hybridisation to probes that contain mismatches; the signal from the mismatch probes are thought to represent cross-hybridisation. In earlier versions of the Affymetrix software (Versions 4 and below), the gene expression is calculated as a combination of the differences between the true probes and corresponding mismatch probes. When the mismatch probes have higher intensity than the complimentary probes, the software generates negative numbers, which are not particularly meaningful. There are four possible approaches to handling genes with negative intensities:

- Discard these genes from the analysis. The reasoning is that if the mismatch probes have higher intensity than the complimentary probes, then the signal is mostly cross-hybridisation and is unreliable.
- Replace the negative numbers with the smallest possible positive number, usually 1. The reasoning behind this is that the genes for which the mismatch signal is less than the perfect probe signal are either not expressed, or expressed at very low level, so we replace the signal with the lowest possible value.
- Use more sophisticated algorithms to estimate the true feature intensity, based on the assumption that the true probe intensity is higher than the mismatch intensity; hence, the effect seen is artifactual and represents some type of experimental error.[3]
- Affymetrix changed their algorithm in version 5 of their software so that it is no longer possible to get negative numbers. Thus this problem is only present in historic Affymetrix data that has not been reanalysed with their latest software.

Taking Logarithms

It is common practise to transform DNA microarray data from the raw intensities into log intensities before proceeding with analysis.[4] There are several objectives of this transformation:

- There should be a reasonably even spread of features across the intensity range.
- The variability should be constant at all intensity levels.
- The distribution of experimental errors should be approximately normal.
- The distribution of intensities should be approximately bell-shaped.

It is usual in microarray data analysis to use logarithms to base 2. The reason is that the ratio of the raw Cy5 and Cy3 intensities is transformed into the difference between the logs of the intensities of the Cy5 and Cy3 channels. Therefore, 2-fold up-regulated genes correspond to a log ratio of $+1$, and 2-fold down-regulated genes correspond to

[3] At the time of writing this book, I am not aware of any published methods for performing such analysis with Affymetrix data. However, the Bayesian methodology applied to background subtraction could be modified and also applied to this problem.

[4] There are a number of alternative transformations that can be applied instead of logarithms, typically aimed at ensuring that the variability is constant at all intensity levels. A reference to one of these methods is given at the end of the chapter.

TABLE 5.1: Conversion from Log (to Base 2) to Raw Intensity and from Raw Intensity to Log (to Base 2) Intensity

Log (to Base 2) Intensity	Raw Intensity	Raw Intensity	Log (to Base 2) Intensity
0	1	1	0
1	2	2	1
2	4	5	2.32
3	8	10	3.32
4	16	20	4.32
5	32	50	5.64
6	64	100	6.64
7	128	200	7.64
8	256	500	8.97
9	512	1,000	9.97
10	1,024	2,000	10.97
11	2,048	5,000	12.29
12	4,096	10,000	13.29
13	8,192	20,000	14.29
14	16,384	50,000	15.61
15	32,768		

a log ratio of −1. Genes that are not differentially expressed have a log ratio of 0. These log ratios have a natural symmetry, which reflects the biology and is not present in the raw fold difference. For example, the fold differences of 2-fold up-regulated genes, genes that are not differentially expressed, and 2-fold down-regulated genes are 2, 1 and 0.5, respectively. The logarithms to base 2 of a range of intensities are shown in Table 5.1, and the log ratios for a range of fold differences are shown in Table 5.2.

Logarithms to base 2 are closely related to natural logarithms favoured by mathematicians. Natural logarithms are implemented in Excel, R and on pocket calculators. You can convert a natural logarithm to a logarithm to base 2 via Equation 5.1:

$$\log\,(\text{to base }2)x = \frac{\log\,(\text{natural})x}{\log\,(\text{natural})\;2} \qquad \text{(Eq. 5.1)}$$

TABLE 5.2: Conversion from Fold Ratios to Log (to Base 2) Ratios

Fold Ratio	Log (to Base 2) Ratio Difference
4-fold down-regulated	−2
3-fold down-regulated	−1.58
2-fold down-regulated	−1
1.5-fold down-regulated	−0.58
No change	0
1.5-fold up-regulated	0.58
2-fold up-regulated	1
3-fold up-regulated	1.58
4-fold up-regulated	2

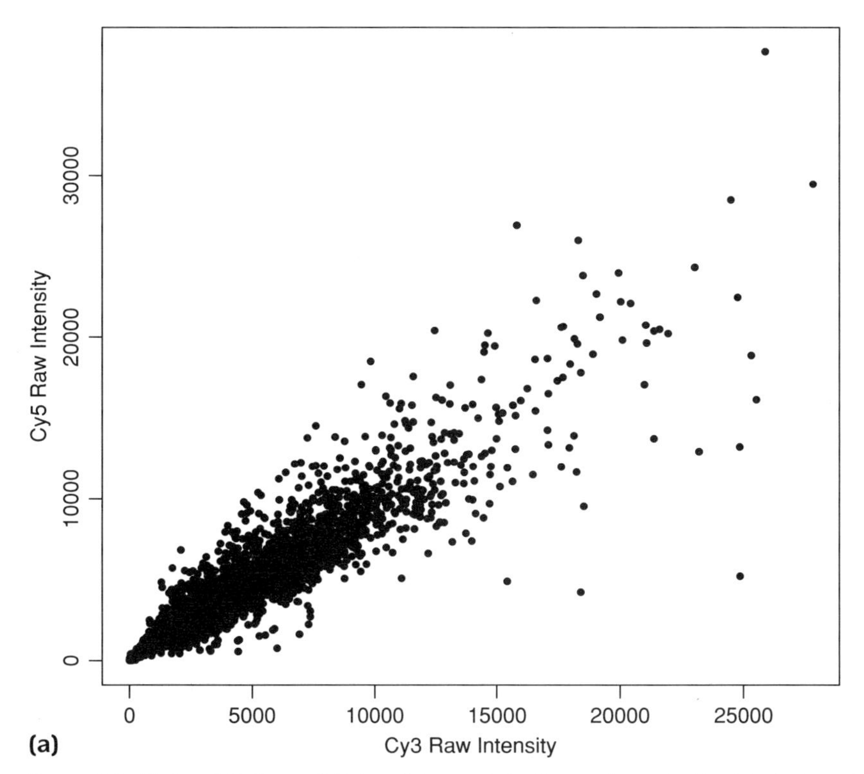

Figure 5.1: Plots of Cy3 vs. Cy5 for data set 5A. Human foreskin fibroblasts have been infected with *Toxoplasma gondii* for a period of 1 hour. A sample has been prepared, labelled with Cy5 (red), and hybridised to a microarray with approximately 23,000 features. The Cy3 (green) channel is a sample prepared from uninfected fibroblasts. Because the infectious period is short, most genes in this experiment are not differentially expressed. **(a)** Scatterplot of the (background-subtracted) raw intensities; each point on the graph represents a feature on the array, with the x coordinate representing the Cy3 intensity, and the y coordinate representing the Cy5 intensity. The graph shows two weaknesses of the raw data that would have a negative impact on further data analysis:

1. Most of the data is bunched in the bottom-left-hand corner, with very little data in the majority of the plot.
2. The variability of the data increases with intensity, so that it is very small when the intensity is small and very large when the intensity is large.

(b) Scatterplot of the log (to base 2) intensities. This plot is better than (a). The data is spread evenly across the intensity range, and the variability of the data is the same at most intensities. The genes with log intensity less than 5 have slightly higher variability, but these genes are very low expressed and are below the detection level of microarray technology.

The straight line is a linear regression through the data. The linear regression is not perfect (the data appears to bend upwards away from the line at high intensities), but is approximately right. The intercept is 1.4, and the gradient is 0.88. If the two channels were behaving identically, the intercept would be 0 and the gradient would be 1. We conclude that the two Cy dyes behave differently at different intensities; this could result from differential dye incorporation or different responses of the dyes to the lasers.

(*continued*)

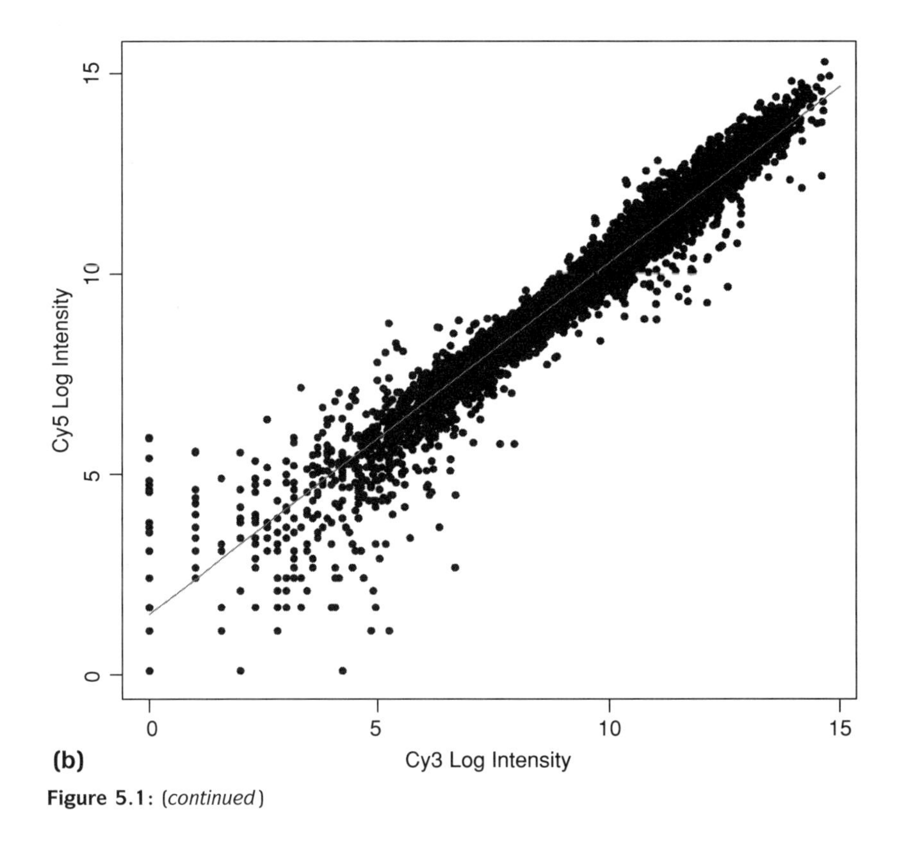

Figure 5.1: (*continued*)

EXAMPLE 5.1 TAKING THE LOG OF *TOXOPLASMA GONDII* DATA (DATA SET 5A)

Fibroblasts taken from human foreskin have been infected with *Toxoplasma gondii*. Samples from uninfected cells and cells treated with *T. gondii* for 1 hour are hybridised to two channels of a microarray with approximately 23,000 features.[5] The researchers want to identify genes that have been differentially expressed.

The raw data (Figure 5.1a) do not satisfy the requirements for effective analysis. Most of the features are in the bottom-left part of the graph; the variability increases with intensity, and the distribution of the intensities is not bell-shaped but very heavily right-skewed (Figures 5.2a and 5.2b).

The logged data (Figure 5.1b), however, do satisfy the requirements. The data are well spread across the range of log intensity values the variability is approximately constant at all intensities and would appear to be normally distributed (with the exception of very low expressed genes, whose intensities are likely to be unreliable); and the distribution of intensities (Figures 5.2c and 5.2d) are closer to being bell-shaped (although these distributions are also slightly right-skewed).

[5] The paper from which this data has been derived is given at the end of the chapter. The data is available from the Stanford Microarray Database.

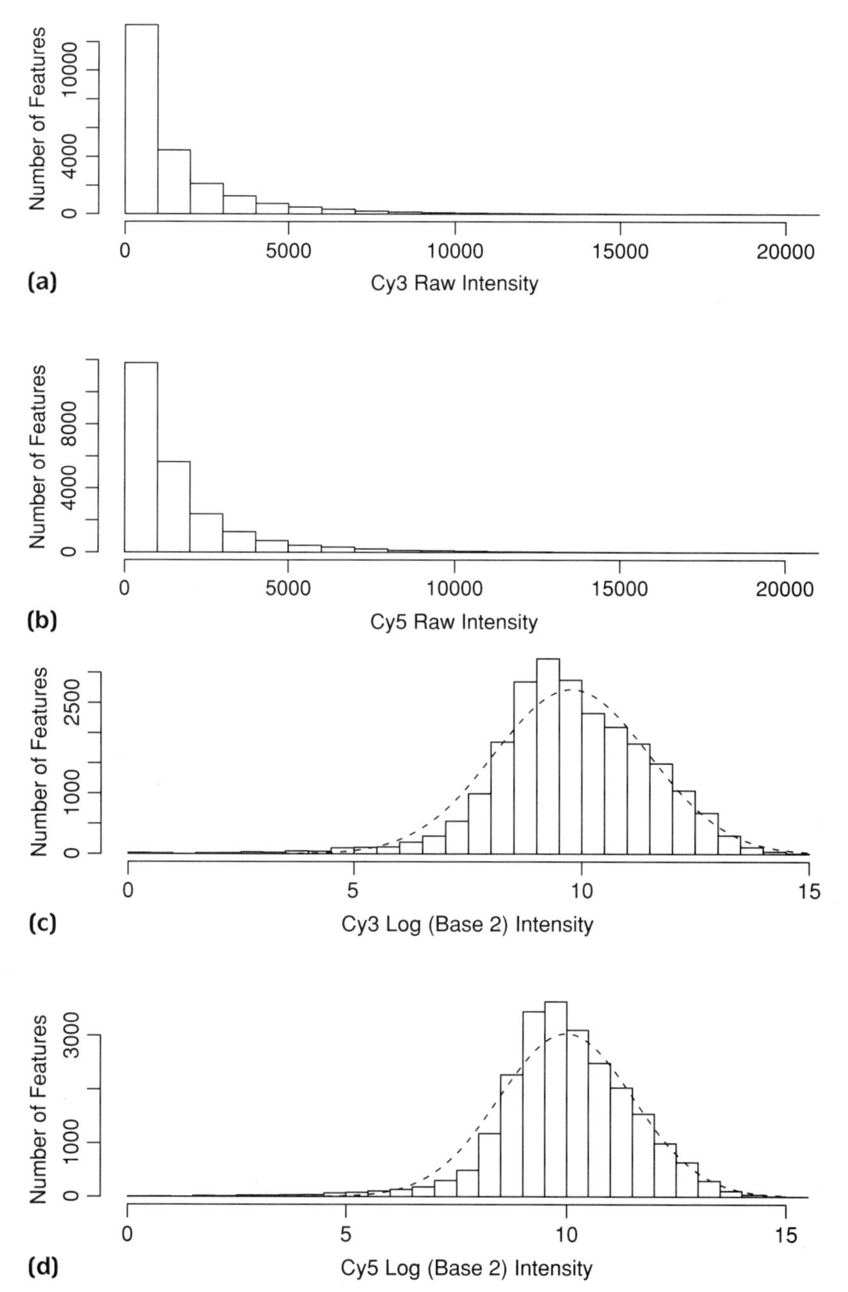

Figure 5.2: Histograms of the raw and log Cy3 and Cy5 intensities. Histograms of the intensities of the features for the human fibroblast data. **(a)** The raw intensities for the Cy3 channel; the data is right-skewed, with the majority of features having low intensity and decreasing numbers of features having higher intensity. **(b)** The raw intensities for the Cy5 channel; the pattern is the same as (a). **(c)** The log intensities for the Cy3 channel; the intensities are closer to a bell-shaped normal curve (shown as a dashed line). There is still a slight right skew, but the logged data is better for data analysis than the raw data. **(d)** The log intensities for the Cy5 channel, along with a normal curve (dashed line). As with (c), the intensities are approximately normal, with a slight right skew.

SECTION 5.3 WITHIN-ARRAY NORMALISATION

Data set 5A is an example of a very typical class of microarray experiment. The experimenters are using the microarray to compare two different samples and identify genes that are differentially expressed.

However, the two samples have been labelled with two different fluorescent dyes in two separate chemical reactions, and their intensity is measured with two different lasers operating at two different wavelengths. In addition, the features on the array are distributed on different parts of the surface of the array. When we measure differential expression between the two samples, we need to ensure that our measurements represent true differential gene expression, and not bias and error introduced by the experimental method. We need to be able to compare the Cy3 and Cy5 intensities on an equal footing – this is achieved by eliminating four sources of systematic bias:

- The Cy3 and Cy5 labels may be differentially incorporated into DNA of different abundance.
- The Cy3 and Cy5 dyes may have different emission responses to the excitation laser at different abundances.
- The Cy3 and Cy5 emissions may be differentially measured by the photomultiplier tube at different intensities.
- The Cy3 and Cy5 intensities measured on different areas on the array may be different because the array is not horizontal and so the focus is different in different parts of the array.

It is not possible to deconfound the first three sources of bias, and so they are combined together. In this section we describe three methods of correcting for different responses of the Cy3 and Cy5 channels:

- Linear regression of Cy5 against Cy3
- Linear regression of log ratio against average intensity
- Non-linear (Loess) regression of log ratio against average intensity

Spatial bias can be corrected separately, and we describe two methods to correct for it:

- Two-dimensional Loess regression
- Block-by-block Loess regression

All of the methods described in this section rely on a core assumption: **the majority of the genes on the microarray are not differentially expressed**. If this assumption is true, then these methods are meaningful. However, if this assumption is not true, then these methods may not be reliable, and a different experimental design and normalisation method, such as using a reference sample, would be more appropriate.

Linear Regression of Cy5 Against Cy3

The first and simplest method to check whether the Cy3 and Cy5 channels are behaving in a comparable manner is via a scatterplot of the two channels (Figure 5.1b). If the Cy3 and Cy5 channels are behaving similarly, then the cloud of points on the scatterplot should approximate a straight line, and the linear regression line through the data should have a gradient of 1 and an intercept of 0. Variations from these values represent different responses of the Cy3 and Cy5 channel:

- A non-zero intercept represents one of the channels being consistently brighter than the other.
- A slope not equal to 1 represents one channel responding more strongly at high intensities than the other.
- Deviations from a straight line represent non-linearities in the intensity responses of the two channels.

EXAMPLE 5.2 LINEAR REGRESSION APPLIED TO DATA SET 5A

A straight line is plotted through the scatterplot of log intensities of the human fibroblast data of data set 5A (Figure 5.1b). The intercept of the line is 1.41 and the gradient is 0.88. This implies that at low intensities, the Cy5 channel gives a stronger response, while at high intensities, the Cy3 channel gives a stronger response. The highest data points are curving away from the straight line, so the relationship between the Cy3 and Cy5 channels is not completely linear.

The straight-line fit can be used to normalise the data. The procedure is straightforward:

1. Plot a Cy3 vs. Cy5 scatterplot.
2. Fit a regression line through the scatterplot and identify the gradient and intercept.
3. Replace the Cy3 values with the fitted values on the regression line.

This method for normalisation works well for data where the linear fit is good and is a reasonable preliminary method for visualising the data. However, there are two disadvantages of this method:

- The human eye and brain are better at perceiving differences from horizontal and vertical lines than from diagonal lines. Therefore it is not always easy to see non-linearities in the data with this type of plot.
- The linear regression treats the Cy3 and Cy5 channels differently and would produce a different result if Cy3 were plotted against Cy5.

Linear Regression of Log Ratio Against Average Intensity

An alternative and very useful approach to visualising and normalising the data is to produce a scatterplot of the log ratio against the average intensity of each feature. Such plots are sometimes called **MA plots** in the microarray literature. In these plots, each

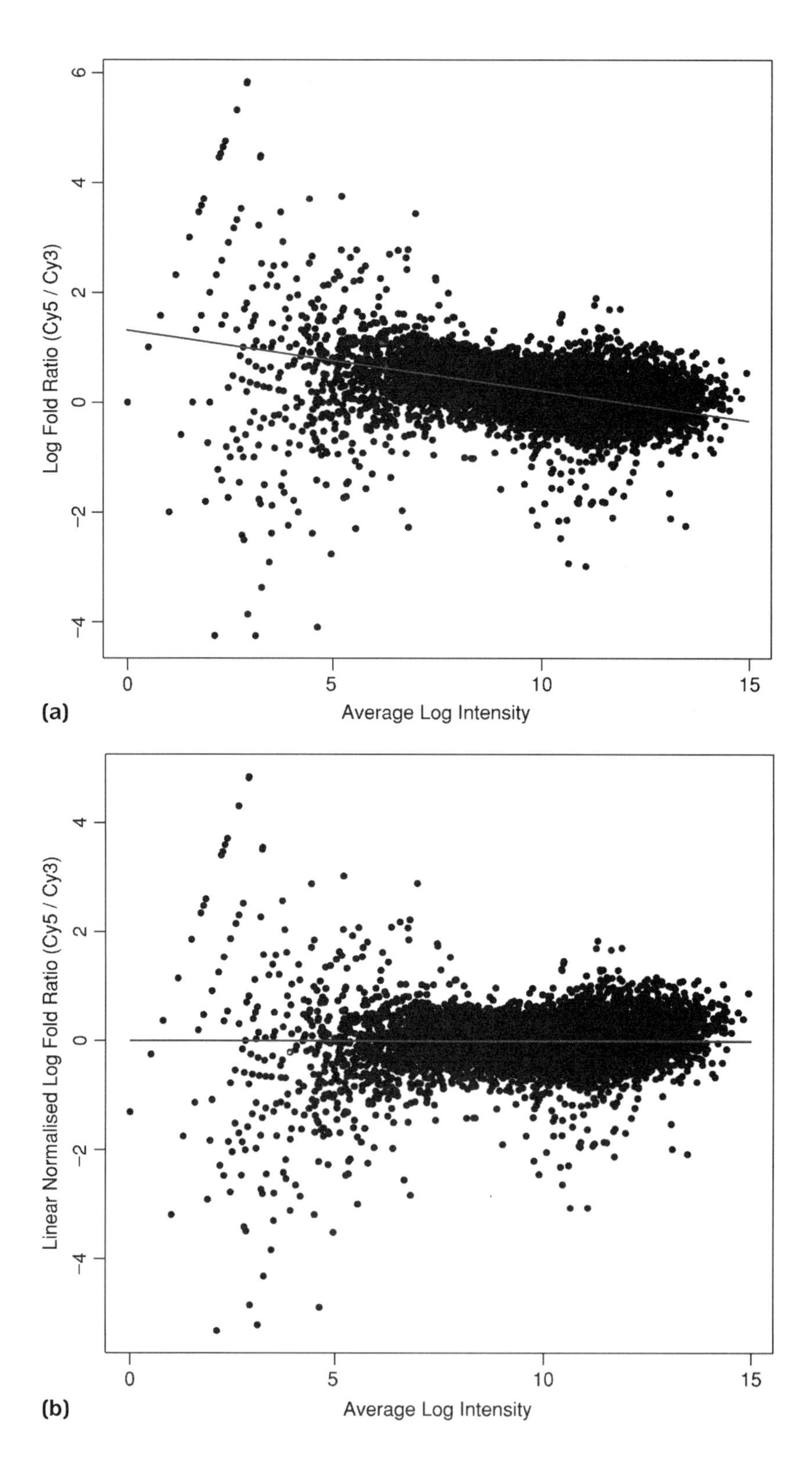

(a)
Log Fold Ratio (Cy5 / Cy3)
Average Log Intensity
(b)
Linear Normalised Log Fold Ratio (Cy5 / Cy3)
Average Log Intensity

point represents one feature, with the x coordinate taking the average value of the Cy3 and Cy5 log intensities, and the y coordinate the difference between the log intensities of the Cy3 and Cy5 channels (i.e., the log ratio). The MA plot derives its name because the average intensity is sometimes called A, and the log ratio is sometimes called M.

The MA plot is related to the scatterplot of the log intensities of the two channels; it can be obtained from the scatterplot by rotating it by 45° and then scaling the two axes appropriately. However, the MA plot is generally a more powerful tool for visualising and quantifying both linear and non-linear differential responses of the Cy3 and Cy5 channels.

First, it is usually clearer if the two channels are responding differently or in a non-linear fashion. If the two channels are behaving similarly, then the data should appear symmetrically about a horizontal line through zero; any deviations from this horizontal line represent different responses of the two channels. The human eye and brain are better at processing horizontal lines than diagonal lines, so it is easier to visualise differences between the channels by using MA plots than by using scatterplots.

Second, any linear or non-linear regression performed on the log ratio against average intensity treats the two channels equally. Thus such regressions are more robust and reproducible than performing regressions of one channel against the other.

EXAMPLE 5.3 PLOT OF LOG RATIO AGAINST AVERAGE INTENSITY FOR DATA SET 5A

The log ratio for each feature is plotted against the average log intensity for all of the features of data set 5A (Figure 5.3). The data are not symmetrical about a horizontal line through zero: the Cy5 channel responds more strongly at low intensities, and the Cy3 channel responds more strongly at high intensities. (This is the same conclusion that we came to through using direct regression.) Because we assume that most genes

Figure 5.3: Plots of log ratio as a function of average intensity and linear normalisation. Scatterplots of the log ratio of the features as a function of the average intensity for the data from data set 5A. Each point on the graph represents a different feature. The x coordinate is the average intensity of the Cy3 and Cy5 channels; the y coordinate is the log of the fold ratio of Cy5 divided by Cy3 (equal to the difference in the log intensities of Cy5 and Cy3). These plots show the average trend of the log ratio as a function of intensity. These plots are sometimes known as MA plots; they are geometrically related to the Cy5 vs. Cy3 scatterplots (Figure 5.1b) obtained by rotating the graph through 45° and then scaling the two axes. **(a)** A straight line has been fitted through the data points, which demonstrates a clear trend in the Cy5 and Cy3 responses. At low intensities, the Cy5 channel is responding more strongly, while at high intensities, the Cy3 channel is responding more strongly. We assume that most genes are not differentially expressed, so this line represents experimental artifact rather than differential expression. The log ratios can be linearly normalised by subtracting the fitted value on the straight line from each log ratio. Although the straight-line fit is very good, it is not an exact fit to the centre of the data. At high intensities, the data appears to flatten out, suggesting that a non-linear fit might give more reliable results. **(b)** The data has been normalised to the regression line in (a) by subtracting the fitted value on the line from the log ratio of each feature. The regression line is transformed to a horizontal line through zero. The points with the highest intensities lie above the line.

are not differentially expressed, we attribute this effect to experimental artifact and normalise the data to remove this effect before testing for differentially expressed genes.

Linear normalisation using the log ratio and average intensity works in a method similar to linear regression on the Cy5 and Cy3 data. There are four steps:

1. Construct the average log intensity and log ratio for each feature.
2. Produce the MA plot.
3. Perform a linear regression of the log ratio on the average log intensity.
4. For each feature, calculate the normalised log ratio by subtracting the fitted value on the regression from the raw log ratio.

EXAMPLE 5.4 LINEAR NORMALISATION OF FIBROBLAST DATA

Linear regression is applied to the MA plot of the fibroblast data of data set 5A (Figure 5.3a). The straight-line fit has intercept 1.31 and gradient −0.11; at low intensities, the Cy5 channel is brighter than the Cy3 channel, while at high intensities, the Cy3 channel is brighter than the Cy5 channel. The fitted values on the straight line are subtracted from the log ratios to produce normalised log ratios that are used for identifying differentially expressed genes.

Nonlinear Regression of Log Ratio Against Average Intensity

If you look carefully at Figure 5.3a, you will see that the straight line does not perfectly fit the cloud of data: at the highest intensities, the line appears to be too low. It is common with microarray data that the relationship between the Cy3 and Cy5 channels is non-linear; when that is the case, linear regression may not produce the best answers, and some form of non-linear regression may be more suitable.

The most commonly used method for non-linear regression with microarray data is called **Loess regression** (and sometimes called lowess regression). Loess stands for *locally weighted polynomial regression*. It works by performing a large number of local regressions in overlapping windows along the length of the data (Figure 5.4a) and then joining the regressions together to form a smooth curve (Figure 5.4b).

Loess regression is a relatively advanced statistical technique and is generally available in advanced statistics packages such as R or Matlab. However, it is straightforward to use with a basic working knowledge of R, and there are also two packages written for R specifically for analysing microarray data that use Loess to perform normalisation.[6] Loess regression has also been implemented in many commercially available microarray data analysis packages.

Loess normalisation is performed in four simple steps:

1. Construct the average log intensity and log ratio for each feature.
2. Produce the MA plot.

[6] URLs for these packages are given at the end of the chapter.

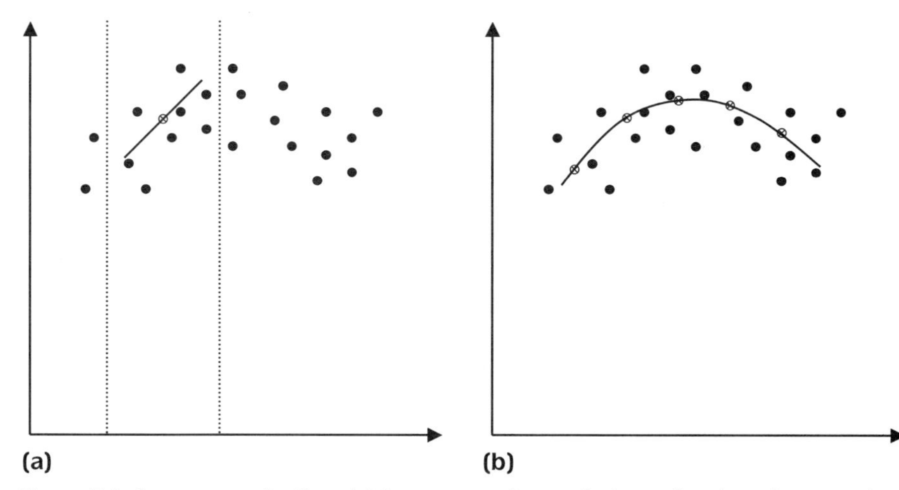

Figure 5.4: Loess normalisation. (a) Loess regression works by performing a large number of local regressions in overlapping windows across the whole range of the data set. The regression curve is usually either a straight line or a quadratic curve. (The default R implementation is a quadratic curve.) Each regression results in a central point and regression line or curve about that point. **(b)** The points and curves from the local regressions are combined to form a smooth curve across the length of the data set.

3. Apply the Loess regression to your data.
4. For each feature, calculate the normalised log ratio by subtracting the fitted value on the Loess regression from the raw log ratio.

EXAMPLE 5.5 NON-LINEAR REGRESSION APPLIED TO DATA SET 5A

Loess regression is applied to the human fibroblast data set 5A (Figure 5.5a).[7] The curve fits the data very well. The normalised data (Figure 5.5b) are balanced about zero and are ready for analysis for differentially expressed genes.

Although Loess is an advanced statistical technique, it is important to remember that it is no more than a computational method for drawing a best-fit curve through a cloud of points. There is no conceptual or theoretical underpinning to the curve produced by Loess; it is only a scaling of the data.

Loess regression has a number of parameters, which can be set by the user, whose values will have an impact on the way in which the curve fits the data. The most important of these is the size of the window, which determines the smoothness of the regression. If the window is too small, the curve will be too sensitive to local ups and

[7] The R statistical software is very well equipped to perform Loess regression. It can be found in the *modreg* package. Suppose the fibroblast data set were in a data frame called *fibroblast*, with variables *average* and *lratio* containing the average log intensity and log ratio. Then the R commands to apply Loess normalisation would be:

```
attach(fibroblast)
lmodel <- loess(lratio~average)
fibroblast$normlratio <- lratio - predict.loess(lmodel,
      average)
```

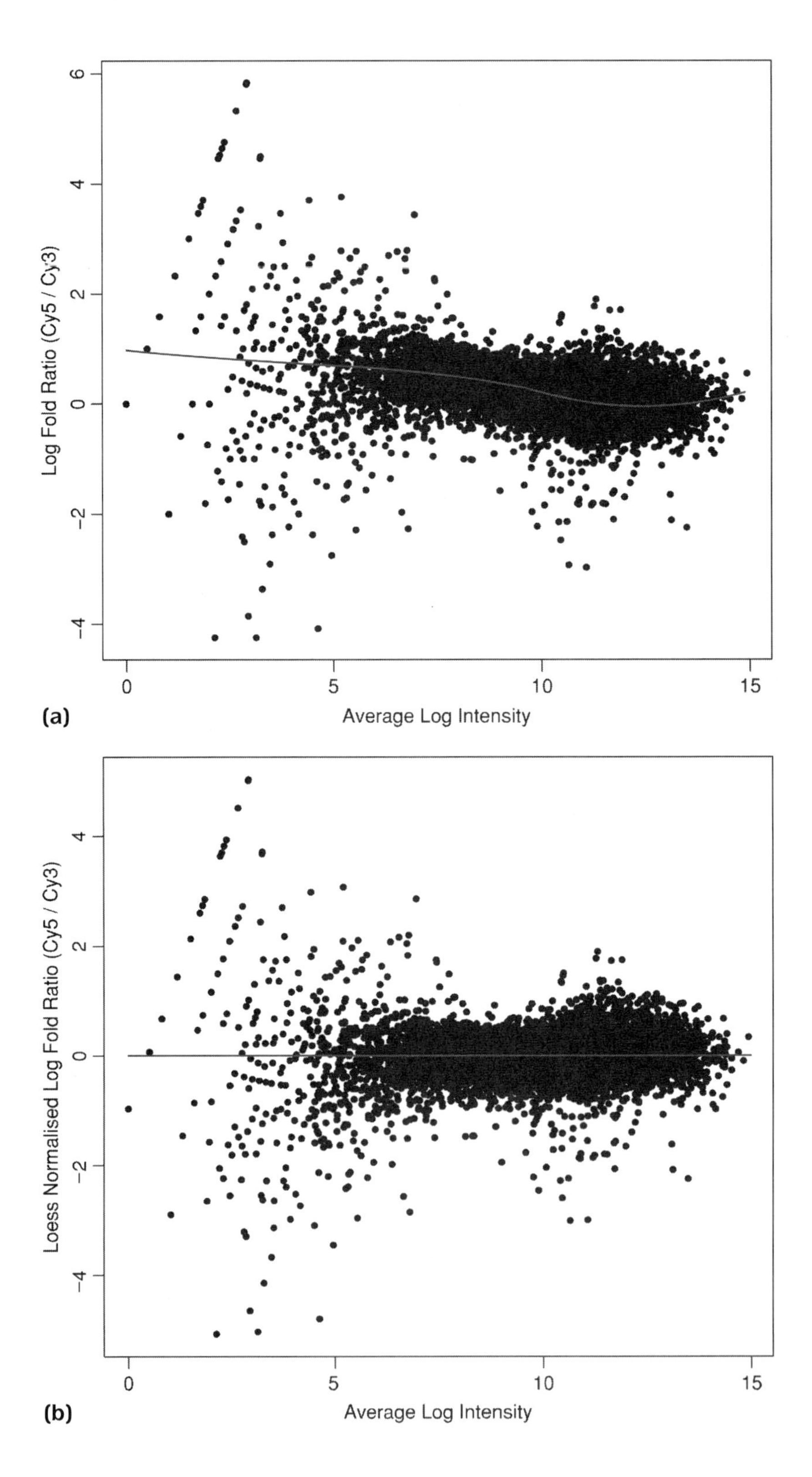
Log Fold Ratio (Cy5 / Cy3)
Average Log Intensity
(a)
Loess Normalised Log Fold Ratio (Cy5 / Cy3)
Average Log Intensity
(b)

downs in the data, and will be very "wiggly" (Figure 5.6a). If the window is too large, the curve will be too "stiff" and will be unable to fit the data effectively (Figure 5.6b).

Correcting for Spatial Effects

In some microarray experiments there is a spatial bias of the two channels: in some regions of the array the Cy3 channel is brighter, and in other regions of the array the Cy5 channel is brighter. This can result from the array not being completely flat or horizontal in the scanner. The depth of focus of the two lasers is different – depth of focus is proportional to wavelength, and so is greater for Cy5 than for Cy3. If the array is not horizontal, then it is possible that in some areas of the array both channels might be in focus, while in other areas of the array the Cy5 channel might be in focus but the Cy3 channel might be slightly out of focus. This can affect the log ratios of the data, with some regions of the array having generally positive log ratios, and other regions having generally negative log ratios.

Where this happens, it is possible to correct for spatial bias using two different normalisation techniques: two-dimensional Loess regression and block-by-block Loess regression.

Two-Dimensional Loess Regression

This is generally the better method for correcting spatial bias on an array. The two-dimensional Loess works in a similar way to one-dimensional Loess, but instead of fitting a curve, it fits a two-dimensional polynomial surface to the data. To perform a two-dimensional Loess regression on microarray data, you would perform the following steps:

1. Calculate the log ratio for each feature on the array.
2. Produce a false-colour plot of the log ratios of the features as a function of the x and y coordinates of the features on the array.
3. Perform a two-dimensional Loess fit of the log ratios as a function of the x and y coordinates of the features.
4. For each feature, calculate the normalised log ratio by subtracting the fitted value on the Loess surface from the raw log ratio.

EXAMPLE 5.6 DATA SET 5B – SPATIAL BIAS ON A KIDNEY–LIVER ARRAY

In a microarray experiment to look at the difference between gene expression in kidney and liver of mice, kidney and liver samples from the same mouse have been

Figure 5.5: Loess normalisation. (a) The scatterplot of the fibroblast data is plotted with a non-linear Loess fit through the data. The non-linear curve appears to fit the shape of the data better than the linear regression. The log ratio can be non-linearly normalised by subtracting the fitted value on the Loess curve from the data. This method works well for data where there is a non-linear relationship between the responses of the Cy3 and Cy5 channels. **(b)** The fibroblast data after Loess normalisation. The horizontal line through zero corresponds to the curved line in (a). The horizontal line appears to go through the centre of the cloud of data very well.

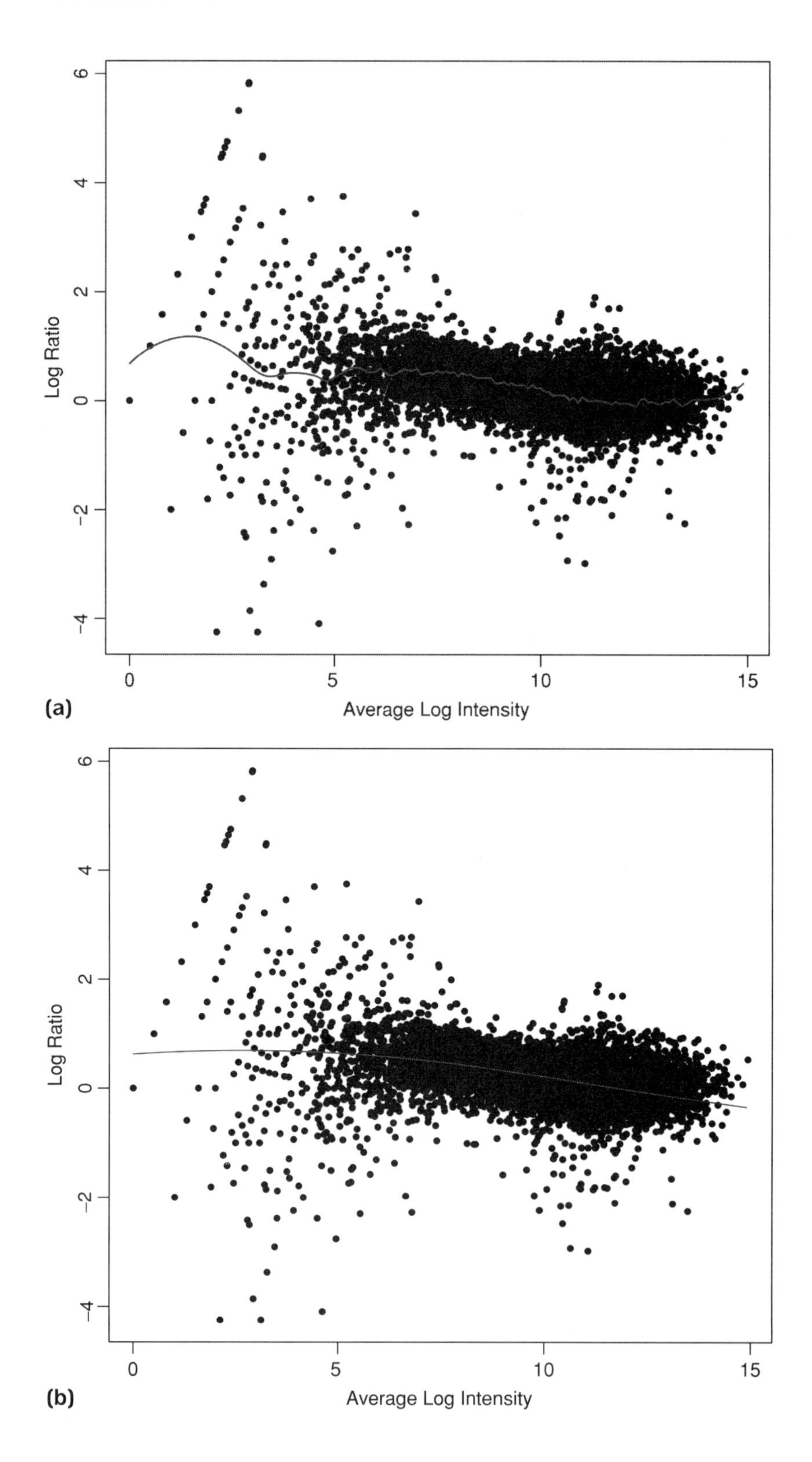

6
4
2
0
-2
-4
Log Ratio
0
5
10
15
(a)
Average Log Intensity
6
4
2
0
-2
-4
Log Ratio
0
5
10
15
(b)
Average Log Intensity

prepared with Cy3 and Cy5 dyes and hybridised to a microarray.[8] The data from the array show Cy3 brighter on the top-left corner and Cy5 brighter in the bottom-right corner (Figure 5.7a)

To correct for this bias, a two-dimensional Loess surface was fitted to that data (Figure 5.7b).[9] From the plot, it is possible to see the contours representing the differential gradient of the Cy3 and Cy5 intensities across the arrays from top left to bottom right. The fitted values from the Loess surface are subtracted from each of the features to produce a normalised data set with no spatial bias (Figure 5.7c).

Block-by-Block Loess Regression

A second method for correcting for spatial bias on the array is to perform one-dimensional Loess regression on the log ratio as a function of average log intensity, but instead of applying this method to the whole array (as done earlier), apply the method to each grid on the array separately. This method works well when bias is introduced from the different pins on the spotting robot, and the Cy3 and Cy5 intensities behave differently for different pins.

EXAMPLE 5.7 BLOCK-BY-BLOCK NORMALISATION ON DATA SET 5B

Block-by-block normalisation is applied to the kidney–liver data set 5B. There are 48 grids on the array (12×4). The log ratio is normalised to the average intensity using a separate Loess regression for each grid (Figures 5.8a and 5.8b). After normalisation, there is no spatial bias on the array (Figure 5.8c). There are two disadvantages with this method. First, the number of data points on each grid can potentially be quite small, and it is possible that the majority of the features within an intensity range could be differentially expressed. This would contravene the requirement that most genes are not differentially expressed. The Loess regression would fit the differentially expressed genes, and so important information would be lost during the normalisation process.

[8] These data were obtained privately from the Microarray Facility at the Mammalian Genetics Unit in the Medical Research Council Laboratories at Harwell in Oxfordshire, UK.

[9] Two-dimensional Loess regression can be performed in R using the same Loess function in the *modreg* package as is used for one-dimensional Loess regression. Suppose the kidney–liver data is held in a data frame called *nonflat*, with variables *x*, *y* and *lratio* corresponding to the *x* coordinate, *y* coordinate and log ratio of each of the features. Then to perform the two-dimensional Loess normalization, use the following commands:

```
attach(nonflat)
lmodel <- loess(lratio~x+y)
nonflat$lrationorm <- lratio - predict.loess(lmodel,nonflat)
```

Figure 5.6: Robustness of Loess regression. Loess regression is applied to the fibroblast data using different window sizes. **(a)** The window for the Loess regression is too small. The Loess curve follows local features of the data too closely and as a result is very "wiggly." **(b)** The window for the Loess regression is too large. The Loess curve is too "stiff" and does not follow the data well; the Loess curve falls below the group of high expressed genes.

Second, it is common for spatial bias to arise from the array not being horizontal in the scanner, which may have no relationship to the variabilities between different pins.

SECTION 5.4 BETWEEN-ARRAY NORMALISATION

Section 5.3 described normalisation methods that can be used to compare the Cy3 and Cy5 channels of a single array. This section looks at normalisation methods that allow you to make comparisons between samples hybridised to different arrays, which could be either two-colour arrays or Affymetrix arrays. In such experiments, each hybridisation reaction may be slightly different, and so the overall intensities of

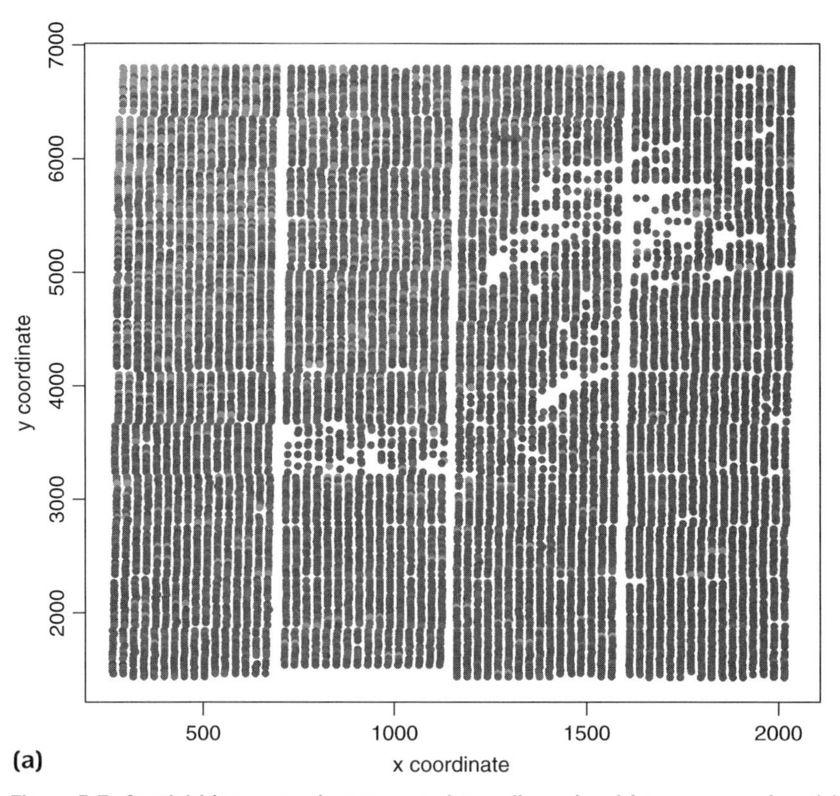

Figure 5.7: Spatial bias on a microarray and two-dimensional Loess regression. (a) False-colour representation of the log ratios of a microarray, with mouse kidney in Cy3 and liver from the same mouse in Cy5 (data set 5B). Each spot represents a feature. The x and y coordinates of each spot correspond to the x and y coordinates of the feature on the array. The colour of the spot represents the log ratio (Cy5/Cy3) of the feature, with red spots having a positive log ratio and green spots having a negative log ratio. There is a strong spatial bias on the array, with green spots in the top-left-hand corner and red spots in the bottom-right-hand corner. The areas of the array with missing spots represent features that have been flagged by the image-processing software, or features with a higher background than signal that have been removed from the data set. **(b)** The same data, but with the fit of a two-dimensional Loess surface to the log ratios superimposed as contours. The contours follow the colour trend, going from negative at top left to positive at bottom right. **(c)** False-colour plot of the normalised log ratio values of the features. These are calculated by subtracting the fitted values of the Loess surface from the raw log ratios. There is no spatial bias on the normalised data. (*Please see also the color section at the middle of the book.*)

(*continued*)

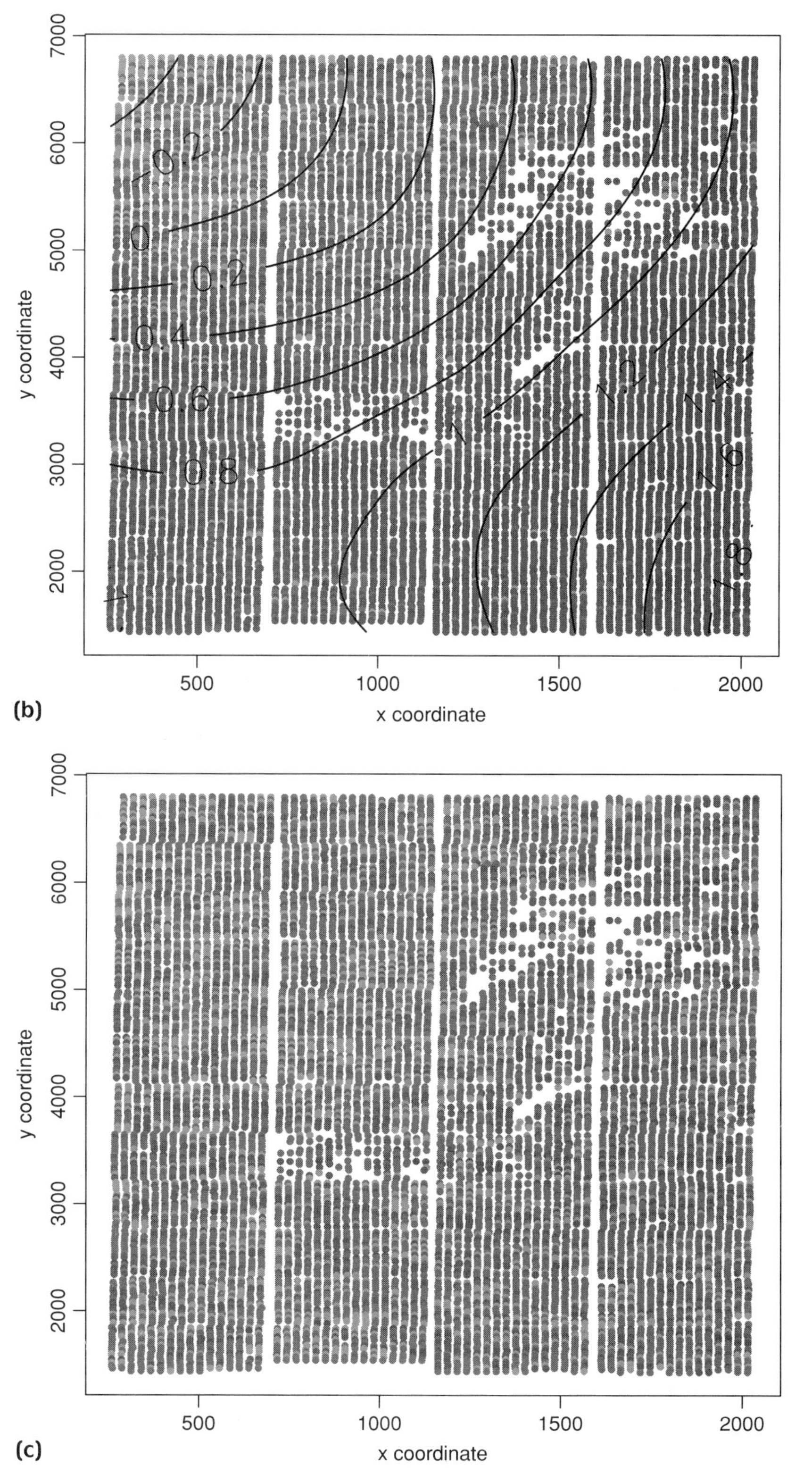

Figure 5.7: (*continued*)

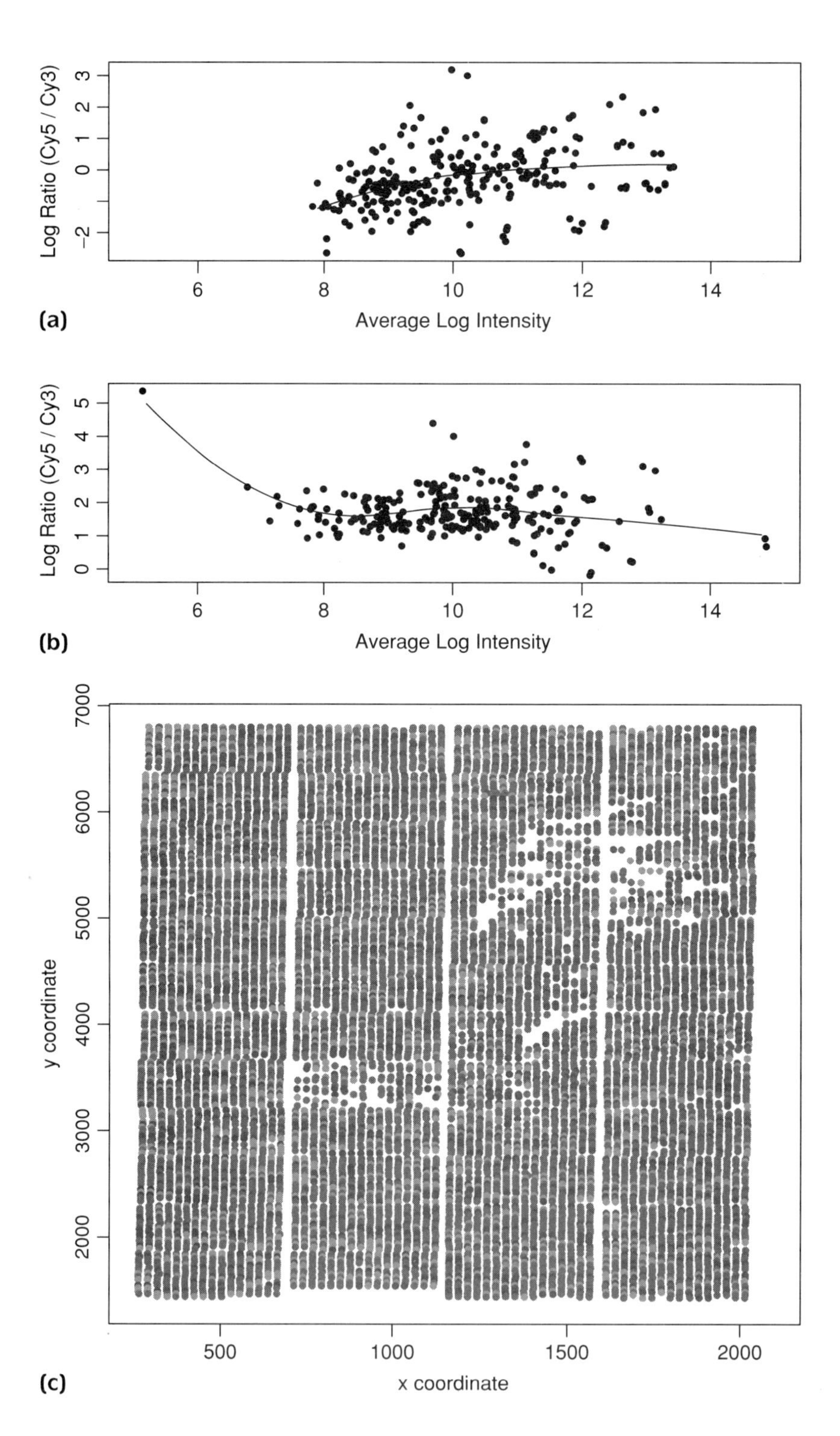

Log Ratio (Cy5 / Cy3)
Average Log Intensity
(a)
Log Ratio (Cy5 / Cy3)
Average Log Intensity
(b)
y coordinate
x coordinate
(c)

different arrays may be different. In order to be able to compare the samples hybridised to different arrays on an equal footing, it is necessary to correct for the variability introduced by using multiple arrays.

Visualising the Data: Box Plots

The box plot is a method for visualising several distributions simultaneously. It is an excellent method for comparing the distributions of log intensities or log ratios of genes on several microarrays. A box plot shows a distribution as a central box bracketed by horizontal lines known as whiskers. The line through the centre of the box represents the mean of the distribution. The box itself represents the standard deviation of the distribution. The horizontal lines bracketing the box represent the extreme values of the distribution.[10]

EXAMPLE 5.8 BOX PLOTS OF DIFFUSE LARGE B-CELL LYMOPHOMA PATIENTS (DATA SET 5C)

Samples have been taken from 39 patients suffering from diffuse large B-cell lymphoma (DLBL) and hybridised to microarrays, with each array containing one sample in the Cy5 channel and a reference sample in the Cy3 channel.[11] Figure 5.9a shows box plots of the log ratios of the patient samples to the reference samples for five of the DLBL patients. The box plot allows you to compare the distributions of the log ratios in the different patients. For example, patient 14 has a lower set of log ratios than the other patients, and patient 13 has a wider range of log ratios than the other patients.

There are three standard methods for normalising data similar to data set 5C so that the arrays can be compared on an even footing. They all make the same central assumption: **the variations in the distributions between arrays are a result of experimental conditions and do not represent biological variability**. If this assumption is not true, then these methods are not appropriate. The three methods are

- Scaling
- Centering
- Distribution normalisation

[10] The box plot function in R is slightly different and plots the median of the distribution at the centre of each box, and the size of the box represents the median absolute deviation from the median. These are robust, non-parametric equivalents of the mean and standard deviation.

[11] A reference to the paper from which these data derive is given at the end of the chapter.

Figure 5.8: Block-by-block regression. Block-by-block regression is performed by applying one-dimensional Loess normalisation to the features in each grid on the array separately. The array in data set 5B has 48 grids. **(a)** MA plot of the features in the top-left-hand grid; most of the log ratios are negative, corresponding to this portion of the array being green. The Loess fit appears to be good, so these features will be normalised well. **(b)** MA plot of the features in the bottom-right-hand grid; most of the log ratios are positive, corresponding to this portion of the array being red. There is a single feature that is low expressed but with very high log ratio. Because there are not enough data points, the Loess curve has fitted this point and so it will appear that it is not differentially expressed. This is a problem with block-by-block Loess normalisation. **(c)** The whole array has been normalised using block-by-block normalisation. The spatial bias has been eliminated. (*Please see also the color section at the middle of the book.*)

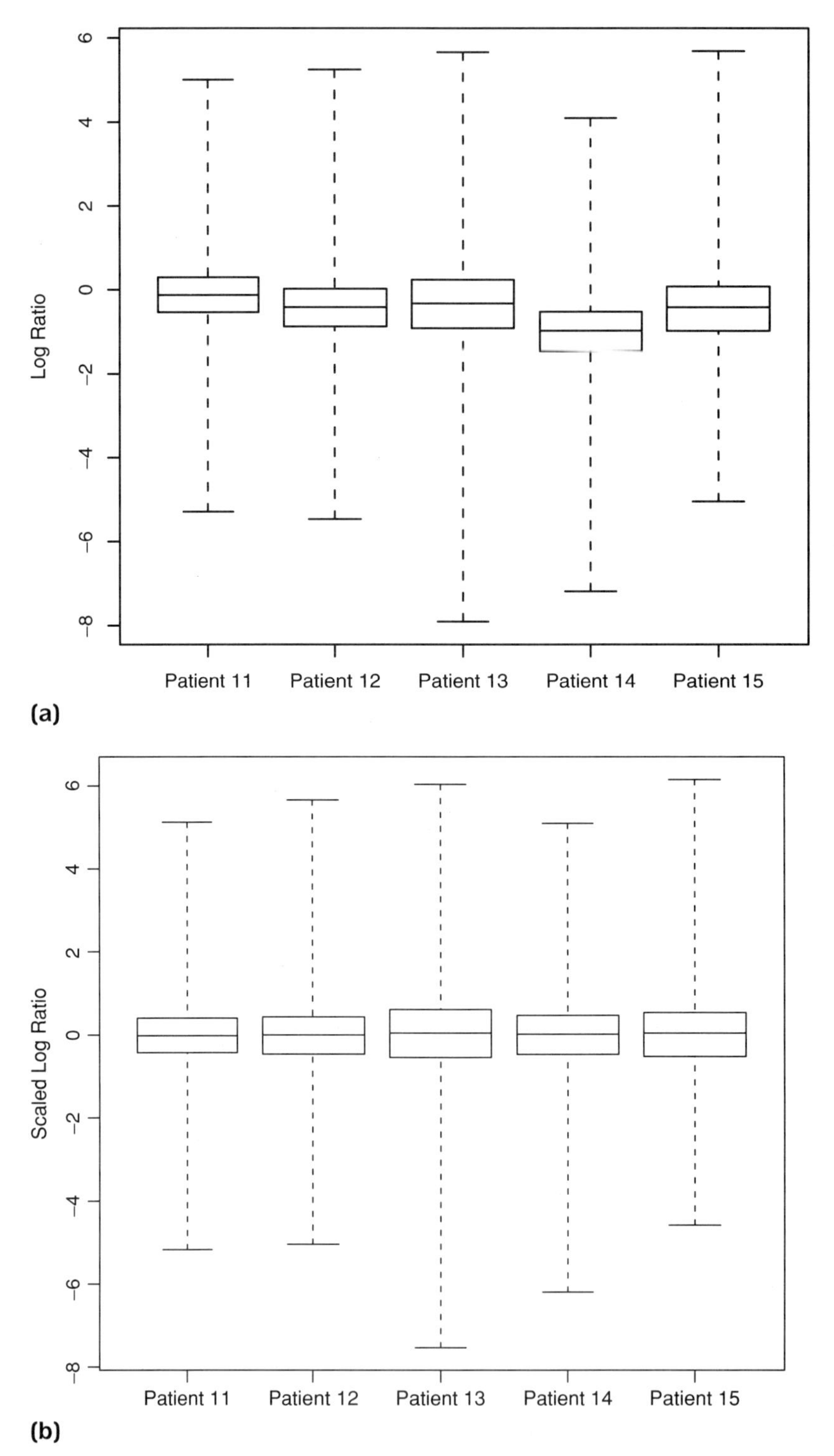

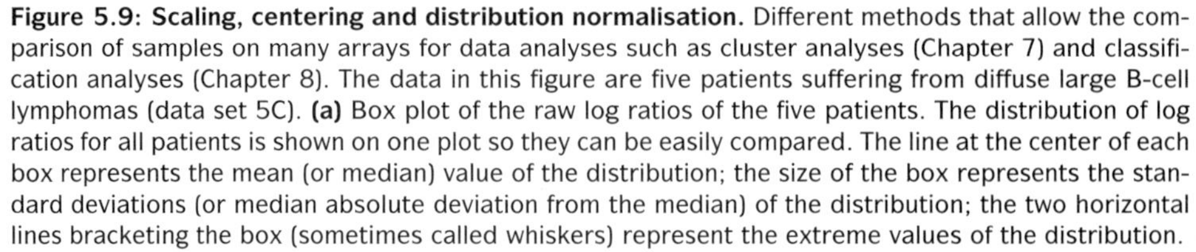

Figure 5.9: Scaling, centering and distribution normalisation. Different methods that allow the comparison of samples on many arrays for data analyses such as cluster analyses (Chapter 7) and classification analyses (Chapter 8). The data in this figure are five patients suffering from diffuse large B-cell lymphomas (data set 5C). **(a)** Box plot of the raw log ratios of the five patients. The distribution of log ratios for all patients is shown on one plot so they can be easily compared. The line at the center of each box represents the mean (or median) value of the distribution; the size of the box represents the standard deviations (or median absolute deviation from the median) of the distribution; the two horizontal lines bracketing the box (sometimes called whiskers) represent the extreme values of the distribution.

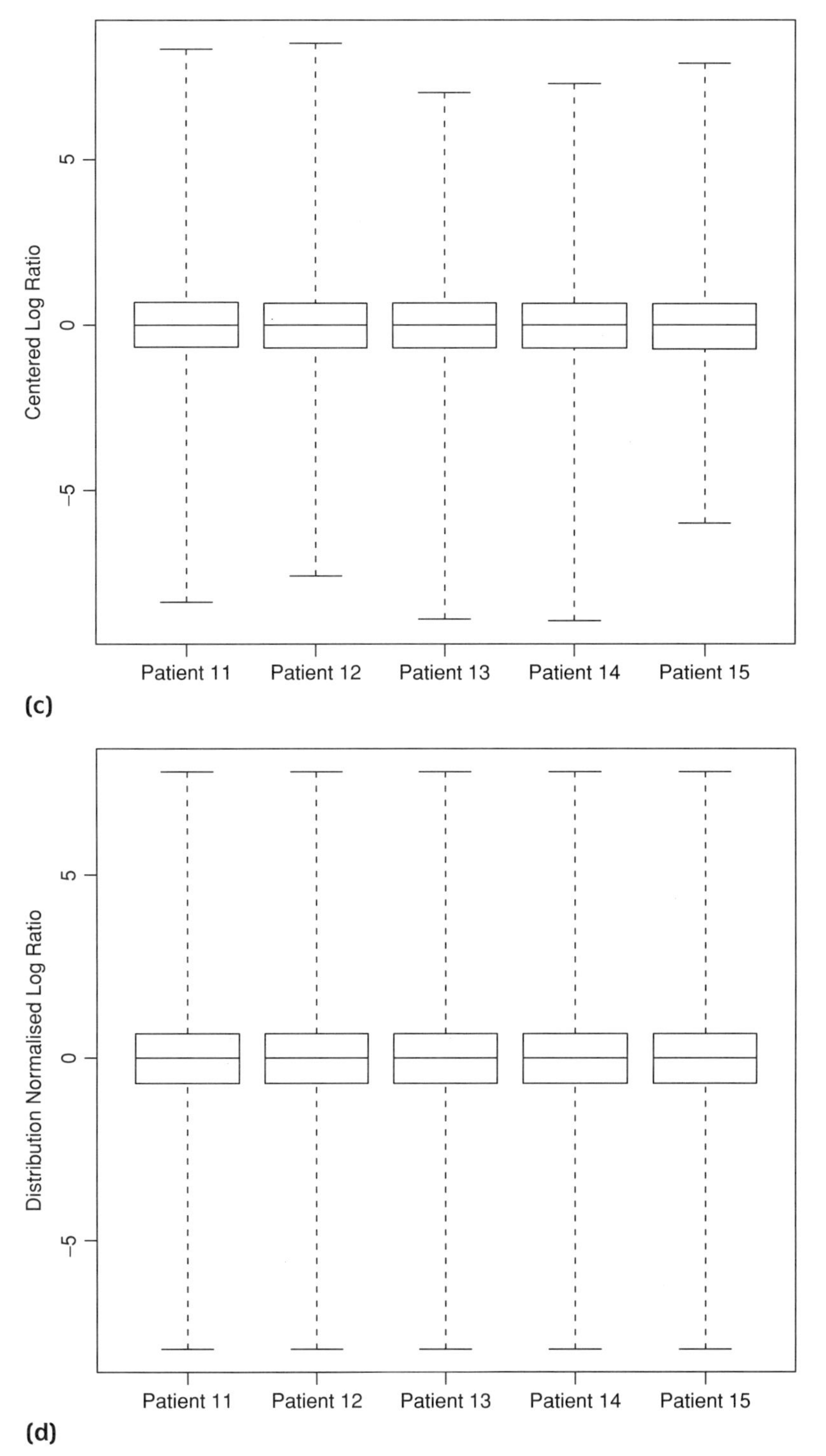

Figure 5.9: (*continued*) In this plot, the five patients all have different means, standard deviations and distributions. **(b)** The data has been scaled by subtracting the mean of the distribution from each of the log ratio values of each patient. The means of the distributions are all equal to zero. **(c)** The data has been centered by subtracting the mean of the distribution and dividing by the standard deviation. The centered distributions for each patient have mean 0 and standard deviation 1. Centering is useful when using correlation as a distance measure (Chapter 7). **(d)** The data has been distribution normalised so that each patient has the same set of measurement values; the distributions for all five patients are identical.

Scaling

Data is scaled to ensure that the **means** of all the distributions are equal (Figure 5.9b). The method is simple: subtract the mean log ratio (or log intensity) of all of the data on the array from each log ratio (or log intensity) measurement on the array. The mean of the measurements on each array will be zero after scale normalisation.

An alternative to using the mean is to use the median; this provides a more robust measure of the average intensity on an array in situations where there are outliers or the intensities are not normally distributed.

Centering

Data is centered to ensure that the means and the standard deviations of all of the distributions are equal (Figure 5.9c). The method is similar to scaling: for each measurement on the array subtract the mean measurement of the array and divide by the standard deviation. Following centering, the mean of the measurements on each array will be zero, and the standard deviation will be 1.

Centering is a very commonly used method for comparing multiple arrays. It is particularly useful when calculating the Pearson correlation coefficient of a large number of data sets prior to cluster analysis, because it ensures that the correlation coefficient can define a distance metric on the data. This is discussed in full in Chapter 7.

An alternative to using the mean and standard deviation is to use the median and median absolute deviation from the median (MAD). This has the advantage of being more robust to outliers than using the mean and standard deviation, but has the disadvantage of not producing a distance metric when using Pearson correlation.

Distribution Normalisation

Data is distribution normalised to ensure that the distributions of the data on each of the arrays are identical. The methodology is slightly more complex:

1. Center the data.
2. For each array, order the centered measurements from lowest to highest.
3. Compute a new distribution whose lowest value is the average of the values of the lowest expressed gene on each of the arrays; whose second-lowest value is the average of the second-lowest values from each of the arrays; and so on until the highest value is the average value of the highest values from each of the arrays.
4. Replace each measurement on each array with the corresponding average in the new distribution. For example, if a particular measurement is the 100th largest value on the array, replace it with the 100th largest value in the new distribution.

Following distribution normalisation, the measurements of each array will have mean 0, standard deviation 1, and identical distributions to all other arrays.

Distribution normalisation is an alternative to centering as a method for normalising data before applying a cluster analysis (Chapter 7) or classification analysis (Chapter 8). It is useful where the different arrays have different distributions of values. However, centering is a simpler method and is the most commonly used method for microarray normalisation.

EXAMPLE 5.9 NORMALISING DATA SET 5C

The DLBL data can be normalised using all three methods. Scaling the data ensures that the means of the log ratios of the five patients are all zero (Figure 5.9b). However, the standard deviations are quite different; for example, patient 13 has a particularly large standard deviation. Centering the data ensures that the standard deviations are all equal to 1 (Figure 5.9c). However, the distributions are not identical; for example, patient 14 has some large negative log ratios. Distribution normalisation ensures that all arrays have identical distributions (Figure 5.9d).

KEY POINTS SUMMARY

- Normalisation can remove unwanted systematic variability from microarray data.
- Visualise the data with scatterplots and MA plots.
- Use within-array normalisation to remove effects of dye bias and spatial bias.
- Use between-array normalisation to enable comparison of multiple arrays.

RESOURCES

The data used in this chapter has come from a number of papers.

Data Set 5A

Blader, I.J., Manger, I.D., and Boothroyd, J.C. 2001. Microarray analysis reveals previously unknown changes in *Toxoplasma gondii*-infected human cells. *Journal of Biological Chemistry* 276: 24223–31.

Data Set 5B

Web site of the Microarray Facility at the Mammalian Genetics Unit in the Medical Research Council Laboratories in Harwell, Oxfordshire. Data set 5B was obtained privately from this laboratory.

http://www.mgu.har.mrc.ac.uk/microarray/

Data Set 5C

Alizadeh, A.A., Eisen, M.B., Davis, R.E., Ma, C., Lossos, I.S., Rosenwald, A., Boldrick, J.C., Sabet, H., Tran, C., Powell, J.I., Yang, L., Marti, G.E., Moore, T., Hudson, J. Jr., Lu, L., Lewis, D.B., Tibshirani, R., Sherlock, G., Chan, W.C., Greiner, T.C., Weisenberger, D.D., Armitage, J.O., Warnke, R., Levy, R., Wilson, W., Grever, M.R., Byrd, J.C., Botstein, D., Brown, P.O., and Staudt, L.M. 2000. Distinct types of diffuse large B-cell lymphoma identified by gene expression profiling. *Nature* 403: 503–11.

Further Reading

Yang, Y.H., Dudoit, S., Luu, P., Lin, D.M., Peng, V., Ngai, J., and Speed, T. 2002. Normalisation for cDNA microarray data: A robust and composite method addressing single and multiple slide systematic variation. *Nucleic Acids Research* 30: e15.

Excellent paper describing normalisation techniques, including the use of MA plots and block-by-block normalisation.

Quackenbush, J. 2002. Microarray data normalization and transformation. *Nature Genetics* 32 Suppl 2: 496–501.

Recent review of normalization techniques.

Kooperberg, C., Fazzio, T.G., Delrow, J.J., and Tsukiyama, T. 2002. Improved background correction for spotted DNA microarrays. *Journal of Computational Biology* 9: 55–66.

Description of a method that allows background subtraction where the background is higher than the feature intensity.

Huber, W., Von Heydebreck, A., Sultmann, H., Poustka, A., and Vingron, M. 2002. Variance stabilization applied to microarray data calibration and to the quantification of differential expression. *Bioinformatics* 18 Suppl 1: S96–S104.

Alternative to log transformation.

Useful Normalisation Resources

http://www.r-project.org/

The homepage for *R*, a free statistics package which is very similar to *S*. This is available for Unix, Windows and Macintosh, and has a wide range of statistics and graphing functionality. Unlike *S+* or *SPSS*, it does not have a graphical user interface and is operated via commands and scripts. A number of groups have written packages for microarray normalization and analysis in R.

http://www.stat.berkeley.edu/users/terry/zarray/Software/smacode.html

Statistics for Microarray Analysis. A package for R written by Terry Speed and co-workers at the University of Berkeley including many methods for microarray normalization.

http://people.cryst.bbk.ac.uk/wernisch/yasma.html

Yasma (Yet Another Statistics for Microarray Analysis). A different package for R written by Lorenz Wernisch and co-workers at Birkbeck College, London.

CHAPTER SIX

Measuring and Quantifying Microarray Variability

SECTION 6.1 INTRODUCTION

Chapter 5 described a number of methods to correct for unwanted systematic variability either within an array or between different arrays. In this chapter, we describe methods to measure and quantify the random variabilities introduced by the microarray experiment. The common sources of variability (Figure 6.1) are

- The variability between replicate features on the same array
- The variability between two separately labelled samples hybridised to the same array
- The variability between samples hybridised to different arrays
- The variability between different individuals in a population hybridised to different arrays

Estimates of these variabilities are essential to gaining an understanding of how well the microarray platform you are using is performing. They are also important parameters for determining the number of replicates required for a microarray experiment – a topic that is discussed in full in Chapter 10.

The first two levels of variability – between replicate features or samples hybridised to the same array – are meaningful only for two-colour arrays. However, the second two levels of variability – between hybridisations to different arrays and between individuals in a population – are meaningful both for two-colour arrays and Affymetrix arrays.

SECTION 6.2 MEASURING AND QUANTIFYING MICROARRAY VARIABILITY

The variabilities between different features on an array, between two samples hybridized to the same array or between samples hybridized to different arrays are all introduced by the microarray experimental process. In contrast, the variation between individuals in the population is independent of the microarray process itself. Experimental variability is measured with **calibration experiments**; population variability is measured with **pilot studies**.

Calibration Experiments

The aim of a calibration experiment is to identify and quantify the sources of variability in your microarray experimental platform. The information might then be used to

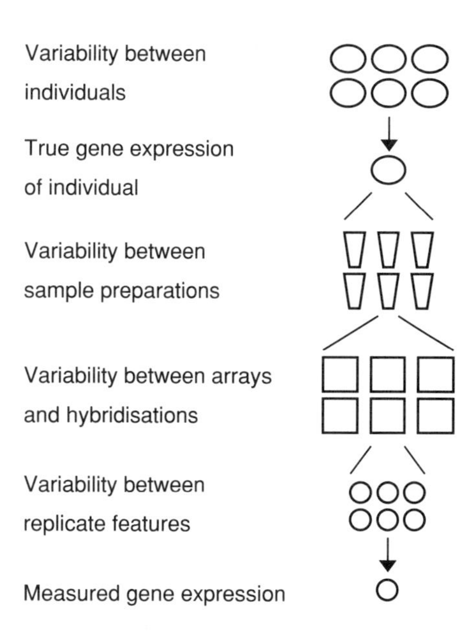

Figure 6.1: Sources of variability in a microarray experiment. There are several levels of variability in the measured gene expression of a feature. At the highest level, there is biological variability in the population from which the samples derive. At an experimental level, there is variability between preparations and labellings of the sample, variability between hybridisations of the same labelled sample to different arrays, and variabilities between the signal on replicate features on the same array. The measured gene expression includes the true gene expression, together with contributions from each of these variabilities.

improve your experimental procedures, or to select appropriate levels of technical replication for quality control. It is important to perform a calibration experiment

- After setting up your laboratory, and before undertaking any microarray projects;
- After any changes in laboratory apparatus, protocols or staff; and/or
- On a regular basis to ensure continued quality.

A typical calibration experiment would include

- An array design with several replicate features for each gene; ideally, these features should be in different locations on the array;
- Producing a sample and labelling it with both Cy3 and Cy5; and/or
- Co-hybridising the sample to several arrays.

In this way, it will be possible to measure the variabilities introduced by all three microarray-specific sources.

Pilot Studies

The aim of a pilot study is to provide an approximate guide to the level of variability in a population prior to performing a large-scale experiment. They are typically used before performing a power analysis to compute the number of biological replicates needed in an experiment (Section 10.4).

EXAMPLE 6.1 PILOT STUDY

Breast cancer patients are treated with a 16-week course of doxorubicin chemotherapy. Samples will be taken before and after chemotherapy and hybridised to microarrays. We want to identify genes that are differentially expressed as a result of this chemotherapy. In order to help design a large-scale experiment, we perform a pilot study on five patients to identify the level of population variability.

Quantifying the Variabilities

In statistics, it is typical to represent the variability of a population as a standard deviation. In microarray experiments, the measured expression (or relative expression) of a gene on a particular feature can be thought of as the true gene expression in the individual or sample, added to which are components for each of the sources of experimental variability (Figure 6.1). Each component of variability can be thought of as a separate distribution with its own standard deviation, and we use the calibration experiments and pilot studies to measure these standard deviations.

For example, if we consider an array with several replicate features for each gene, then we can calculate the standard deviation of the population of replicate measurements from the differences between the replicate measurements of each gene and the average of the replicates. The collection of these differences for all genes can be thought of as a sample from a random variable representing the differences between replicate features. The sample standard deviation would thus be an estimate of the standard deviation of this random variable and would be a measure of the variability between replicates. In this example, we are assuming that the variability between replicate features is the same for all genes. Later, we will see examples where this is not the case.

Log-Normal Distribution

In microarray experiments, it is common to model the distribution of errors at each of the four levels of variability using a **log-normal** distribution. One advantage of making a log-normal assumption is that it allows the different levels of variability to be expressed as a percentage known as the **coefficient of variability**. The coefficient of variability is equal to the standard deviation of a set of measurements divided by the mean. If the variability in a microarray data set follows a log-normal distribution, then the standard deviation of the errors in the raw intensities is proportional to their mean (i.e., the raw gene expression), and so the coefficient of variability is well defined.

In the log-normal model, the errors in the logs of the intensities – and hence also the errors of the log ratios – follow a normal distribution. This model is only approximately correct for real microarray data. We will see in the examples that follow that the distribution of errors tends to have a sharper peak and *heavier tails* than a normal distribution. This means that there are many features with smaller errors, and a small number of features with much larger errors than predicted by the log-normal model.

The use of a log-normal distribution also makes the assumption that the magnitude of the errors in the log intensities is approximately the same for features of all intensities. This is approximately true for some microarray data, but we will also show several examples where the errors are larger for low-intensity features than for high-intensity features. In such cases, it is possible to resolve this problem by partitioning the data into low- and high-intensity features and to provide more than one measure of variability for features with different ranges of expression.

If we use the log-normal model, then the coefficient of variability (v) relates to the standard deviation of the normally distributed errors in the logged measurements (σ) via the following formula:

$$v = \sqrt{(\exp(\sigma^2) - 1)} \qquad \text{(Eq. 6.1)}$$

This equation depends on the use of natural logarithms, so if you are using logs to base 2, the standard deviation must be multiplied by ln(2) (approximately 0.69) to obtain the correct value of σ for Equation 6.1.

Method for Measuring Variability

We discussed four levels of variability: between features, hybridizations, arrays and individuals. We will now describe a method for calculating the standard deviations for each of these levels of variability. There will be slightly different considerations in each case; the general method is as follows:

1. For each set of replicates (features, hybridisations or individuals), calculate the mean of the replicates.
2. For each replicate, calculate the deviation from the mean by computing the difference between the intensity of the replicate and the mean of the set of replicates.
3. Produce MA plots of the deviations against the mean to check that the variability is independent of intensity.
4. If desired, an appropriate linear or non-linear normalisation can be applied to the deviations. (See Section 5.3).
5. Calculate the standard deviation of the error distribution using all of the replicates.[1] If the variability depends on the intensity, you may wish to partition the data into different intensity ranges and calculate a standard deviation for each partition.
6. If the log-normal assumption is true, then the deviates should be distributed as a normal distribution. This can be checked by plotting a histogram of the deviates.
7. Convert the standard deviation to a percentage coefficient of variability by multiplying by ln(2) and applying Equation 6.1.

[1] Although the full set of replicate deviates are not independent random variables, the sample standard deviation is still an unbiased estimator of the population standard deviation and so this procedure is statistically meaningful. However, it would not be meaningful to perform a statistical test for normality of the distribution of all replicate deviates.

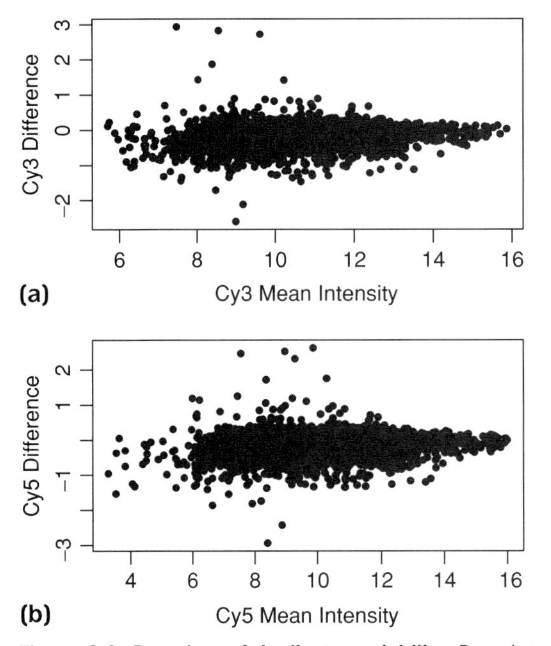

Figure 6.2: Dot plots of duplicate variability. Dot plots that show the relationship between the replicate features on an array from data set 6A. The array has been made with 6,000 genes spotted in duplicate. A sample of mouse kidney was prepared and labelled twice, once with Cy3 and once with Cy5, and both labelled samples were hybridised to the array. This allows for the estimation of variability between duplicate spots, and between the two labelled samples. Each point on the figure represents a pair of duplicate features for a different gene. The x coordinate is the average log intensity of the duplicate features (to base 2). The y coordinate is the difference between the log intensity of the first replicate and the average intensity. **(a)** The duplicates from the Cy3 channel on the array. There are two phenomena to observe. First, at low intensities, the cloud of points is generally negative, implying that the first replicate is less intense than the second replicate of the same gene. There are three reasons why this could be the case: (i) spatial bias on the array; (ii) if the replicates are spotted with different pins, there could be pin-to-pin variability; (iii) if the replicates are spotted with the same pin, more fluid could be released from the pin the second time it is applied to the glass. These effects can be normalised using Loess regression (Section 5.3). Second, although the variability of the duplicates is reasonably constant, it decreases with intensity. It might be meaningful to report two different coefficients of variability between the duplicates, one for low expressed genes [log(2) intensity less than 13], and one for high expressed genes [log(2) intensity greater than 13]. **(b)** A similar plot but for the Cy5 channel on the array. This plot is very similar to (a), but with slightly greater variability.

Variation Between Replicate Features on an Array

The variation between replicate features on an array can be measured in any experiment for which replicate features have been used.

EXAMPLE 6.2 CALCULATING THE VARIABILITY BETWEEN REPLICATE FEATURES USING A SELF–SELF-HYBRIDISATION (DATA SET 6A)

In an experiment to determine the quality of a microarray facility, RNA was extracted from mouse kidney and labelled twice, once with Cy3 and once with Cy5. The two labelled samples were hybridised to an array with 6,000 genes spotted in

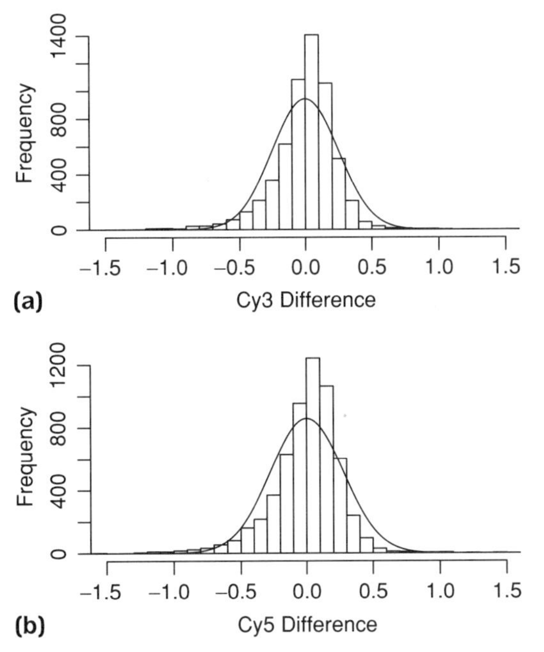

Figure 6.3: Distributions of duplicate variability. Distributions of the variability between the duplicates from data set 6A are shown as histograms, with the curve showing a fitted normal distribution with the same mean and standard deviation. **(a)** Histogram for the Cy3 channel. The variability is bell-shaped. The normal approximation is reasonable but not exact: there are more features with smaller variability than with a normal distribution and, correspondingly, there are more outlier features with greater variability than a normal distribution (this is difficult to see on this plot). It is common to refer to this phenomenon as the distribution having heavy tails. **(b)** Histogram for the Cy5 channel, showing the same properties as (a).

duplicate.[2] The researchers want to calculate the variability between the duplicate spots.

The variability between the duplicate spots is calculated separately for the two channels. The MA plots (Figure 6.2) show that while the variability is approximately constant, the variability is in fact smaller for the highest expressed genes. In addition, the plot is not centered on zero, indicating a systematic bias between the two replicates; this can be corrected using Loess normalisation (Section 5.3). The normalised error distributions are approximately but not exactly normal, with the errors having a sharper peak and longer tails than a normal distribution (Figure 6.3).

In this example, we could calculate standard deviations and coefficients of variabilities either for the whole data set, or partition the data into low-expressed and high-expressed genes and calculate separate standard deviations for the two partitions. When using all of the data, the standard deviations are 0.25 for the Cy3

[2] This data was obtained privately from the Microarray Facility at the Mammalian Genetics Unit in the Medical Research Council Laboratories at Harwell in Oxfordshire, UK.

channel and 0.27 for the Cy5 channel. These can be converted to coefficients of variability by multiplying each standard deviation by ln(2) and applying Equation 6.1; they correspond to coefficients of variability of 17 and 19%, respectively. If we partition the data into low expressed genes with log (to base 2) intensity less than 13, and high expressed genes with log intensity greater than 13, then the coefficients of variability for the low expressed genes are 18 and 19% in the Cy3 and Cy5 channels, respectively, and for the high expressed genes are 14% in both channels.

Variability Between the Cy3 and Cy5 Channels

The variability between two samples hybridised to the same array is best measured with a self–self-hybridisation. The same biological sample is labelled twice, once with Cy3 and once with Cy5, and we measure the variability between the two sets of measurements.

This variability is different from the systematic bias between the Cy3 and Cy5 channels discussed in Section 5.3. The systematic bias represents a consistent difference introduced by the experimental apparatus and is removed using the normalisation methods of Section 5.3. But after this bias has been removed, there remains random variability between the two sets of measurements in the two channels. This is the variability we measure here.

EXAMPLE 6.3 CALCULATING THE VARIABILITY BETWEEN THE CY3 AND CY5 CHANNELS

The self–self-hybridisation of data set 6A can be used to calculate the variability between the two labelled samples hybridised to the array. The same procedure is applied to the average measurements of the duplicate features in each of the Cy3 and Cy5 channels, which are then normalised with Loess regression. In this example, the variability is approximately constant and approximately normally distributed. The standard deviation of the error distribution is approximately 0.18 (in log to base 2), which corresponds to a coefficient of variability of 12%.

Variability Between Hybridisations to Different Arrays

The variability of hybridisations to different arrays contains two components: a variability among the arrays themselves (in relation to their production), and a variability among different hybridisation reactions. It is not possible to measure the components separately (they are what is known as confounded variables; this is discussed in depth in Chapter 10), so the two variabilities are combined into a single measurement.

In order to calculate this variability, you will need to hybridise the same labelled sample to a number of different arrays. In experiments that have used a reference sample, the reference sample on the different arrays can serve to estimate this variability.

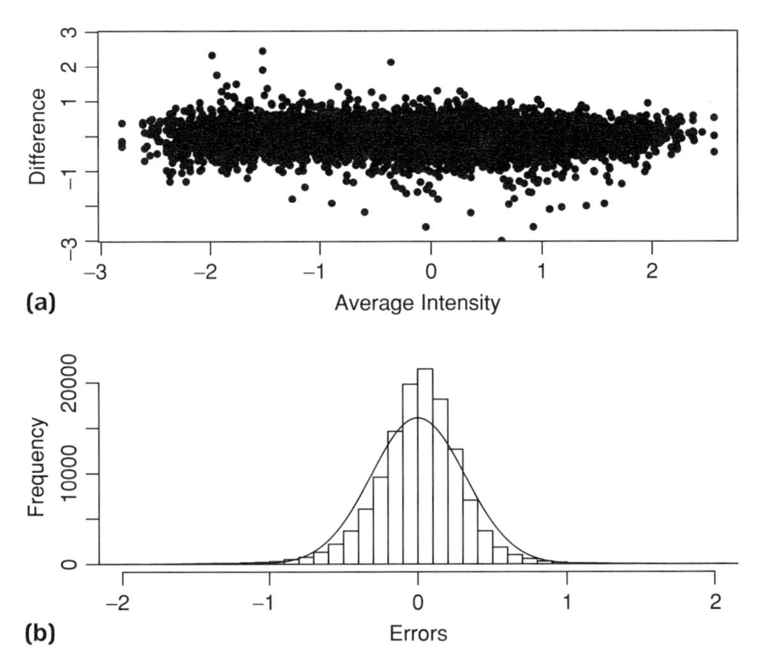

Figure 6.4: Variability between hybridisations. Figures to visualise the variability between hybridisations of the same reference sample to 20 different arrays in data set 6B. The reference sample intensities on each array have been centered so that they have mean 0 and standard deviation 1. **(a)** Dot plot of the variabilities. Because of the large volume of data in this example, the dot plot shows a randomly selected sample of replicates representing 10% of the original data. The variability is fairly constant across the range of intensities. **(b)** Histogram of the error distribution. The distribution is closer to a normal distribution than the example of duplicate features, but there are duplicates with smaller variability than that predicted by a normal distribution, indicating that there are also heavy tails of outlier replicates.

EXAMPLE 6.4 CALCULATING THE VARIABILITY BETWEEN REFERENCE SAMPLE HYBRIDISATIONS TO MULTIPLE ARRAYS (DATA SET 6B)

In an investigation of breast cancer chemotherapy, samples were taken from 20 patients before and after treatment with doxorubicin chemotherapy.[3] The 20 samples before treatment were labelled with Cy5 and hybridised to 20 different arrays; each array was hybridised with the same reference sample that was labelled with Cy3. We want to calculate the variability between hybridisations of the reference sample to different arrays.

We restrict the analysis to 6,350 genes for which all data is present in the data set. The intensities of the reference samples on each array are centered (Section 5.3) to allow for comparison between the arrays. We follow the aforementioned procedure and produce an MA plot for all of the data (Figure 6.4a). The standard deviation is fairly constant at all intensity levels, so it is meaningful to talk about a single coefficient of variability. The histogram of the error distribution (Figure 6.4b) shows that the distribution is approximately normal.

[3] A reference to the paper from which this data set derives is given at the end of the chapter.

The standard deviation of the error distribution is 0.31. We multiply this by ln(2) and apply Equation 6.1 to obtain a coefficient of variability of 22%. So in this experiment, there is 22% variability between the same labelled samples hybridised to different arrays.

Variability Between Individuals

The variability between individuals in a population is fundamentally different from the three other sources of variability described earlier. The other sources of variability are all introduced as part of the experimental process. There is no inherent interest in those variabilities, and any improvements in experimental practises that reduce those variabilities would be an advantage. Population variability, on the other hand, comes from the biological system being studied. In many experiments, we are explicitly interested in the variability between individuals, because it could be important in disease and treatment outcomes. However, the measurement of variability between individuals is performed in the same way as for the other variabilities and can also be expressed as a coefficient of variability.

EXAMPLE 6.5 CALCULATING THE VARIABILITY BETWEEN INDIVIDUALS IN DATA SET 6B

Using data set 6B we can calculate the variability of gene expression in the population of breast cancer patients. We perform the same procedure, but instead of looking at the reference sample, we look at the log ratio of each gene relative to the reference sample as a measurement of gene expression. The log ratios on each array are centered; for each gene, calculate the average log ratio of the 20 patients. For each gene in each patient, subtract the centered log ratio from the average log ratio. The standard deviation of this distribution is 0.60. This corresponds to a coefficient of variability of 44%.

It is very common for the differences between individuals to be the largest source of variability. This is one of the reasons why it is essential to replicate experiments with several individuals. The number of individuals needed depends on the type of experiment being performed as well as the level of population variability. Methods to estimate how many replicates to use are discussed in full in Chapter 10.

KEY POINTS SUMMARY

- Microarray experiments have several sources of variability.
- Variability can be measured and quantified via calibration experiments and pilot studies.
- With a log-normal assumption, variability can be expressed as a percentage coefficient of variability.

RESOURCES AND FURTHER READING

Data Set 6A

http://www.mgu.har.mrc.ac.uk/microarray/

Web site of the Microarray Facility at the Mammalian Genetics Unit in the Medical Research Council laboratories in Harwell, Oxfordshire. Data set 6A was obtained privately from this laboratory.

Data Set 6B

Perou, C.M., Sorlie, T., Eisen, M.B., van de Rijn, M., Jeffrey, S.S., Rees, C.A., Pollack, J.R., Ross, D.T., Johnsen, H., Akslen, L.A., Fluge, O., Pergamenschikov, A., Williams, C., Zhu, S.X., Lonning, P.E., Borresen-Dale, A., Brown, P.O., and Botstein, D. 2000. Molecular portraits of human breast tumours. *Nature* 406: 747–52.

CHAPTER SEVEN

Analysis of Differentially Expressed Genes

SECTION 7.1 INTRODUCTION

Data analysis is seen as the largest and possibly the most important area of microarray bioinformatics. Reflecting this, there are three chapters in this book describing data analysis methods, which themselves answer three sets of scientific questions that are asked of microarray data:

1. Which genes are differentially expressed in one set of samples relative to another?
2. What are the relationships between the genes or samples being measured?
3. Is it possible to classify samples based on gene expression measurements?

In this chapter, we describe the methods for the first of these questions: the search for up- or down-regulated genes; Chapters 8 and 9 answer the other two questions. This chapter covers a variety of techniques, drawn from both classical statistics and more modern theory, to give a detailed account of how to analyze DNA microarray data for differentially expressed genes. We start the chapter with three examples to illustrate what we mean by the identification of differentially expressed genes.

EXAMPLE 7.1 DATA SET 7A

Samples are taken from 20 breast cancer patients, before and after a 16-week course of doxorubicin chemotherapy, and analyzed using microarrays. We wish to identify genes that are up- or down-regulated in breast cancer following that treatment.[1]

EXAMPLE 7.2 DATA SET 7B

Bone marrow samples are taken from 27 patients suffering from acute lymphoblastic leukemia (ALL) and 11 patients suffering from acute myeloid leukemia (AML) and analyzed using Affymetrix arrays. We wish to identify the genes that are up- or down-regulated in ALL relative to AML.[2]

[1] Data are from the paper of Perou et al. (2000) and are available from the Stanford Microarray Database. All references are given at the end of the chapter.

[2] Data are from the paper of Golub et al. (1999) and are available from the Stanford Microarray Database.

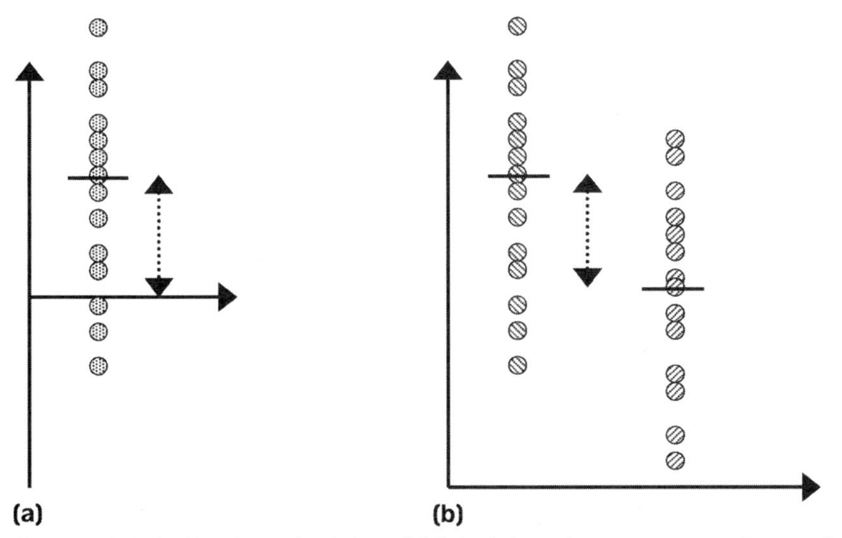

Figure 7.1: Paired and unpaired data. (a) Paired data; the measurements from each individual are subtracted to produce a single measure. We want to know whether the mean (or median) of the measures is different from 0. **(b)** Unpaired data; there is a gene expression measurement from each individual in each of the two groups. We want to know whether the mean (or median) of the two groups is different.

EXAMPLE 7.3 DATA SET 7C

There are four types of small round blue cell tumours of childhood: neuroblastoma (NB), non-Hodgkin lymphoma (NHL), rhabdomyosarcoma (RMS) and Ewing tumours (EWS). Sixty-three samples from these tumours, 12, 8, 20 and 23 in each of the groups, respectively, have been hybridised to microarrays.[3] We want to identify genes that are differentially expressed in one or more of these four groups.

In all of these examples, we are interested in identifying differentially expressed genes. The methods we describe in this chapter are designed for looking at one gene at a time to determine whether or not it is differentially expressed. The method would then be applied to every gene on the microarray in order to identify those genes that are differentially expressed. Thus the microarray is being used as a tool to study many individual genes in parallel. This approach contrasts with the analysis methods of Chapters 8 and 9, which are specifically interested in the interactions between the genes on the microarray.

Although these examples are similar, they are each examples of paired, unpaired or more complex data.

Data set 7A is an example of paired data (Figure 7.1a). There are two measurements from each patient, one before treatment and one after treatment. These two measurements relate to one another; indeed, we are really interested in the difference

[3] The data are from the paper of Khan et al. (2001) and are available from the Stanford Microarray Database.

between the two measurements (the log ratio) to determine whether a gene has been up-regulated or down-regulated following treatment.

Data set 7B is an example of unpaired data (Figure 7.1b). There are two groups of patients, and we are interested in seeing if a gene is differentially expressed between the two groups. There is no inherent relationship between the patients in one group and the patients in the other group.

Paired and unpaired data require slightly different analyses; these will be elucidated during the chapter. Data set 7C has four groups and requires more complex analysis. In this chapter, we will only give a brief introduction to more complex types of data, and the analysis of variance (ANOVA) analyses that are required for them.

The remainder of the chapter is organized into the following six sections:

Section 7.2: Fundamental Concepts, introduces the ideas behind all of the methods in this chapter: statistical inference, hypothesis tests, *p*-values and independence.

Section 7.3: Classical Parametric Statistics – *t*-Tests, discusses the traditional statistical approach to this type of data.

Section 7.4: Non-parametric Statistics, looks at methods that allow robust analysis of data that is not normally distributed. We will examine both traditional non-parametric statistics, and the more modern approach of bootstrapping, which combines the power of *t*-tests with the robustness of traditional non-parametric statistics.

Section 7.5: Multiplicity of Testing, describes the statistical problems associated with applying analyses to many genes and a simple method to resolve these problems.

Section 7.6: ANOVA and General Linear Models, gives a brief introduction to analysis of more complex data, such as in data set 7C where there may be more than two groups of patients, or in data sets where there are several factors that determine the gene expression measurements.

SECTION 7.2 FUNDAMENTAL CONCEPTS

This section describes four concepts that underlie all the methods described in this chapter:

- Statistical inference
- Hypothesis tests
- *p*-Values
- Independence

Statistical Inference

Statistical inference lies at the core of both science and classical statistics. Consider again data set 7A. We are interested in identifying genes that are up- or down-regulated in breast cancer following chemotherapy. Suppose we were to perform an experiment

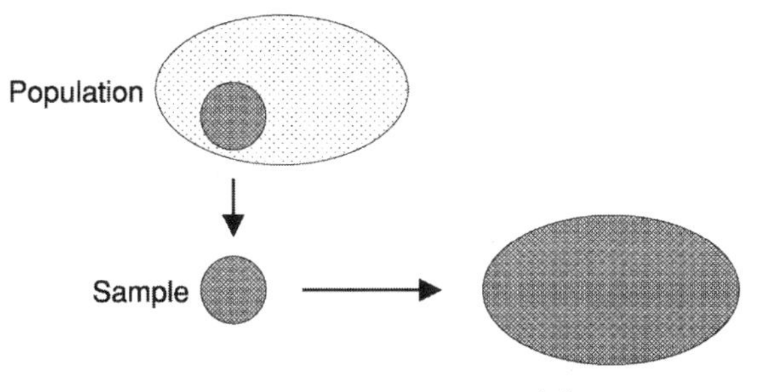

Figure 7.2: The idea of a statistical inference is central to classical statistics and underlies the scientific method. We are interested in describing a population of individuals; however, it is not practical to measure every individual in the population. Instead, we choose a representative sample of individuals from the population and make our measurements on these individuals. We then seek to extrapolate from the measurements of the individuals in the sample to make assertions about the population from which the sample derived.

on all breast cancer patients in the world who have received this treatment. This would completely describe the results of this therapy, but would be a slow and expensive experiment, and impossible in practical terms. Instead, we choose a **sample**[4] of 20 patients, which we hope is representative of the breast cancer population, and perform an analysis on these patients. We then seek to generalise our results from the 20 patients and make scientific assertions about changes in gene expression following breast cancer chemotherapy in the whole population of breast cancer patients (Figure 7.2).

Where we are trying to make a statistical inference, we are explicitly interested in the variabilities between individuals in the population to which we are extrapolating. We want to capture as much of this variability in our experiment and statistical analyses as we can: therefore, we seek to maximise the number of biological replicates we use, subject to budgetary and practical constraints. We may also seek to include population subtypes, for example, age or genetic factors, both in the experimental design and the statistical analyses.

Hypothesis Tests and *P*-values

For each gene in data set 7A, we have a measurement of expression from before and after treatment in each patient. These would be normalized, logged and combined into a log ratio for each patient, which describes numerically the extent to which the gene is differentially expressed, and whether it is up-regulated or down-regulated.

We want to identify those genes that are consistently differentially regulated across all 20 patients, with a view to asserting that these genes are differentially expressed following this chemotherapy. In early microarray experiments, researchers would have

[4] The word sample here is a reference to a statistical sample from a population, and not a biological sample that will be hybridized to a microarray. Unfortunately, the statistics community and the biological community use the same word to mean different things, which can sometimes lead to confusion.

chosen a threshold, for example, 2-fold differential expression, and selected those genes whose average differential expression is greater than that threshold.

From a statistical perspective, however, this is not a good approach, for two reasons:

- The average fold ratio does not take into account the extent to which the measurements of differential gene expression vary between the individuals being studied. The level of population variability is critical if we are trying to use the experimental sample to infer a general scientific statement about the overall population.
- The average fold ratio does not take into account the number of patients in the study, which statisticians refer to as the **sample size**. Intuitively, we would expect that the larger the sample size, the more confident we could be about determining genes that are differentially expressed.

For these reasons, statisticians determine whether or not a gene is differentially expressed via methodologies known as **hypothesis tests**. A hypothesis test builds a probabilistic model for the observed data based on what is known as a **null hypothesis**.

The null hypothesis is that there is no biological effect. For a gene in data set 7A, it would be that this gene is not differentially expressed following doxorubicin chemotherapy; for a gene in data set 7B, it would be that this gene is not differentially expressed between ALL and AML patients. If the null hypothesis were true, then the variability in the data does not represent the biological effect under study, but instead results from differences between individuals or measurement error.

Each hypothesis test in this chapter builds a probabilistic model under the null hypothesis. Using this model, it is possible to calculate the probability of observing a statistic, for example, an average differential gene expression, that is at least as extreme as the observed statistic in the data. This probability is known as a ***p*-value**. The smaller the *p*-value, the less likely it is that the observed data have occurred by chance, and the more significant the result. For example, a *p*-value of 0.01 would mean there is a 1% chance of observing at least this level of differential gene expression by random chance.

We then select differentially expressed genes not on the basis of their fold ratio, but on the basis of their *p*-values. We hypothesize that the differential expression observed in those genes with very small *p*-values is unlikely to have occurred by chance, and therefore has resulted from the biological effect being tested. In traditional applications, a *p*-value less than 0.01 would be deemed a significant result. In microarray applications, we need to use more stringent *p*-values; this is discussed in detail in Section 7.5.

Independence

Two measurements are independent if knowing the value of one measurement does not give information about the value of the other. All of the statistical tests we describe in this chapter require that the measurements being analysed are independent.

For example, for any gene, the measurements of expression in two different patients are independent. On the other hand, replicate measurements from the same patient, for example replicate features on an array, are not independent and could not be included as separate variables in a hypothesis test.

The simplest method for ensuring that all data points in the analysis are independent is to combine non-independent measurements into single variables. In data set 7A, we combine the two points from the same patient by subtracting one from the other in order to create a single data point for each patient and perform an analysis on these values. We could not treat the 40 samples as independent variables.

SECTION 7.3 CLASSICAL PARAMETRIC STATISTICS – *t*-TESTS

These hypothesis tests are historically the standard approach to analyze data in the forms of data sets 7A or 7B. There are two versions of the test, depending on whether the data is paired or unpaired.

Paired or One-Sample *t*-Test

The **paired *t*-test**, also referred to by statisticians as the **one-sample *t*-test**,[5] is applicable to data in the form of data set 7A. In that data set, there is a pair of measurements for each patient, one before treatment and the other after treatment, and these are combined to generate a single log ratio for each patient. Thus the data to be analysed look like a single column of numbers, one for each patient. From this data, one would calculate the ***t*-statistic**, using the following formula:

$$t = \frac{\bar{x}}{s/\sqrt{n}} \qquad \text{(Eq. 7.1)}$$

where $\bar{x}$ is the average of the log ratios of each of the patients; s is the standard deviation of the sample of patients; and n is the number of patients in the experiment. A p-value is then calculated from the t-statistic by comparing it to a t-distribution with an appropriate number of **degrees of freedom**. The degrees of freedom is the number of independent variables in the analysis; in the case of paired t-tests, it is the number of patients minus one.

The method of comparing the average log ratio with a threshold to determine the differentially expressed genes would use exactly the quantity $\bar{x}$ from Equation 7.1. However, we observe that the t-test is more sophisticated than this method. The significance of differentially expressed genes depends not only on the average log ratio, but also on both the population variability and the number of individuals in the

[5] The confusion over the two uses of the word sample is even greater when it comes to the names of t-tests: one-sample t-tests are used when two biological samples are taken from each patient (where there is one statistical sample of patients); two-sample t-tests are used when one biological sample is taken from each patient (when there are two statistical samples of patients). I think this really highlights the benefit of having biologists and statisticians working very closely together in order to learn each other's language.

TABLE 7.1: Data for ACAT2 from Data Set 7A

Patient	Before Treatment	After Treatment	Log Ratio	Fold Difference
7	−0.86	−2.17	−1.30	−2.47
10	−1.97	−1.93	0.04	+1.03
12	−2.07	−1.28	0.79	+1.73
14	−1.91	−2.32	−0.41	−1.33
15	−0.94	−2.00	−1.06	−2.09
18	−1.29	−1.74	−0.45	−1.37
26	−1.09	−1.54	−0.44	−1.36
27	−0.65	−0.60	0.06	+1.04
39	−1.69	−2.06	−0.37	−1.30
41	−0.79	−1.22	−0.43	−1.35
47	−1.19	−2.11	−0.91	−1.88
48	−1.36	−1.40	−0.04	−1.03
53	−1.11	−1.59	−0.48	−1.40
61	−1.82	−1.72	0.10	+1.07
100	−2.22	−2.13	0.10	+1.07
101	−1.76	−1.94	−0.18	−1.14
102	−1.51	−2.37	−0.86	−1.81
104	−1.65	−1.98	−0.33	−1.25
109	−0.78	−1.49	−0.71	−1.63
112	−1.80	−1.82	−0.03	−1.02
Average	**−1.42**	**−1.77**	**−0.35**	**−1.21**
Sample SD	**0.48**	**0.43**	**0.48**	

Note: In this experiment, the samples from before and after treatment have been hybridised to two separate arrays, with a common reference sample in the second channel. The measurements before and after treatment are the log ratios of the experimental sample to the reference sample. The log ratio is the difference between these two values; the logs are taken to base 2, so a value of 1 represents a 2-fold up-regulation, and −1 represents a 2-fold down-regulation. The sample standard deviations have been calculated with a denominator of $n - 1 = 19$ to ensure that they are unbiased estimators of the population standard deviation.

study. So a gene could be only 1.5-fold differentially expressed, but detected as such if the population variability is small. Similarly, the more individuals in the experiment, the easier it is to determine differentially expressed genes.

Paired t-tests are widely implemented in computer software, including Excel, SPSS, SAS, S+, R and GeneSpring. They are also straightforward to implement in code, and there are libraries available for all major programming languages that include functions for t-distributions. Most users will use one of these software packages to apply a t-test: for example, in Excel this would be achieved via the `TTEST` function, and in R this would be achieved via the `t.test` function.

EXAMPLE 7.4 PAIRED t-TEST APPLIED TO A GENE FROM DATA SET 7A

The gene acetyl-Coenzyme A acetyltransferase 2 (ACAT2) is on the microarray used for the breast cancer data, data set 7A. We can use a paired t-test to determine whether or not the gene is differentially expressed following doxorubicin chemotherapy (Table 7.1). In this particular experiment, the samples from before and after chemotherapy have been hybridized on separate arrays, with a reference sample in the other channel. There are three steps:

1. Normalise the data using one of the methods described in Chapter 5; this is not part of the *t*-test but is usually required before microarray data can be analysed. Because this is a reference sample experiment, we calculate the log ratio of the experimental sample relative to the reference sample for before and after treatment in each patient.
2. Calculate a single log ratio for each patient that represents the difference in gene expression due to treatment by subtracting the log ratio for the gene before treatment from the log ratio of the gene after treatment.
3. Perform the *t*-test, either using a software package or by calculating the average and sample standard deviation of the log ratios and applying Equation 7.1. The *t*-statistic is 3.22; this is compared with a *t*-distribution with 19 degrees of freedom (20 patients minus 1). The *p*-value for a two-tailed one-sample *t*-test is 0.0045, which is significant at a 1% confidence level.

We would therefore conclude (for the moment) that this gene has been significantly down-regulated following chemotherapy at the 1% level. There are some caveats to this statement that will become evident in later parts of this chapter.

Unpaired or Two-Sample *t*-Test

The **unpaired *t*-test** is very similar to the paired *t*-test, the difference being experimental design. In data set 7A, the data are paired: there are two values from each patient, which are first subtracted from each other, and the paired *t*-test is applied to the difference. In data set 7B, the data are unpaired: there are two groups of patients with no relationship to each other, and we test to see if the difference between the means of the two groups is zero. The unpaired *t*-test, also called the **two-sample *t*-test,** uses a formula similar to the paired *t*-test:

$$t = \frac{\bar{x}_1 - \bar{x}_2}{\left(\frac{s_1^2}{n_1} + \frac{s_2^2}{n_2}\right)} \qquad \text{(Eq. 7.2)}$$

where $\bar{x}_1$ and $\bar{x}_2$ are the means of the two groups; s_1 and s_2 are the sample standard deviations of the two groups; and n_1 and n_2 are the sizes of the two groups. There are actually two forms of the unpaired *t*-test: the version here allows the standard deviation between the two groups to be different, which is a better version to use for microarrays. There is also a version that calculates a single standard deviation for all the data. The *t*-statistic is compared with a *t*-distribution with an appropriate number of degrees of freedom to obtain a *p*-value.[6]

[6] The number of degrees of freedom for a two-sample *t*-test with unequal variance is given by a complicated formula; the interested reader is referred to one of the two statistics books referenced at the end of this chapter.

TABLE 7.2: Data for Metallothionein IB from Data Set 7B

Patient	ALL Log	Patient	AML Log
1	8.60	28	8.42
2	7.85	29	8.35
3	8.85	30	9.58
4	8.20	31	9.18
5	7.60	32	9.41
6	8.21	33	8.96
7	8.47	34	8.81
8	8.51	35	9.55
9	8.75	36	8.18
10	6.75	37	8.71
11	7.93	38	9.46
12	7.71		
13	7.88		
14	7.55		
15	6.61		
16	8.75		
17	9.32		
18	8.40		
19	7.16		
20	8.41		
21	4.75		
22	7.92		
23	7.82		
24	8.42		
25	7.08		
26	7.38		
27	9.29		
Average	**7.93**		**8.97**
Sample s.d.	**0.94**		**0.51**
Fold Ratio	**−1.84**		**+1.84**

Note: This data came from Affymetrix arrays; the values have been logged (to base 2) to ensure that the data are normally distributed.

The unpaired t-test is implemented in the same range as the paired t-test in the software. Because of the similarity between the two tests, the software frequently uses the same formula, for example, Excel the `TTEST` formula in Excel or the `t.test` formula in R.

EXAMPLE 7.5 UNPAIRED t-TEST APPLIED TO A GENE FROM DATA SET 7B

The gene metallothionein IB is on the Affymetrix array used for the leukaemia data, data set 7B. We want to identify whether or not this gene is differentially expressed between the AML and ALL patients. We want to identify genes which are up- or down-regulated in AML relative to ALL (Table 7.2). There are three steps:

1. The data is log transformed.
2. The average and sample standard deviations are computed for each set of patients: one average and standard deviation for the ALL patients, and a separate

average and standard deviation for the AML patients. The statistical test determines whether these two means are equal.

3. The *t*-test is performed using one of the software packages, or via Equation 7.2. In this case, the *t*-statistic is 4.35. This is compared with a *t*-distribution with 33 degrees of freedom and produces a *p*-value of 0.00012.

We conclude that the expression of metallothionein IB is significantly higher in AML than in ALL at the 1% level.

Requirements of *t*-Tests

The *t*-tests just described are very commonly used in statistics and appear frequently in the biological and medical literature. However, *t*-tests require that the distribution of the data being tested is normal. This has slightly different meaning for the paired and unpaired tests:

- For paired *t*-tests, it is the distribution of the subtracted data that must be normal.
- For unpaired *t*-tests, the distribution of both data sets must be normal.

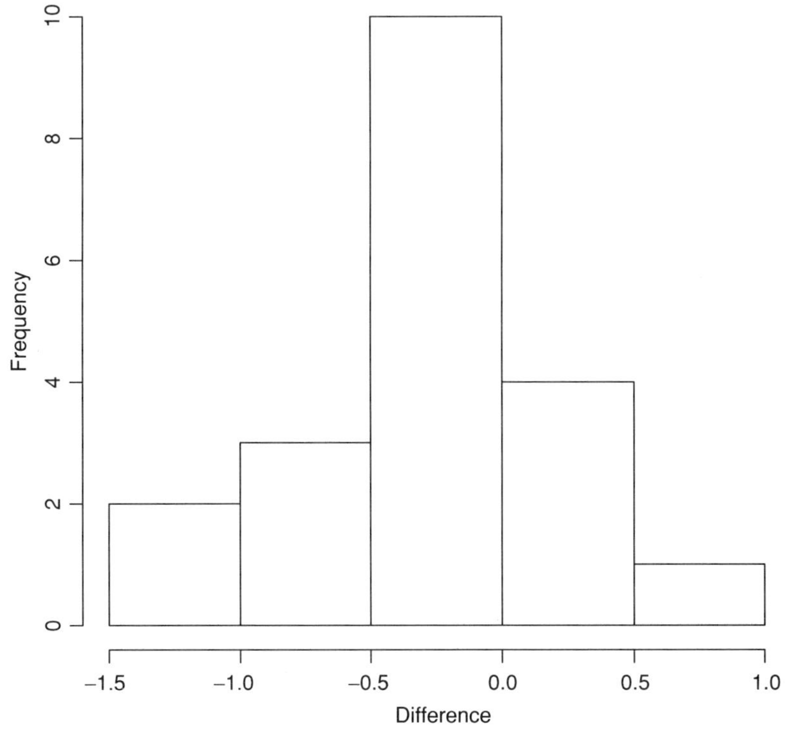

Figure 7.3: Histogram of the difference between the log ratios of the expression of ACAT2 in 20 breast cancer patients before and after a 16-week course of chemotherapy with doxorubicin. The data are approximately normal; the mean of the distribution appears to be less than zero, suggesting that this gene might be down-regulated.

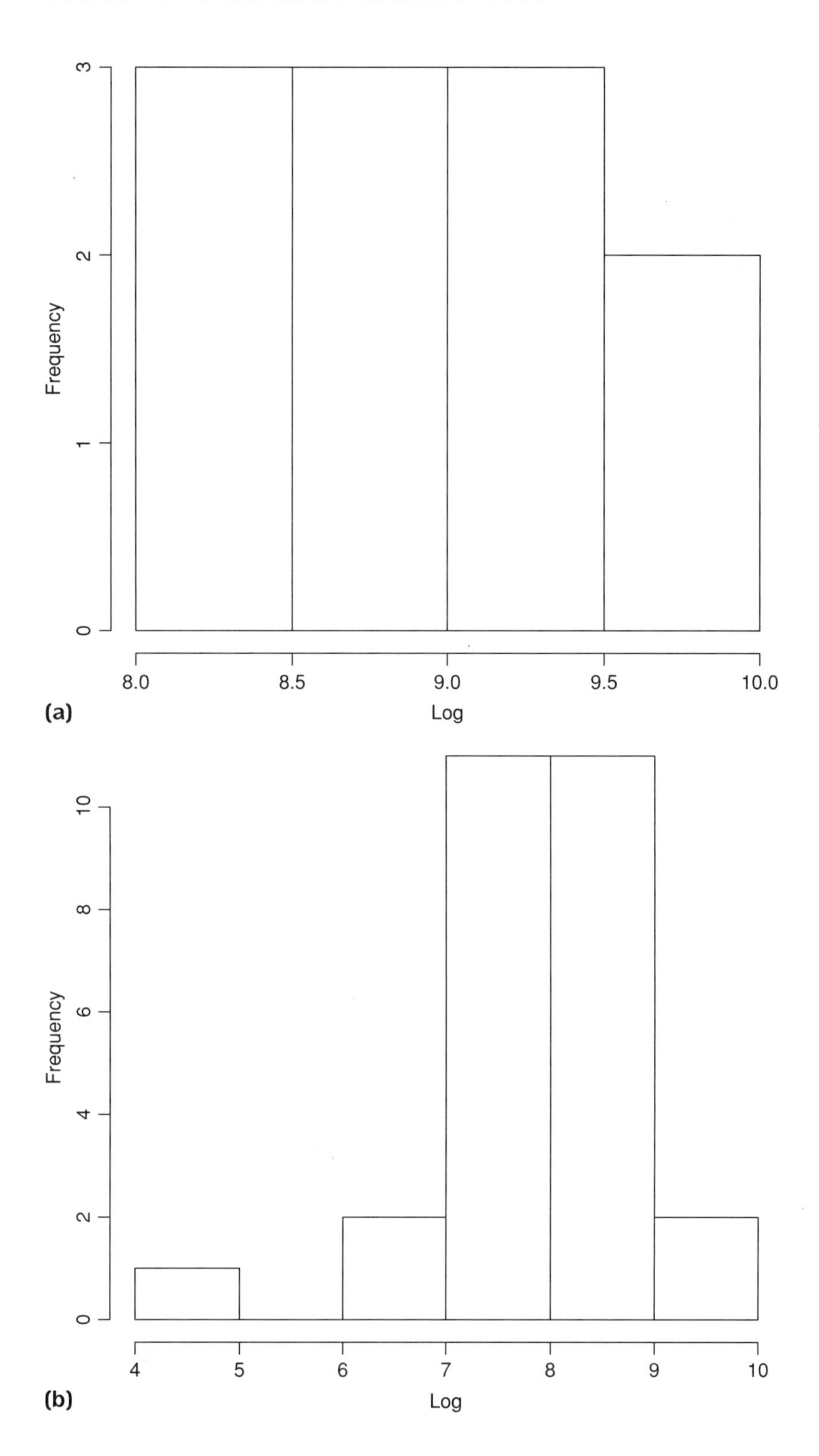
Frequency
0
1
2
3
8.0
8.5
9.0
9.5
10.0
(a)
Log
Frequency
0
2
4
6
8
10
4
5
6
7
8
9
10
(b)
Log

TABLE 7.3: Data for Diubiquitin from Data Set 7A

Patient	Unlogged Difference	Log Ratio	Fold Change
7	−10.08	−2.91	−7.54
10	0.85	0.62	+1.54
12	−0.28	−0.11	−1.08
14	0.04	0.08	+1.06
15	−0.68	−0.42	−1.34
18	0.17	0.12	+1.09
26	−4.93	−0.99	−1.99
27	−0.12	−0.16	−1.12
39	−1.67	−0.44	−1.35
41	−27.98	−1.64	−3.12
47	−0.92	−0.55	−1.46
48	−2.00	−0.99	−1.99
53	−3.04	−1.37	−2.58
61	−3.80	−2.05	−4.14
100	−3.53	−3.20	−9.18
101	−1.44	−1.12	−2.17
102	−0.62	−0.72	−1.64
104	−4.50	−1.19	−2.27
109	−0.23	−0.34	−1.27
112	0.10	0.12	+1.09

Note: The unlogged difference includes two outliers, patient 7 and patient 41. These have a detrimental effect on the data analysis, and provide unreliable results in the t-test applied to the raw data. These data points are not outliers in the logged data, and so the t-test applied to the logged data is more reliable.

EXAMPLE 7.6 EXAMPLES OF NORMALLY DISTRIBUTED DATA

Figure 7.3 shows a histogram of the data from Example 7.4, the gene ACAT2 from data set 7A. Although the sample is small, the data look approximately normal.

Figures 7.4 shows histograms of the data from Example 7.5, the gene metallothionein from data set 7B. Again, the samples are small, particularly the AML sample which only has 11 patients, but the data look approximately normal.

Both of these data sets meet the normality requirement, so t-tests are appropriate analyses of these data.

EXAMPLE 7.7 EXAMPLES OF DATA THAT ARE NOT NORMALLY DISTRIBUTED

The raw data for the gene diubiquitin from data set 7A are not normally distributed (Figure 7.5a; Table 7.3). There are two outliers, with values of approximately −10 and −28. When the t-test is applied to the unlogged data, the p-value is 0.03, which is not significant at the 1% level.

Figure 7.4: Histograms of the log of the gene expression of metallothionein in (a) 11 AML patients and (b) 27 ALL patients. Both distributions are approximately normal. The mean of the histogram for the ALL patients appears to be lower than the mean for the AML patients, suggesting that this gene might be differentially regulated in these two diseases.

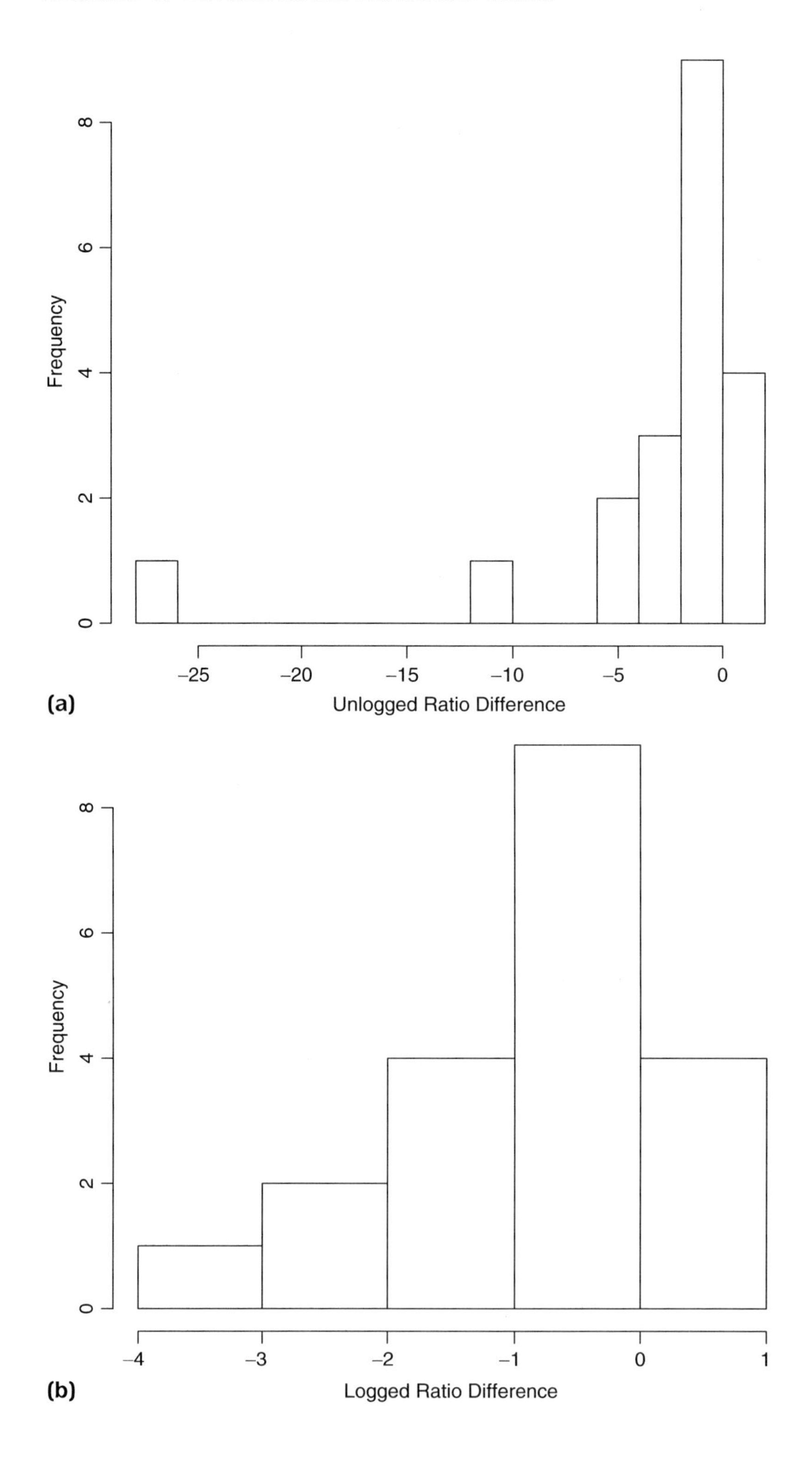

Frequency
0
2
4
6
8
−25
−20
−15
−10
−5
0
(a)
Unlogged Ratio Difference
Frequency
0
2
4
6
8
−4
−3
−2
−1
0
1
(b)
Logged Ratio Difference

TABLE 7.4: Summary Statistics for Diubiquitin Gene from Data Set 7A

Data	Mean	Standard Deviation	Standard Error	Standard Error/Mean	t-statistic	p-value
Unlogged	−3.23	6.36	1.46	−0.45	−2.27	0.03
Logged	−0.86	1.00	0.23	−0.27	−3.86	0.001

Note: The standard error of the unlogged data is relatively large relative to the mean; this is because of the two outlier values. Because of this, the p-value is not significant. The logged data are closer to a normal distribution; the outliers do not have extreme values, and the standard error is small relative to the mean. This is reflected in the significant p-value.

After the data have been log transformed, the outliers are no longer extreme, and the distribution looks normal (Figure 7.5b). When the t-test is applied to the logged data, the p-value is 0.001, which is significant at the 1% level.

When we use the correct analysis, we conclude that the gene is significantly down-regulated. This is in line with what we would expect from inspection of Figure 7.5b; in most patients, this gene is down-regulated. We would not have reached this conclusion with the incorrect analysis.

This example is quite counterintuitive. You might have thought the unlogged data, with its two large negative scores, would show clearer evidence of down-regulation than the logged data. In fact, the reverse is true. We can see why this is the case from the summary statistics (Table 7.4): the two outliers increase the standard error of the raw data. The high standard error decreases the certainty of the test, which returns a less significant result. In contrast, the standard error of the logged data is much smaller relative to the mean. As a result, the variability of the data is smaller, and the t-test returns a significant result.

SECTION 7.4 NON-PARAMETRIC STATISTICS

This section discusses methods that do not assume that the data is normally distributed. There are two good reasons to use these methods in preference to t-tests for the analysis of microarray data.

- **Microarray data is noisy.** There are many sources of variability in a microarray experiment, and outliers are frequent. Thus the distribution of intensities of many genes may not be normal. Non-parametric methods are robust to outliers and noisy data.
- **Microarray data analysis is high throughput.** When performing a t-test on a single set of data, it is straightforward to check the distributions for normality.

Figure 7.5: Histograms of the difference of expression of diubiquitin in 20 breast cancer patients. **(a)** The data have not been logged. The distribution is not normal: there are two outliers, with values of approximately −28 and −11. **(b)** The data have been logged. The distribution is normal; the outliers have been pulled in. Note that in both cases the mean difference is less than zero. However, with the unlogged data, a t-test gives a not-significant result because the standard error of the mean is so high, while for the logged data, a t-test is significant because the standard error is much lower.

However, when analysing the many thousands of genes on a microarray, we would need to check the normality of every gene in order to ensure that a *t*-test is appropriate. Those genes with outliers or which were not normally distributed would then need a different analysis. It makes more sense to apply a test that is distribution free and thus can be applied to all genes in a single pass.

There are two types of non-parametric test we will discuss in this section:

- **Classical non-parametric tests** are equivalent to parametric tests but do not assume that the data are normally distributed.
- **Bootstrap tests** are more modern and applicable to a wide range of analyses.

Classical Non-parametric Statistics

These are simple methods to apply and are implemented in all statistics packages, including SPSS, SAS, S+ and R, but not in Excel. There are non-parametric equivalents of both the paired and unpaired *t*-tests described in Section 7.3. The non-parametric equivalent of the paired *t*-test is called the **Wilcoxon sign-rank test**. The non-parametric equivalent of the unpaired *t*-test is called the **Mann–Whitney test**, or sometimes the **Wilcoxon rank-sum test**.

The Wilcoxon sign-rank test works by replacing the true value of the log-ratio data with ranks according to the magnitude of the log ratio: 1 for the smallest, 2 for the second smallest, and so on. The sum of the ranks for the "positive" (up-regulated) values is calculated and compared against a precomputed table to obtain a *p*-value.

The Mann–Whitney test is similar. The data from the two groups are combined and given ranks: 1 for the smallest, 2 for the second smallest, and so on. The ranks for the larger group are summed and that number is compared against a precalculated table to obtain a *p*-value.

These tests have the advantage of not requiring that the data are normally distributed, although the sign-rank test does require that the data are symmetric. The disadvantage of these tests is that they are less powerful than their parametric or bootstrap equivalents. The **power** of a statistical test is defined as the probability of seeing a positive result when there really is a positive result to be seen. We discuss power in greater detail in Chapter 10. Because of the loss of power, classical non-parametric statistics have not become popular for use with microarray data, and instead bootstrap methods tend to be preferred.

EXAMPLE 7.8 WILCOXON SIGN-RANK TEST ON DIUBIQUITIN FROM DATA SET 7A

The Wilcoxon sign-rank test is applied to both the unlogged and logged version of the diubiquitin data from Example 7.7. The results are as follow:

Unlogged: *p*-value is 0.00032.
Logged: *p*-value is 0.00048.

TABLE 7.5: Data for the Gene RYK from the Leukemia Data Set

Patient	ALL Log	Patient	AML Log
1	7.22	28	6.78
2	5.25	29	4.95
3	6.58	30	6.52
4	6.19	31	4.81
5	3.00	32	6.19
6	5.61	33	6.38
7	0.00	34	6.67
8	0.00	35	7.34
9	7.21	36	3.81
10	0.00	37	6.09
11	0.00	38	6.02
12	0.00		
13	6.81		
14	6.29		
15	1.00		
16	1.58		
17	7.29		
18	4.25		
19	5.32		
20	5.91		
21	4.58		
22	6.02		
23	6.00		
24	2.81		
25	5.78		
26	4.46		
27	0.00		
Average	**4.60**	**Average**	**5.96**

Note: These data were generated from Affymetrix arrays using version 4 of their software. The genes for which the hybridisation intensity of the mismatch probes was greater than the true probes had negative scores; these have been replaced with zeros in the logged data set (see Section 5.2). Later versions of the Affymetrix software do not generate negative numbers.

In both cases, the test gives a significant result, in line with the t-test analysis on the logged data. The Wilcoxon test is robust to outliers and so gives a significant result even on the unlogged data.

EXAMPLE 7.9 MANN–WHITNEY TEST ON RECEPTOR-LIKE TYROSINE KINASE FROM DATA SET 7B

The gene receptor-like tyrosine kinase (RYK) appears on the Affymetrix arrays used for data set 7B. These have been created using an early version of Affymetrix's microarray analysis suite. A number of values have negative scores and have been replaced with zeros (see Section 5.2) in the logged data set (Table 7.5).

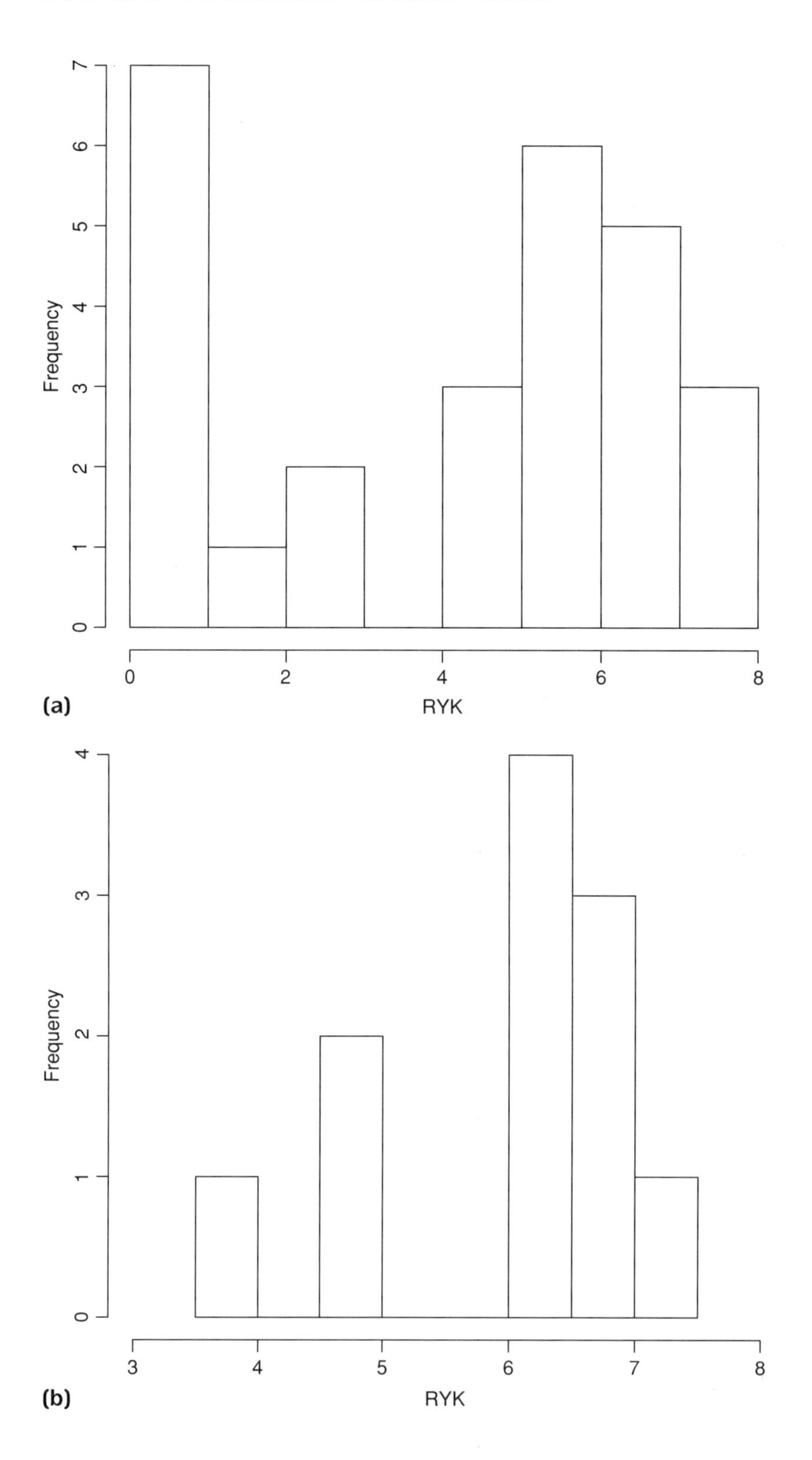
Frequency
0
1
2
3
4
5
6
7
0
2
4
6
8
(a)
RYK
Frequency
0
1
2
3
4
3
4
5
6
7
8
(b)
RYK

TABLE 7.6: Advantages and Disadvantages of Different Statistical Analyses

t-Tests	Non-parametric Tests	Bootstrap Analyses
✓ Easy	✓ Easy	✓ Robust
✓ Powerful	✓ Robust	✓ Powerful
✓ Widely implemented	✓ Widely implemented	× Requires use of specialist packages or programming
× Not appropriate for data with outliers	× Less powerful	

The p-value from the Mann–Whitney test is 0.039, which is not significant at a 1% confidence level. The two-sample t-test applied to this data gives a p-value of 0.0032, which is significant at the 1% confidence level.

The answers are different because neither the ALL data nor the AML data are normally distributed (Figure 7.6). Worse, the ALL data (Figure 7.6A) is bimodal, with 10 patients having very low or no expression, and the remaining patients showing relatively high expression. The ALL sample is very small, so it is difficult to be conclusive, but it also does not appear to be normally distributed.

Therefore, the t-test is not an appropriate analysis, and we should not believe the significant result from the t-test. We would conclude from the non-parametric analysis that this gene is not significantly differentially expressed between these two disease types. However, we must remember that the Mann–Whitney test is a less powerful test and is more likely to lead to a false negative result. In the next section we will show that the bootstrap test applied to the same data gives a significant result.

Bootstrap Analyses

As with classical non-parametric tests, bootstrap analyses do not require that the data are normally distributed and are thus robust to noise and experimental artifacts. They are also more powerful than the classical non-parametric tests. Therefore, bootstrap analyses are more appropriate for microarray analysis than either t-tests or classical non-parametric tests (Table 7.6). The disadvantage of bootstraps is that they are computationally intensive, so it is only since the advent of modern computer technology that bootstrapping has become widespread.

There are bootstrap equivalents for both the paired and unpaired analyses described earlier. It is also possible to use bootstraps for more complex analyses, such as ANOVA models (Section 7.6) and cluster analysis (Chapter 8). In this section, we will describe how the bootstrap works for unpaired data, because this is the simplest analysis to understand. At the end of the chapter, we reference an excellent book on bootstrapping should you wish to study these methods further.

Figure 7.6: Histograms of the log of the gene expression of RYK in (a) 27 ALL patients and **(b)** 11 AML patients. Negative scores in the raw data have been replaced with zeros. Neither distribution is normal. The ALL data is bimodal, which means it has two peaks: 10 patients have little or no expression and 17 patients have gene expression. The AML is a small data set, but is also not normal.

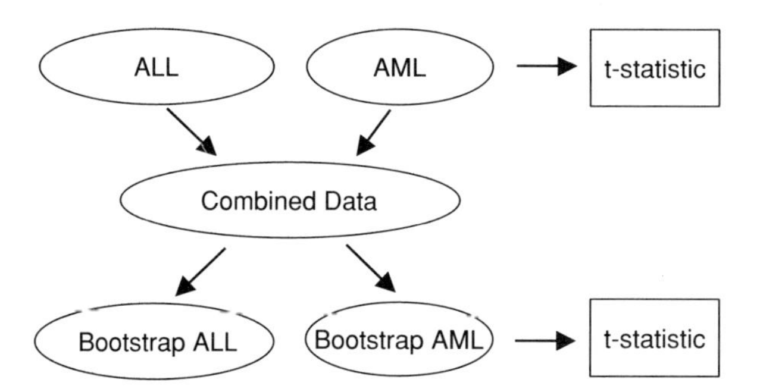

Figure 7.7: The methodology of the bootstrap applied to Data Set 7B. A t-statistic is calculated using the real data, as a measure of differential gene expression, but a p-value is not computed directly from this statistic. Instead, the data are combined, and then bootstrap data sets are constructed from the combined data. The bootstrap data sets also have 27 ALL patients and 11 AML patients, but with each patient having a measurement chosen at random from the combined 38 values from the original data. A t-statistic is computed for each of the bootstrap data sets to produce a population of t-statistics representing randomized data with measurements similar to the real data. The real t-statistic is compared with this distribution to generate a p-value.

With the unpaired analysis, there are two groups, and we seek to determine whether the means of the two groups are different. For example, with the gene RYK from data set 7B, there are 27 measurements from ALL patients and 11 measurements from AML patients; we want to know whether the gene is differentially expressed between the two groups of patients.

Under the null hypothesis, there is no difference in gene expression between the two groups. If that were the case, then any of the measurements in the data could have been observed in any of the individuals; in the example, any of the AML patients could have had any of the 38 measurements in Table 7.5 associated with both the AML and ALL patients. The bootstrap works by constructing a large number of random data sets by resampling from the original data, in which each individual is randomly allocated one of the measurements from the data, which could be from either of the groups (Figure 7.7).[7] Thus the bootstrap data sets look like the real data, in that they have similar values, but are biologically nonsense because the values have been randomized.

The aim of the test is to compare some property of the real data with a distribution of the same property in random data sets. The most commonly used property to use is the t-statistic (Equation 7.2); this is a good measure because it relates the difference in means (fold ratio) to the population variability and the number of individuals in the experiment. However, we do not use the t-distribution to calculate a p-value. Instead, we generate an empirical distribution using the t-statistics calculated from the randomized bootstrap data.

[7] There are in fact two ways of performing a bootstrap: with or without *replacement*. When a bootstrap is performed with replacement, different individuals in the bootstrap data could have the same value from the real data. When a bootstrap is performed without replacement, each of the real values is only used once in the bootstrap data. In this chapter, we describe the method with replacement, while the significance analysis of microarrays (SAM) software mentioned later performs the bootstrap without replacement. Although there is some debate about which is better, in practical terms the two methods are fairly equivalent and produce very similar results.

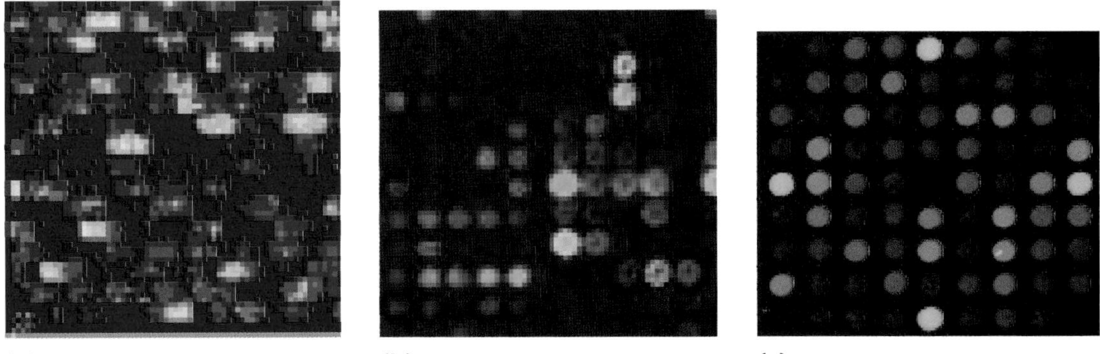

(a) (b) (c)

Figure 1.7: Array quality. (a) On Affymetrix arrays the features are rectangular regions. The masks refract light, so there is leakage of signal from one feature to the next. The Affymetrix image-processing software compensates for this by using only the interior portions of the features. **(b)** Spotted arrays produce spots of variable size and quality. This image shows some of this variation; we cover image processing of spotted arrays in detail in Chapter 4. **(c)** Inkjet arrays tend to be of the highest quality, with regular, even spots.

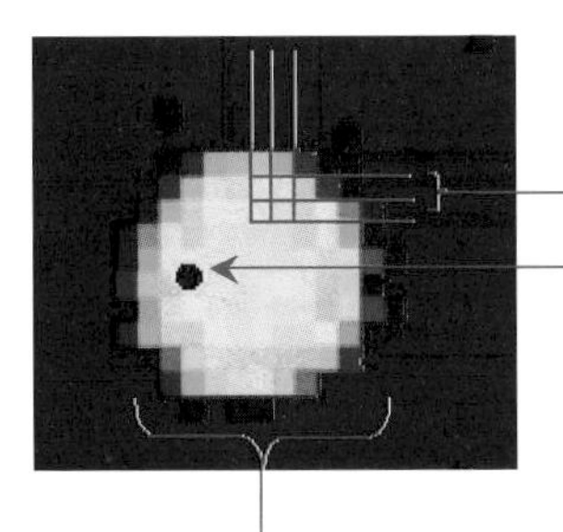

(a)

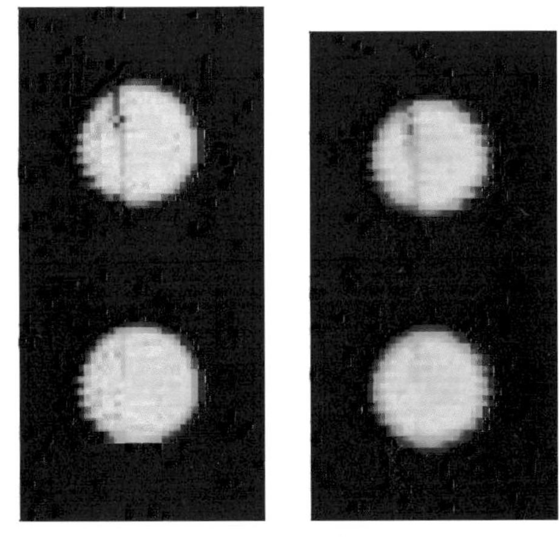

(d) (e)

Figure 1.12: The pixels comprising a feature. (a) A false-colour image of the pixels from a single scan of a 100-μm microarray feature. The size of the laser spot is 5 μm. The pixel size has been set to 5 μm so that each pixel represents the area from the size of the laser spot. **(b)** and **(c)** See pp. 9 and 10. **(d)** Two neighbouring features on an array with a streak through them, measured with a laser spot size of 5 μm and a pixel size of 5 μm. The streak is clear on both spots and so the spot can be identified as problematic. **(e)** The same features scanned with laser spot size of 10 μm and a pixel size of 5 μm. The streak has become blurred.

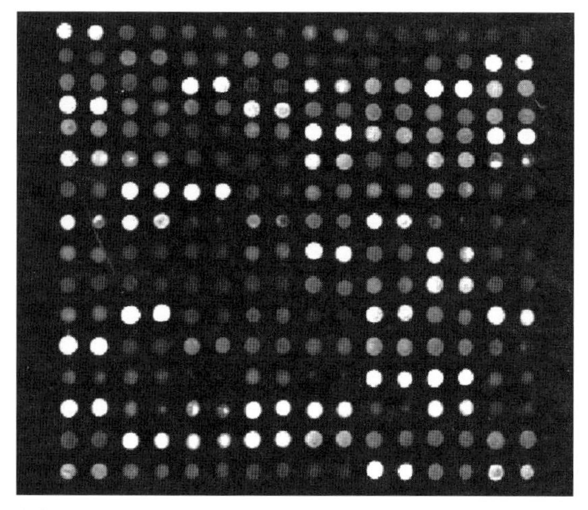

(a)

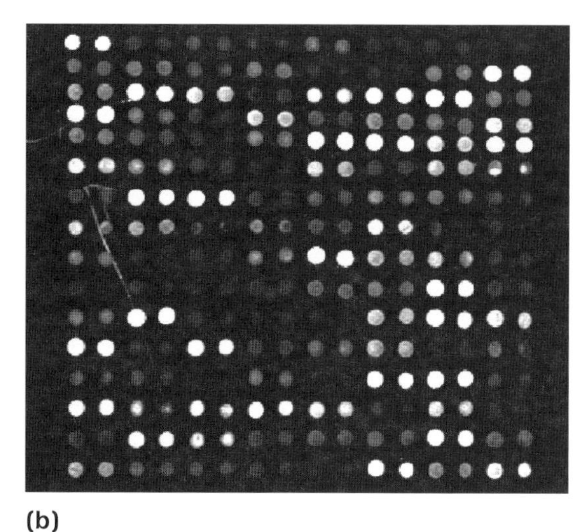

(b)

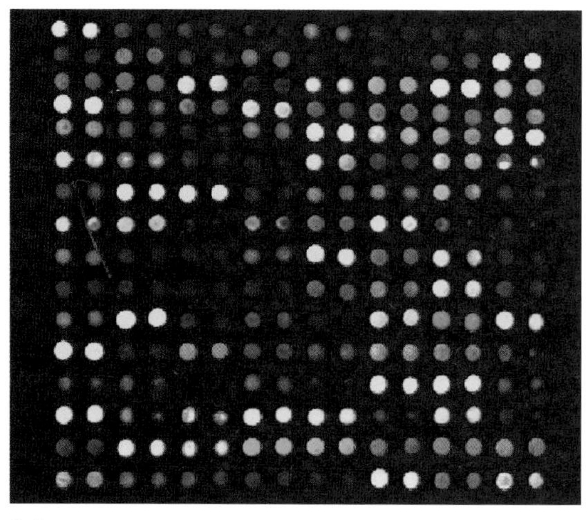

(c)

Figure 1.13: Output of scanners. (a) This is the scanner output for a part of a microarray – in this case one of twelve 16×16 blocks of features. This is the monochrome image of the Cy3 (green) channel. **(b)** The scanner output for the same part of the array but using the Cy5 (red) channel. **(c)** It is usual to combine the two monochrome images into a composite false-colour image of the array. Green features correspond to features that are expressed more in the sample labelled with Cy3 than the sample labelled with Cy5, and so will be bright in (a) and dark in (b). Similarly, red spots will be bright in (b) and dark in (a). Yellow features have a similar level of expression in both samples. Dark features are low expressed in both samples.

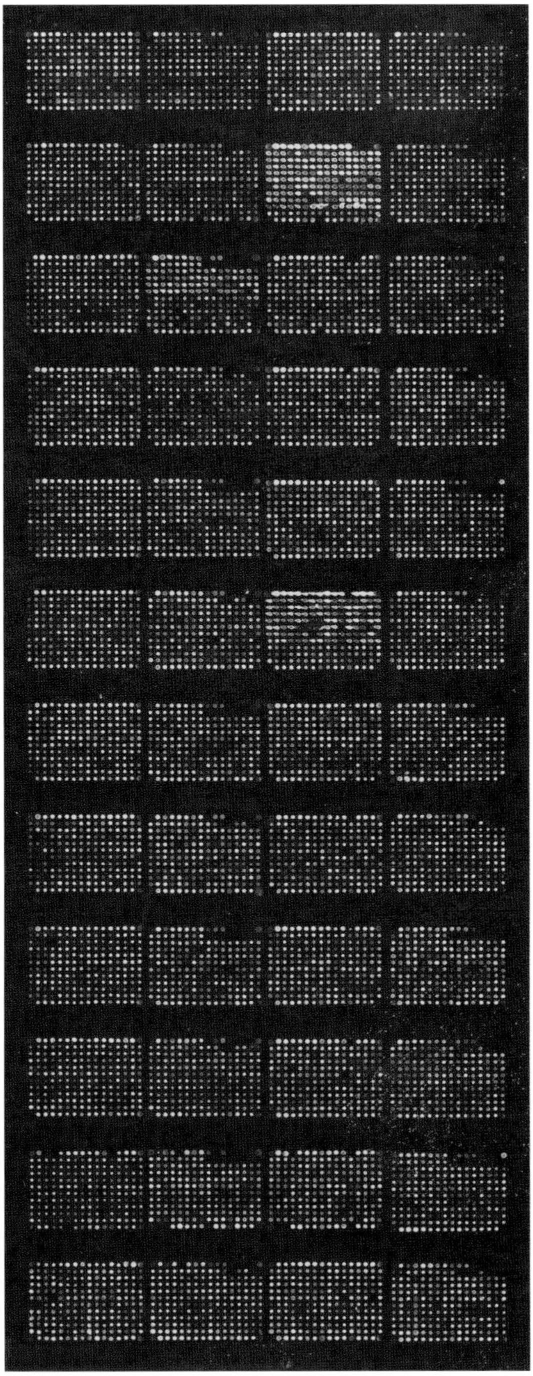

(a)

Figure 4.1: (a) An example image of a complete microarray. In this case, there are 48 grids in a 12×4 pattern, and each grid has 12×16 features. Therefore, there are a total of 9,216 features on this array. **(b)** See p. 64.

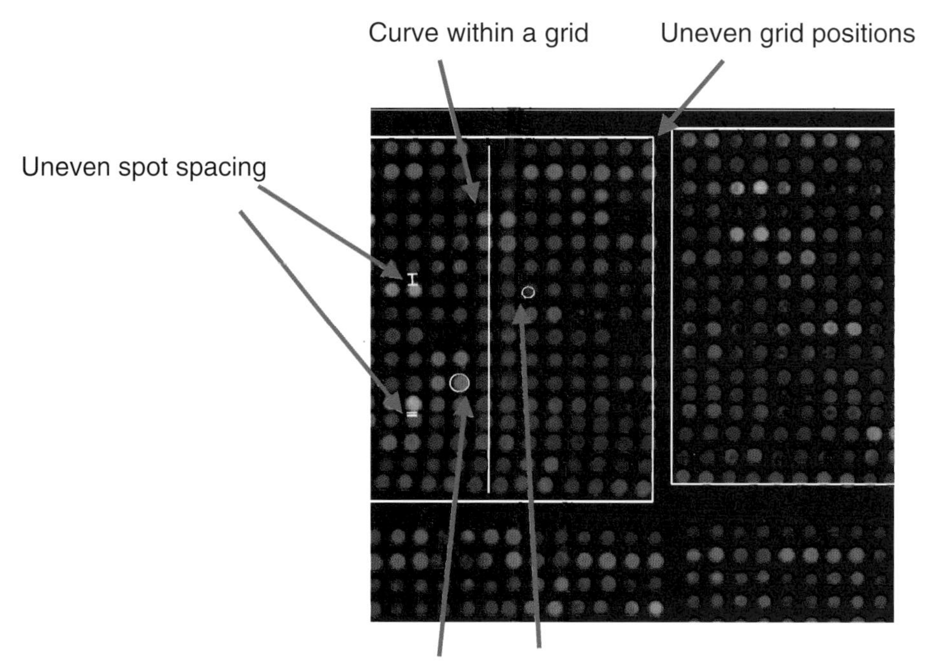

Figure 4.2: Problems with microarray images from pin-spotted arrays: (1) Uneven grid positions. The two grids are not aligned. This occurs because the pins are not perfectly aligned in the cassette. (2) Curve within the grid. Note that the centers of the features at the top of the vertical line lie on the line, but that the centers of the features at the bottom of the line are to the left of the line. This can happen if the array is not horizontal during array manufacture, or because of movement of the pins during manufacture. (3) Uneven spacing between features. This occurs because of pins moving during manufacture; this itself could result from the glass slide not being perfectly flat. (4) Uneven feature sizes. Different features can have different sizes as a result of different volumes of liquid being deposited on the array. This can also result from uneven drying of the features, so it is important to maintain constant temperature and humidity of the array during the manufacture process.

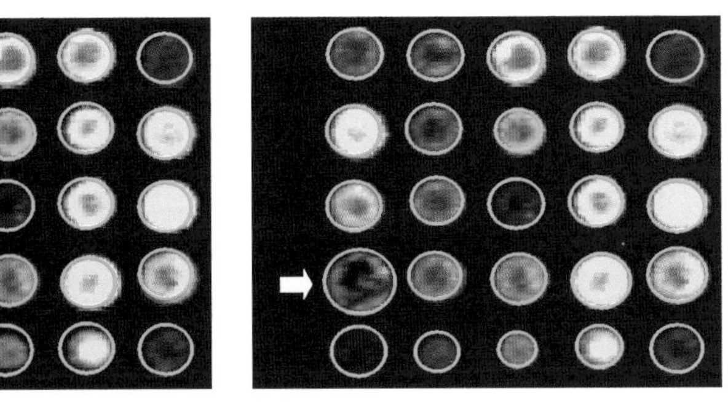

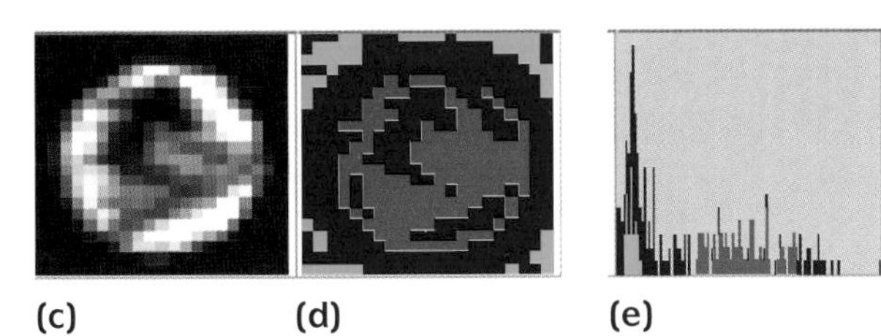

Figure 4.3: (a) Fixed circle segmentation. A circle of the same size is placed on every feature on the array and the pixels inside the circle are used to determine the intensity of the feature. This is not a good method because the circle will be too large for some features and too small for others. **(b)** Variable circle segmentation. A circle of different size is applied to each feature and the pixels inside the circle are used to determine the intensity of the feature. This performs better on different size features but does not perform so well on features with irregular shapes, for example, the irregular red feature that is marked with an arrow. **(c)** Zoom in on the red channel of the irregularly shaped feature marked with the arrow in (b). Note the black region where there is no hybridisation, probably because there is no probe attached to the glass in that area. **(d)** Histogram method applied to that feature. The red pixels are the ones that have been used to calculate the feature signal; the green pixels have been used to calculate the feature background. The black pixels are unused. The area corresponding to the black region in (c) is not used for calculating the feature intensity. The brightest features have also been excluded. The red-to-green ratio of this feature calculated by fixed circle segmentation is 1.8, variable circle segmentation is 1.9, and histogram segmentation is 2.6; so the measured differential gene expression between the samples is different with the different algorithms. Because of the irregular shape of the feature, the histogram method probably gives the most realistic measurement. **(e)** Histogram of the intensities of the pixels in the irregularly shaped feature. The red bars represent pixels used for the signal intensity; the green bars represent pixels used for the background intensity; the black bars are unused pixels. The brightest and darkest pixels are not used, thus giving a better measurement of hybridisation intensity.

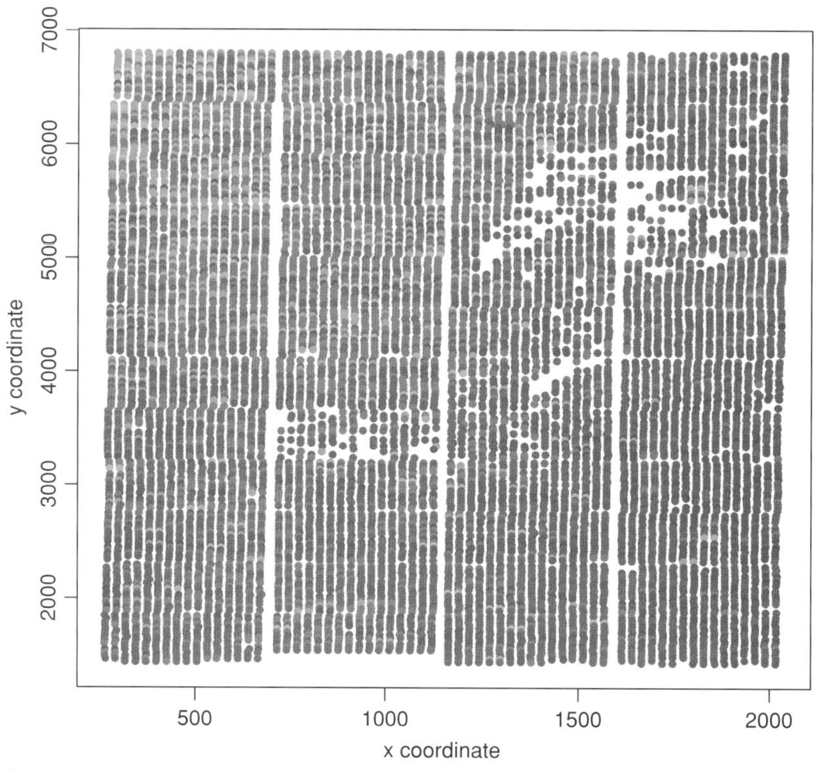

Figure 5.7: Spatial bias on a microarray and two-dimensional Loess regression. (a) False-colour representation of the log ratios of a microarray, with mouse kidney in Cy3 and liver from the same mouse in Cy5 (data set 5B). Each spot represents a feature. The *x* and *y* coordinates of each spot correspond to the *x* and *y* coordinates of the feature on the array. The colour of the spot represents the log ratio (Cy5/Cy3) of the feature, with red spots having a positive log ratio and green spots having a negative log ratio. There is a strong spatial bias on the array, with green spots in the top-left-hand corner and red spots in the bottom-right-hand corner. The areas of the array with missing spots represent features that have been flagged by the image-processing software, or features with a higher background than signal that have been removed from the data set. **(b)** The same data, but with the fit of a two-dimensional Loess surface to the log ratios superimposed as contours. The contours follow the colour trend, going from negative at top left to positive at bottom right. **(c)** False-colour plot of the normalised log ratio values of the features. These are calculated by subtracting the fitted values of the Loess surface from the raw log ratios. There is no spatial bias on the normalised data.

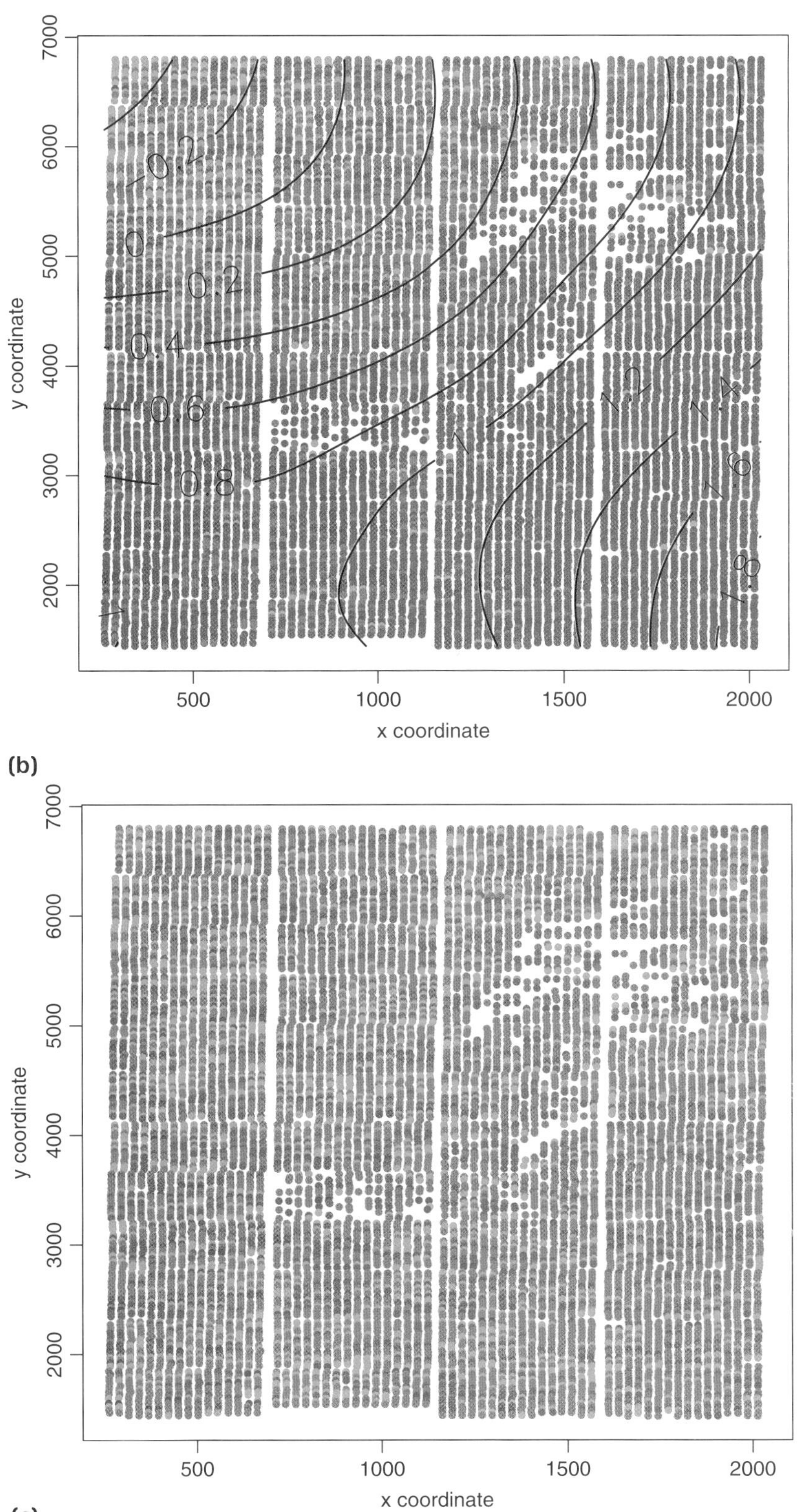
y coordinate
x coordinate
7000
6000
5000
4000
3000
2000
500
1000
1500
2000
y coordinate
x coordinate
7000
6000
5000
4000
3000
2000
500
1000
1500
2000

(b)

(c)

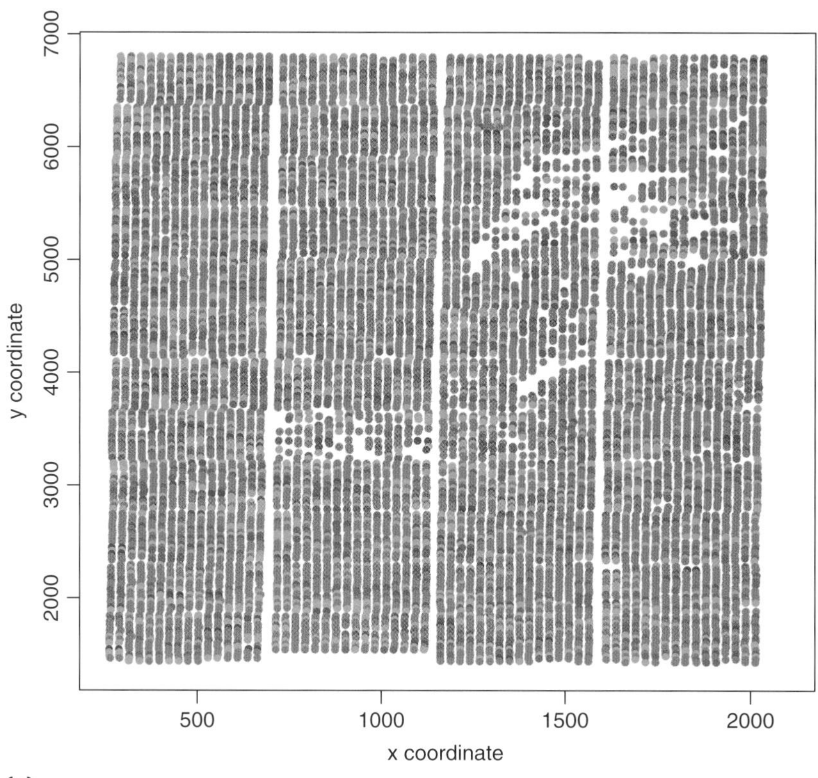

(c)

Figure 5.8: Block-by-block regression. Block-by-block regression is performed by applying one-dimensional Loess normalisation to the features in each grid on the array separately. The array in data set 5B has 48 grids. **(a)** and **(b)** See p. 92. **(c)** The whole array has been normalised using block-by-block normalisation. The spatial bias has been eliminated.

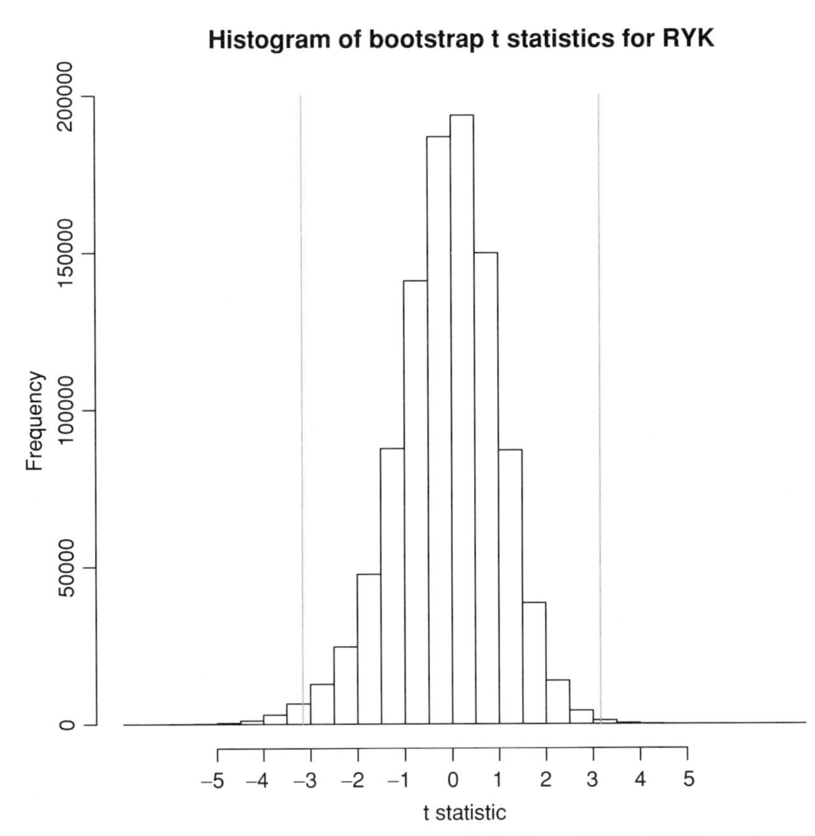

Figure 7.8: Histogram of the results of t-statistics from 1,000,000 bootstrap resamples for the gene RYK from the leukemia data. The t-statistic of the real data is 3.1596. We have marked $+3.1596$ and -3.1596 on the histogram to demonstrate that the majority of the bootstrap statistics are less extreme: 9,750 of the bootstrap t-statistics lie outside these lines. Thus, the p-value is just under 0.001, and we conclude that this gene is significantly differentially expressed at the 1% level.

The t-statistic from the real data is compared with the distribution of t-statistics from the bootstrap data (Figure 7.8). We calculate an empirical p-value by computing the proportion of bootstrap statistics that have a more extreme value than the t-statistic from the real data. If the real t-statistic is in the belly of the distribution, then it is indistinguishable from t-statistics generated from randomized data. We would conclude that the gene is not significantly differentially expressed. If, on the other hand, the statistic from the real data is towards the edge of the bootstrap distribution, then it is unlikely that the experimental result can have arisen by chance, and we would conclude that the gene is significantly differentially expressed.

EXAMPLE 7.10 BOOSTRAPPING RYK FROM DATA SET 7B

Consider again the gene RYK from the previous example. There are many arrays on which the control signal is greater than the gene signal, so the Affymetrix software calls the gene absent. We replaced these readings with zeros in the logged data. We construct a t-statistic for the real data, using Equation 7.2. The t-statistic is 3.1596, but we do not use the t-distribution to calculate a p-value from this statistic.

TABLE 7.7: Advantages of Different Bootstrap Implementations

SAM	BOOT Package in R	Writing Your Own Program
✓ Easy to use Excel plug-in	✓ Easy to use if you can use R or S+	✓ Maximum speed and efficiency
✓ Can handle many types of data	✓ Faster than SAM	✓ Can run in background on server
✓ Good false discovery rate algorithm		✓ Good for high-throughput analyses

Next, we create bootstrap data sets, each of which also consists of 27 ALL patients and 11 AML patients. For each patient, we choose a measurement at random from the 38 observed values in Table 7.6 and assign that value to that patient. For each bootstrap data set we construct a t-statistic, also using Equation 7.2, and record the value. We repeated this procedure 1,000,000 times to generate a bootstrap distribution of the t-statistics (Figure 7.8). Of the 1,000,000 values, 9,750 had an absolute value greater than 3.1596, the t-statistic from the real data. Thus the bootstrap p-value for RYK is just under 0.001, which is significant at the 1% level. This contrasts with the result for the classical non-parametric test.

In Example 7.10, we used 1,000,000 bootstrap data sets to generate a distribution. This is more than is usually necessary. In general, we recommend performing the bootstrap at 10 times the number of genes on the array being analysed. So on a microarray with 10,000 genes, we would recommend 100,000 bootstrap data sets. The reason for this is that the number of replicates determines the granularity of the p-values; we will see in Section 7.5 that we may need to use p-values equal to one over the number of genes on the array. The use of this recommended number of bootstrap data sets enables you to do so with reasonable accuracy.

Significance Analysis of Microarrays

It is straightforward for a reasonably proficient programmer to write code to perform bootstrap analyses. However, there is a package available called SAM that performs bootstrap analyses on microarray data. It also provides a false discovery rate estimate using a method that is more sophisticated than the method described in Section 7.5.

SAM is available as an Excel plug-in from the URL given at the end of the chapter. It is very easy to use, but can be slow on large data sets. I was not able to perform more than a few hundred replicates on the two data sets I have used as examples in this chapter using SAM on my desktop PC. However, it is useful for smaller data sets and exploratory work. We summarise the advantages of using SAM or other bootstrap implementations in Table 7.7.

SECTION 7.5 MULTIPLICITY OF TESTING

In Sections 7.3 and 7.4, we have been performing statistical tests on different genes and concluding that these genes may be up- or down-regulated based on these tests.

When we analyse a microarray experiment, we want to apply these tests to many genes in parallel. In data set 7A, the breast cancer data, there are 9,216 genes. In data set 7B, the leukemia data, there are 7,070 genes. Microarrays today typically have between 10,000 and 30,000 genes, and it is probable that in the future there will be arrays for each splice variant of every human gene, with perhaps as many as 300,000 features.

There is a serious consequence of performing statistical tests on many genes in parallel, which is known as **multiplity of *p*-values.** Suppose we perform a thought experiment, in which we had microarrays with 10,000 genes. We take a large supply of reference sample, label it with Cy3 and Cy5, and co-hybridise the reference sample to a number of arrays. We take the data, perform appropriate normalisation to remove dye bias, and perform our statistical analysis of choice to the 10,000 genes: this could be a t-test, a Wilcoxon test, or a bootstrap test.

Since every sample hybridised to the arrays is the same reference sample, we know that no genes are differentially expressed: all measured differences in expression are experimental error. But our statistical analysis will tell us a different story. By the very definition of a p-value, each gene would have a 1% chance of having a p-value of less than 0.01, and thus be significant at the 1% level. Because there are 10,000 genes on this imaginary microarray, we would expect to find 100 significant genes at this level. Similarly, we would expect to find 10 genes with a p-value less than 0.001, and 1 gene with p-value less than 0.0001.

Now consider data set 7A, with 9,216 genes. Even if the chemotherapy had no effect whatsoever, we would expect to find 92 "differentially expressed" genes with p-values less than 0.01, simply because of the large number of genes being analyzed. This leads to an important question: how do we know that the genes that appear to be differentially expressed are truly differentially expressed and are not just artifact introduced because we are analyzing a large number of genes? More concretely, how would we interpret our result for ACAT2 (Example 7.4)? Is this gene truly differentially expressed, or could it be a false positive result?

This is a deep problem in statistics, which affects many applications, and not just microarray analysis. We will describe a simple method that can be used to estimate the percentage of genes that have been called up- or down-regulated that are likely to be false positives. A more sophisticated method is implemented in the SAM software. We will also describe the Bonferroni correction and demonstrate that this is not appropriate for microarray analysis.

Estimation of False Positive Rate

We describe a simple but effective analysis to estimate the false positive rate of a statistical test and thereby choose an appropriate p-value threshold for significantly differentially expressed genes that gives an acceptable false positive rate. There are five steps:

1. Perform the statistical analysis of choice on every gene being analyzed and record the p-value for each gene.

TABLE 7.8: Number and Percentage of False Positives in the Breast Cancer Analysis

For each p-value threshold, we count the number of genes observed in the data with p-values less than that threshold. This is compared with the expected number of false positives, which is the number of genes being tested multiplied by the p-value threshold. The percentage of false positives is the expected number of false positives divided by the observed number of genes. The smaller the threshold used, the fewer false positives, and the better the false positive rate. However, this is at the cost of introducing greater numbers of false negative results.

p-Value Less Than or Equal To	Observed Number of Genes	Expected Number of False Positives	Percentage of False Positives
10^{-2}	184	64	35
10^{-3}	35	6	18
10^{-4}	15	0.6	4
10^{-5}	6	0.06	1

2. For an appropriate range of significance thresholds, identify the number of genes with p-values less than that threshold.
3. For the same significance thresholds, calculate the expected number of false positives by mutliplying the p-value by the number of genes being analyzed.
4. For each threshold, the percentage of false positives is the expected number of false positives divided by the number of genes identified as expressed at that threshold.
5. Choose a threshold that gives an acceptable false positive rate.

EXAMPLE 7.11 MULTIPLICITY OF p-VALUES IN DATA SET 7A

We performed a bootstrap test on 6,350 genes for which there was data from all 20 patients from data set 7A, using 100,000 bootstrap data sets. In Table 7.8, we show number of genes from the breast cancer data with different ranges of p-values, alongside the number of genes expected to have those p-values from the multiple testing.

We can see from Table 7.8 that using a traditional significance threshold of 1%, which would be stringent in classical statistics, we would expect to see 64 genes significantly differentially regulated; with the real data, we see 184 differentially expressed genes. Thus we estimate that 35% of these are false positive results.

On the other hand, the expected number of false positives with p-value less than 0.0001 is 0.6, so it is likely that 14 or 15 out of the 15 observed differentially expressed genes are true positive results. However, at this more stringent level, we will be missing at least 100 truly differentially expressed genes: our false negative rate has increased.

This illustrates the trade-off that is always played between controlling false positive and false negative results: a more stringent p-value threshold may lead to fewer false positives, but will give more false negatives; a less stringent p-value may give fewer false negatives, but will give more false positives. The only way to improve both rates is to increase the number of individuals in the study; this is discussed in Chapter 10.

TABLE 7.9: Significant Genes from the Breast Cancer Data Set

The unadjusted p-values are the proportion of the 100,000 bootstrap data sets that had t-statistics more extreme than the t-statistic from the real data. Thus the smallest possible p-value is 1/100,000 (or 10^{-5}). Because of the number of genes in the analysis, the Bonferroni corrected p-values are all too large to be significant, illustrating that this method is not applicable to most microarray data.

Accession	Description	p-Value	Bonferroni Adjusted p-Value
AA598794	connective tissue growth factor	10^{-5}	0.064
N23941	cyclin-dependent kinase inhibitor 1A	10^{-5}	0.064
AA478553	dopachrome tautomerase	10^{-5}	0.064
W96134	v-jun avian sarcoma virus 17 oncogene homolog	10^{-5}	0.064
AA044993	connective tissue growth factor	10^{-5}	0.064
AA040944	v-fos FBJ murine osteosarcoma viral oncogene homolog	10^{-5}	0.064
N95402	copine V	2×10^{-5}	0.13
R12840	v-fos FBJ murine osteosarcoma viral oncogene homolog	3×10^{-5}	0.19
AA442853	cyclin-dependent kinase 5, regulatory subunit 1 (p35)	$4.\times10^{-5}$	0.25
AA418077	GTP-binding protein overexpressed in skeletal muscle	5×10^{-5}	0.32
AA133129	transcription elongation factor B (SIII), polypeptide 3	5×10^{-5}	0.32
AA485377	v-fos FBJ murine osteosarcoma viral oncogene homolog	6×10^{-5}	0.38
AA134757	fibulin 1	6×10^{-5}	0.38
AI831083	dihydropyrimidinase-like 3	7×10^{-5}	0.45
AA004637	ESTs	9×10^{-5}	0.57
No Annotation		1.2×10^{-4}	0.76
AA025939	CD4 antigen (p55)	2×10^{-4}	1.3
H21041	activating transcription factor 3	2.3×10^{-4}	1.5
AA449463	KIAA0220 protein	2.6×10^{-4}	1.7
H05099	KIAA0182 protein	3.8×10^{-4}	2.4

EXAMPLE 7.12 CHOOSING DIFFERENTIALLY EXPRESSED GENES FROM DATA SET 7A

In an example analysis of data set 7A, we decided that it was important not to have false positive results. Therefore, we set a p-value threshold so that the expected number of false positive results was 1; this threshold is equal to the reciprocal of the number of genes tested, or $1/6350 = 1.6 \times 10^{-4}$. The top 20 genes from the analysis are listed in Table 7.9. We can see from the table that there are 16 genes that pass this threshold; the false positive rate is thus approximately 1 in 16, or about 6%.

Bonferroni Adjustment

The final column in Table 7.9 gives Bonferroni adjusted p-values. The Bonferroni correction is a traditional approach for modifying the p-values when performing many statistical tests in parallel. It is very similar to the approach we described, but more stringent. p-Values are calculated in the normal way via the statistical test of choice, and the p-values are then all multiplied by the number of tests being performed. In the case of gene expression analysis, the p-values would all be multiplied by the number of genes in the analysis.

The problem with the Bonferroni adjustment is that it is usually too stringent for microarray analysis. The p-values obtained are frequently so large that no genes are deemed differentially expressed. The next example illustrates this.

EXAMPLE 7.13 BONFERRONI ADJUSTED p-VALUES FOR DATA SET 7A

The Bonferroni correction is applied to the breast cancer data of Example 7.12. There are 6,350 genes tested, so each p-value is multiplied by 6,350 (Table 7.8). Even if we use a liberal significance threshold of 5%, not a single gene would be listed as significant, so this would not be a good method for analyzing this data.

SECTION 7.6 ANOVA AND GENERAL LINEAR MODELS

Up to this point, we have described methods for analyzing very simple experiments for differentially expressed genes, in which the data is either paired, with two biological samples derived from the same individual, or unpaired, with two groups of individuals being compared.

Increasingly, microarrays are being used to perform more complex experiments, in which there may be more than two groups, or in which the response to more than one variable is being measured. These types of experiments require more sophisticated analyses known as ANOVA and general linear models. We will introduce these ideas very briefly; there are many intermediate-level and advanced-level statistics text books that the interested reader can consult for greater detail. These methods are also implemented in all statistics software, for example, SPSS, SAS, S+ and R, which all have documentation to describe how to use them.

The One-Way ANOVA

Data set 7B had two groups of patients, and we were interested in comparing gene expression in the two groups to identify differentially expressed genes. Suppose instead there were three groups of patients, and we were interested in identifying genes that were differentially expressed on one or more of the groups relative to the others. There are two ways we could perform the analysis:

- Naively, we could apply an unpaired t-test three times, to each pair of groups in turn, and select genes that are significant in one or more of the t-tests.
- Instead, we could use a statistical test that compares all three groups simultaneously and reports a single p-value.

There are two problems with the first method. The first is multiplicity: by performing three tests, we increase the likelihood of seeing a significant difference between two of the groups as a result of measurement errors. This problem gets worse as the number of groups increases: for example, with 7 groups, there would be 21 separate comparisons. The second problem is that each of these comparisons is not independent of the other, so it becomes very difficult to interpret the results.

The approach taken by statisticians is the **one-way ANOVA**. This performs an analysis of this type of data, where we are comparing two or more groups, and returns a single p-value that is significant if one or more groups is different from the others.

EXAMPLE 7.14 DATA SET 7C AND ANOVA ANALYSIS

In data set 7C, samples were taken from four groups of patients suffering from four different types of cancer: neuroblastoma (NB), non-Hodgkin lymphoma (NHL), rhabdomyosarcoma (RMS) and Ewing tumours (EWS). If we wanted to identify genes that are differentially expressed in one or more of these four groups, we use a one-way ANOVA. This is much better than performing six separate tests to compare all of the groups with the others.

Multifactor ANOVAs

In the preceding example, there is only one variable that affects gene expression: the group to which the individual belongs, which in data set 7B or 7C, is the type of cancer the patient is suffering from. Suppose, however, we are analysing data in which the patients are suffering from two types of leukemia, and we also know the sex of the patients, who may be male or female. In this case, gene expression could depend either on the type of disease, on the sex of the patient, or both. More general ANOVA models can be built that include two or more factors and that will return a p-value for each of the factors separately. In this example, there would be one p-value for whether or not the gene is differentially expressed because of disease type, and another p-value for whether or not the gene is differentially expressed because of the sex of the patient.

With multifactor ANOVAs, it is possible for two factors to behave together in an additive or multiplicative fashion. If the response to the two factors is additive, then the effect of one factor does not influence the effect of the other. However, suppose there is a gene that is differentially expressed in male ALL patients relative to male AML patients, but which is not differentially expressed in women. In that case, the factors are behaving multiplicatively, and statisticians talk about an **interaction** between the factors. The statistics packages allow for the user to build interactions into the ANOVA models.

Frequently in microarray analysis, there may be factors in the experiment that are of no intrinsic scientific interest, but which can influence the observed gene expression. For example, there may be two or more scientists performing the hybridisations. In such cases, we may want to include these factors into the statistical model; such factors are called **random effects** and have to be handled slightly differently.

General Linear Models

ANOVA analyses are appropriate where the factors on which gene expression depends are all **categorical variables**, such as type of disease or sex of the individual. Sometimes, however, we may have factors that are continuous variables, for example, the dose of compound added to a sample. If it is thought that the gene expression responds in a linear fashion to such a variable, then statisticians will use **general linear models** to analyse the data.

General linear models can combine both categorical variables and continuous variables; thus, they are generalizations of both ANOVAs and linear regressions. As with ANOVA models, the general linear model returns a separate *p*-value for each of the factors being tested. It is also possible for general linear models to include interactions. Suppose we have an experiment in which different doses of a compound are given to male and female mice: if the dose response is different in females vs. males, then there is an interaction between these factors.

Both ANOVA models and general linear models are similar to *t*-tests in that they require that variability in the data is normally distributed. However, it is possible to apply bootstrap analyses to these more sophisticated tests, which makes them applicable to microarray analysis. This approach has been taken in many of the papers describing the applications of ANOVA to microarrays.

KEY POINTS SUMMARY

- Statistical analyses for differentially expressed genes are best carried out via hypothesis tests rather than using a simple fold ratio threshold.
- The structure of your data might be paired, unpaired, or more complex.
- The traditional *t*-tests may not be appropriate for microarray data because they require that the data are normally distributed.
- Non-parametric tests are robust to experimental noise and bootstrap tests are the most powerful versions available.
- The large number of genes being tested introduces the problem of multiplicity, and so it is important to perform an analysis of the false positive rate.
- More complex data may require analysis via ANOVA or general linear models and may also include bootstrapping.

SUPPLEMENTARY INFORMATION AND REFERENCES

Publications from Which We Have Used Data

Data Set 7A

Perou, C.M., Sorlie, T., Eisen, M.B., van de Rijn, M., Jeffrey, S.S., Rees, C.A., Pollack, J.R., Ross, D.T., Johnsen, H., Akslen, L.A., Fluge, O., Pergamenschikov, A., Williams,

C., Zhu, S.X., Lonning, P.E., Borresen-Dale, A., Brown, P.O., and Botstein, D. 2000. Molecular portraits of human breast tumours. *Nature* 406: 747–52.

Data Set 7B

Golub, T.R., Slonim, D.K., Tamayo, P., Huard, C., Gaasenbek, M., Mesirov, J.P., Coller, H., Loh, M.L., Downing, J.R., Caligiuri, M.A., Bloomfield, C.D., and Lander, E.S. 1999. Molecular classification and class prediction by gene expression monitoring. *Science* 286: 531–37.

Data Set 7C

Khan, J., Wei, J.S., Ringner, M., Saal, L.H., Ladanyi, M., Westermann, F., Berthold, F., Schwab, M., Antonescu, C.R., Peterson, C., and Meltzer, P.S. 2001. Classification and diagnostic prediction of cancers using gene expression profiling and artificial neural networks. *Nature Medicine* 7: 673–79.

Statistics Textbooks

Witner, J.A. and Samuels, M.L. 2002. *Statistics for the Life Sciences* (3rd Edition). Prentice-Hall: Englewood Cliffs, New Jersey.

An excellent introduction to statistics that I have found very helpful for teaching statistics to biology undergraduates.

Kanji, G.K. 1993. *100 Statistical Tests*. SAGE Publications: London.

A very useful summary of statistical tests for a wide range of data. Will be understandable to someone with basic statistics background (e.g., who has worked through the material in Witner and Samuels).

Efron, B. and Tibshirani, R.J. 1998. *An Introduction to the Bootstrap*. Chapman & Hall/CRC: New York.

Seminal text covering all aspects of bootstrap analyses. Required reading for anyone who wants to write their own bootstrap routines, but does require good numeracy skills.

Publications for Microarray Analysis Methods

Tusher, V.G., Tibshirani, R., and Chu, G. 2001. Significance analysis of microarrays applied to ionizing radiation response. *PNAS* 98(9): 5116–21.

The paper for SAM.

Kerr, M.K., Martin, M., and Churchill, G.A. 2000. Analysis of variance for gene expression microarray data. *Journal of Computational Biology* 7(6): 819–37.

ANOVA analysis of microarray data. Uses an ANOVA model to fit the data with bootstrapping to deduce p-values. An excellent method, but requires specialist code to apply.

Ideker, T., Thorsson, V., Siegel, A.F., and Hood, L.E. 2000. Testing for differentially expressed genes by maximum likelihood analysis of microarray data. *Journal of Computational Biology* 7(6): 805–17.

A very similar ANOVA analysis that computes parameters using maximum likelihood estimation.

Thomas, J.G., Olson, J.M., Tapscott, S.J., and Zhao, L.P. 2001. An efficient and robust statistical modeling approach to discover differentially expressed genes using genomic expression profiles. *Genome Research* 11: 1227–36.

Another similar ANOVA analysis.

Internet Resources

http://genome-www5.stanford.edu/MicroArray/SMD/

The Stanford Microarray Database. All the data sets we refer to in this chapter are available for download from this database. There are also links to the papers from which the data derive.

http://www-stat.stanford.edu/~tibs/SAM/

The homepage for the SAM software. Academic users can download the SAM software for free after obtaining a license code.

http://www.r-project.org/

The homepage for R, a free statistics package, which is very similar to S. This is available for Unix, Windows and Macintosh, and has a wide range of statistics and graphing functionality. Unlike S+ or SPSS, it does not have a graphical user interface and is operated via commands and scripts.

CHAPTER EIGHT

Analysis of Relationships Between Genes, Tissues or Treatments

SECTION 8.1 INTRODUCTION

Microarrays are a genomic technology. Genomics is different from genetics in that it looks not at genes in isolation, but at how many genes work together to produce phenotypic effects. In Chapter 7 we saw how microarrays can be used to study genes in isolation. But much of the real power of microarrays is their ability to be used to study the relationships between genes and to identify genes or samples that behave in a similar or coordinated manner. This chapter looks at a number of analysis techniques to find and verify such relationships.

We will use two example data sets to examine the ideas of this chapter.

EXAMPLE 8.1 YEAST SPORULATION DATA (DATA SET 8A)

Budding yeast can reproduce sexually by producing haploid cells through a process called sporulation. Yeast was placed in a sporulating medium, and samples were taken at six time points following the start of sporulation and hybridised to microarrays. We want to identify groups of genes that behave in a coordinated manner in this time series.[1]

EXAMPLE 8.2 DIFFUSE B-CELL LYMPHOMA SUBTYPES (DATA SET 8B)

Samples were taken from 39 patients suffering from diffuse large B-cell lymphomas and hybridized to microarrays. We want to identify genes that are co-regulated in this disease. We are also interested in whether there are groups of patients with similar gene expression profiles.[2]

This chapter discusses methods that can be used to answer such questions; it is organised into the following five sections:

Section 8.2: Similarity of Gene or Sample Profiles, looks at different methods for quantifying the similarity or dissimilarity of gene expression profiles. We show how the different methods can give different results and, hence, the need to think carefully about choosing the method you use.

[1] The data are taken from the work of Chu et al. (1998). The full reference is given at the end of the chapter, and the data are available from the Stanford Microarray Database.

[2] The data are taken from the work of Alizadeh et al. (2000). The full reference is given at the end of the chapter, and the data are available from the Stanford Microarray Database.

Section 8.3: Dimensionality Reduction, describes two methods for reducing the dimensionality of data: principal component analysis and multidimensional scaling. One of the problems of microarray data analysis is that it is very difficult for the human brain to conceptualise the large number of genes and samples typically involved. For example, in data set 8B, there are 40 samples hybridised to microarrays with nearly 18,000 genes. These methods can reduce the data to two or three dimensions, allowing the human user to easily visualise it; these are also useful tools for using the classification methods described in Chapter 9.

Section 8.4: Hierarchical Clustering, introduces the most commonly used method for identifying groups of closely related genes or tissues. Hierarchical clustering is a method that successively links genes or samples with similar profiles to form a tree structure – much like a phylogenetic tree. We describe different versions of the algorithm and show how they can give different results from the same data.

Section 8.5: The Reliability and Robustness of Hierarchical Clustering, describes methods for validating hierarchical clustering. We focus on bootstrapping and consensus tree construction, which can be used both to validate clusters and to place a numerical measure of confidence on them.

Section 8.6: Machine-Learning Methods for Cluster Analysis, describes two further methods for clustering data, both of which derive from the machine-learning community. K-means clustering is a method of non-hierarchical clustering that requires the analyst to supply the number of clusters in advance and then allocates genes or samples to clusters appropriately. The self-organised map is a related method that allocates genes or samples to a predefined number of clusters that relate to each other on a spatial grid. Both methods are implemented in a wide range of gene expression analysis software packages.

SECTION 8.2 SIMILARITY OF GENE OR SAMPLE PROFILES

When we use microarrays as a genomic research tool, we want to identify genes or samples that have similar expression profiles. Given graphs or charts of such profiles, most people would have a good intuition about what this means. When we perform computational data analysis, we need to be able to transform this intuition into quantitative measures that can be computed by software, and which reflect this intuition in a reliable and robust way.

In this chapter, we have deliberately chosen two quite different data sets: data set 8A is a time series, and data set 8B looks at samples from a cohort of patients. We will describe a number of measures of similarity between profiles and show examples of these measures applied to them. We will then show how two profiles can be similar under one measure, but different under another – hence, the importance of choosing an appropriate similarity measure for the analysis techniques described later in the chapter.

There are two ways to analyse microarray data: either we are interested in the similarity of genes, or we are interested in the similarity of samples. In the first case,

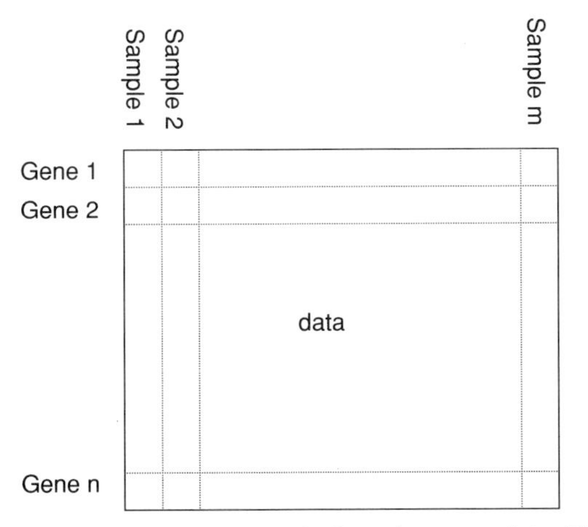

Figure 8.1: Two ways to look at the same data. The data matrix of m samples and n genes can be analysed in two ways. Either we look for relationships between the genes, using the expressions in each of the samples as measurements of the genes, or we look for relationships between the samples, using the expressions in each of the genes as measurements of the samples.

each gene is measured by the samples; in the second case, each sample is measured by the genes (Figure 8.1).

From a scientific perspective, these are very different analyses. From the perspective of the data analysis methods in this chapter, they are essentially the same. Therefore, throughout this chapter, we shall refer to *gene profiles* with the understanding that these might be gene or sample profiles.

Features of a Distance Measure

It is common to describe the similarity between two profiles in terms of the distance between them in the high-dimensional space of gene expression or sample measurements. Before we describe any specific measures of distance, we first set out some theoretical properties that a measure of similarity (or dissimilarity) between two genes should have. These may appear obvious, but they are very important if one is to use a measure of similarity as part of a successful data analysis.

- The distance between any two profiles must be greater than or equal to zero – distances cannot be negative.
- The distance between a profile and itself must be zero.
- Conversely, if the difference between two profiles is zero, then the profiles must be identical.
- The distance between profile A and profile B must be the same as the distance between profile B and profile A.
- This distance between profile A and profile C must be less than or equal to the sum of the distances between profiles A and B and profiles B and C.

The first four rules are very intuitive; the last rule is what is known as the **triangle inequality** (Figure 8.2).

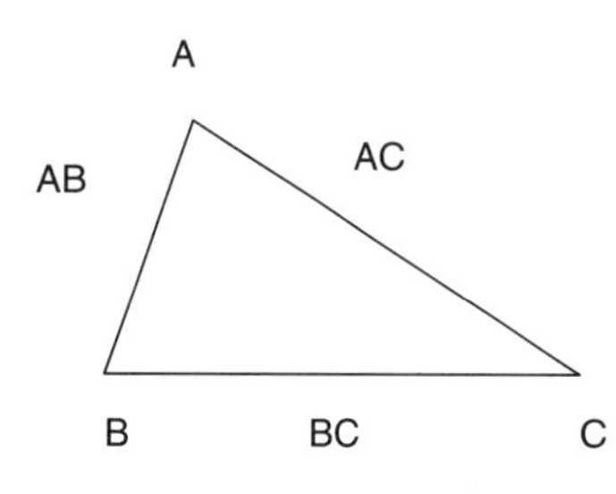

Figure 8.2: The triangle inequality. The triangle ABC has distances AB, AC and BC. These distances satisfy the following equations:

$$AB \leq AC + BC$$
$$AC \leq AB + BC$$
$$BC \leq AB + BC$$

These equations mean that the length of each side is less than or equal to the sum of the lengths of the other two sides. If the lengths are equal, then all three points are on a straight line and the triangle is completely flat.

Correlation Coefficient

The first measure of similarity we describe is the correlation coefficient. This is a statistical concept that quantifies the level of relationship between two sets of measurements (Figure 8.3).

If we denote the two sets of measurements with the notation (x_i) and (y_i), where i is an index from 1 to n, then the correlation coefficient r is given by the following formula:

$$r = \frac{n\sum_{i=1}^{n} x_i y_i - \sum_{i=1}^{n} x_i \sum_{i=1}^{n} y_i}{\sqrt{\left(n\sum_{i=1}^{n} x_i^2 - \left(\sum_{i=1}^{n} x_i\right)^2\right)\left(n\sum_{i=1}^{n} y_i^2 - \left(\sum_{i=1}^{n} y_i\right)^2\right)}} \qquad \text{(Eq. 8.1)}$$

The correlation coefficient takes a value from between -1 and $+1$. A value of -1 represents strong negative correlation: when one variable is high, the other is low. A value of $+1$ represents strong positive correlation: when one variable is high, the other is high. A value of 0 represents uncorrelated variables.

It is common to centre gene expression profiles (Section 5.4) to ensure that they have mean equal to 0 and standard deviation equal to 1 before calculating the correlation coefficient. When the profiles have been centred, the correlation coefficient is given by a much simpler formula[3]:

$$r = \sum_{i=1}^{n} x_i y_i \qquad \text{(Eq. 8.2)}$$

Centering data is an excellent method for data similar to those in data set 8B, where we are looking at the expression of genes in patients relative to a reference sample. However, when we are analysing time series data, as in data set 8A, where the data is relative to gene expression at time zero, centering can lose the natural notion of genes that are up- or down-regulated in the time series. This can be a disadvantage.

The correlation coefficient is a measure of similarity and needs to be converted into a distance measure with the properties listed earlier. There are a number of formulae

[3] This formula is the dot product of the two sets of measurements (x_i) and (y_i) when thought of as vectors in n-dimensional space. The dot product measures the angle between the two vectors, and so the correlation coefficient has a geometric interpretation: parallel vectors have correlation $+1$ or -1, whereas orthogonal vectors have correlation coefficient 0. To convert this to a distance, we need to apply Equation 8.3 or 8.4 so that parallel vectors have distance 0 and orthogonal vectors have distance 1.

that can be used to achieve this, including the following:

$$d\,(X,Y) = 1 - \text{abs}(r\,(X,Y)) \quad \text{(Eq. 8.3)}$$

$$d\,(X,Y) = 1 - r\,(X,Y)^2 \quad \text{(Eq. 8.4)}$$

Equation 8.3 states that the distance between profile X and profile Y is equal to one minus the absolute value of the correlation coefficient between X and Y. So if X and Y are perfectly negatively or positively correlated ($r = -1$ or $r = 1$, respectively), the distance between them is zero, and if they are perfectly uncorrelated, the distance between them is 1. Equation 8.4 is very similar, but subtracts the square of the correlation coefficient instead of its absolute value.

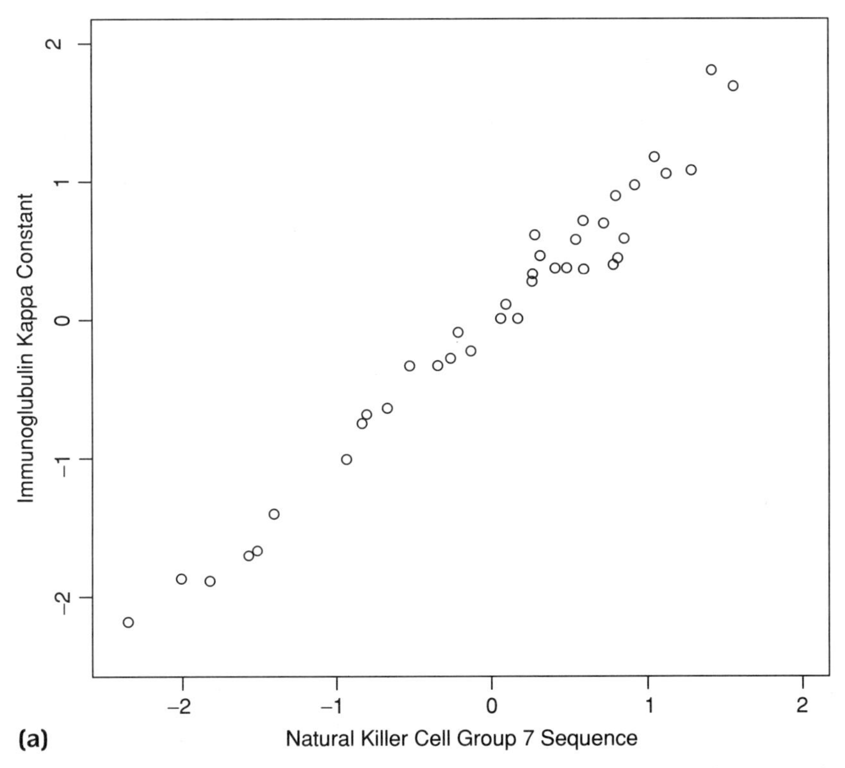

Figure 8.3: Correlated and uncorrelated variables. These three figures show examples of correlated and uncorrelated variables using data from data set 8B. For each gene, we have computed the log ratio of the intensity of that gene relative to the reference sample. Each gene has then been centred so that the values for each gene have mean 0 and standard deviation 1. Each point on the graphs represents one of the 38 patients, where the *x*-axis value is the log ratio of one gene and the *y*-axis value is the log ratio of the other gene. **(a)** The genes Natural Kill Cell Group 7 (NKG7) and Immunoglobulin Kappa Constant (IGKC) are strongly positively correlated ($r = 0.97$). The 38 points lie on a straight line from bottom left to top right. **(b)** The two unannotated genes FLJ13207 and MGC10771 are weakly negatively correlated ($r = -0.47$). The 38 points lie approximately on a line from top left to bottom right – but this is largely the effect of the three measurements at the top left of the plot, and the three measurements at the bottom right. The majority of points are in a cloud in the middle. **(c)** The genes Stromal Antigen 3 (STAG3) and Src-like-adapter are not correlated ($r = 0.054$). The 38 points lie in a cloud around zero with no discernable trend.

(*continued*)

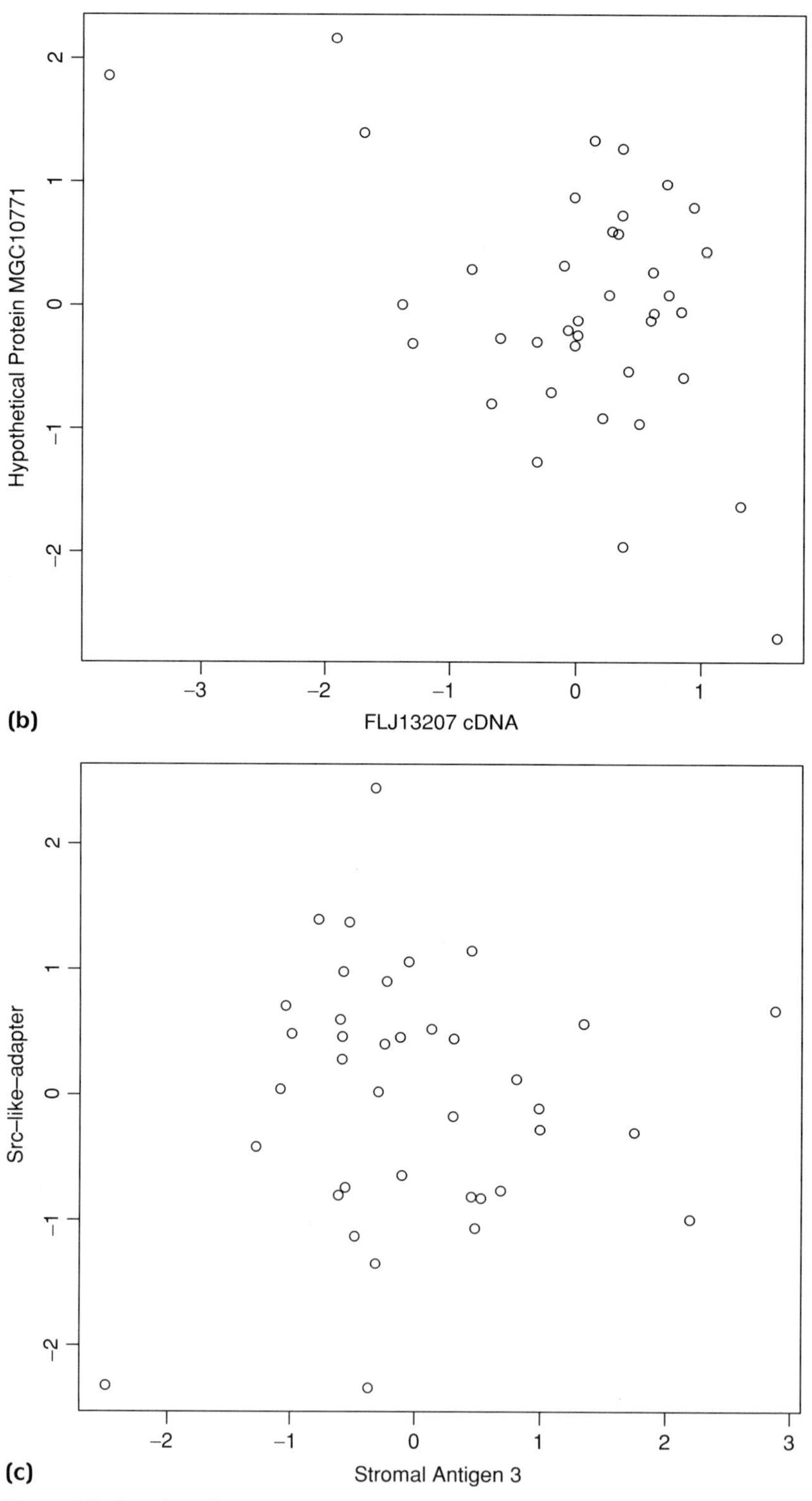

Figure 8.3: *(continued)*

Spearman Correlation

One of the problems with using the standard correlation coefficient just given is that it is susceptible to being skewed by outliers: a single data point can result in two genes appearing to be correlated, even when all of the other data points suggest that they are not. Spearman correlation is a non-parametric measure of correlation that is robust to outliers and, because of this, it is often more appropriate for microarray analysis.

EXAMPLE 8.3 FALSELY CORRELATED TIME SERIES

Figure 8.4 shows two genes from the time series data set 8A, ENB1 and NPR2, which appear to be reasonably correlated, with a Pearson correlation coefficient of 0.63.

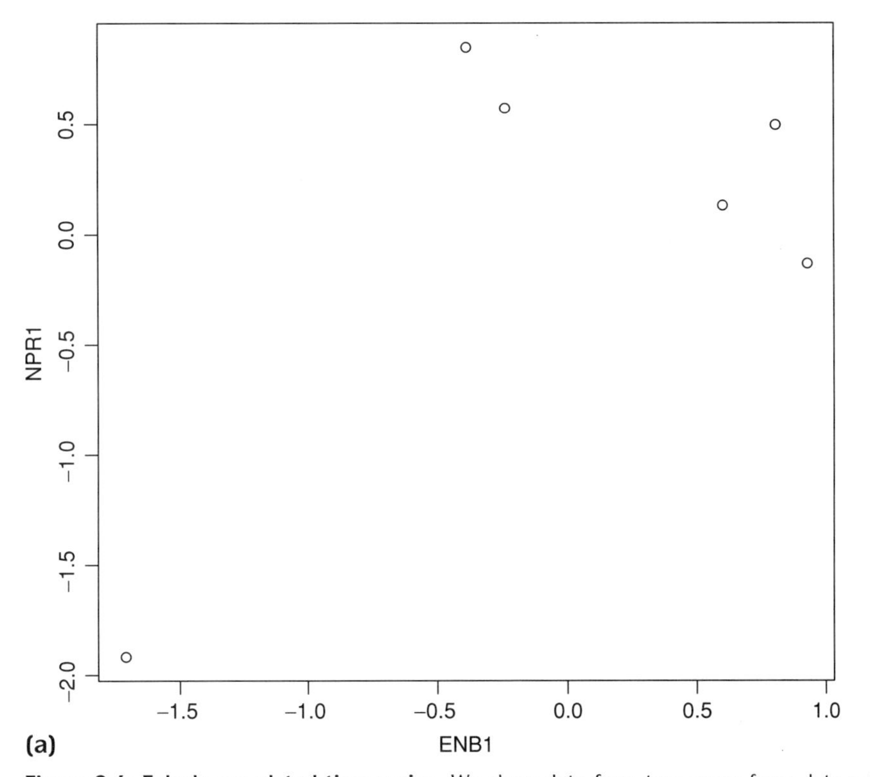

Figure 8.4: Falsely correlated time series. We show data from two genes from data set 8A: ENB1 and NPR1. **(a)** Correlation plot. For this plot, the values of each time series have been centred so that they have mean 0 and standard deviation 1. The correlation coefficient is 0.63, a medium-size positive correlation. The positive correlation is a result of the single outlier at the bottom left of the figure. The other data points are not positively correlated. **(b)** Time series plot. On this graph, we have plotted the log ratio (to base 2) of the signal at each time point relative to the signal at time 0. The two graphs show no relationship at all. The "outlier" that results in the strong correlation is at 30 minutes, where both genes are down-regulated. However, the behaviour of the two genes after the initial down-regulation is completely different, with ENB1 showing decreasing expression and NPR2 showing an increase in expression. **(c)** Ranked plot. The values on the time series have been ranked so that the lowest value in each time series has rank 1, and the highest rank 6, and then plotted. There does not appear to be a correlation between the ranked time series.

(continued)

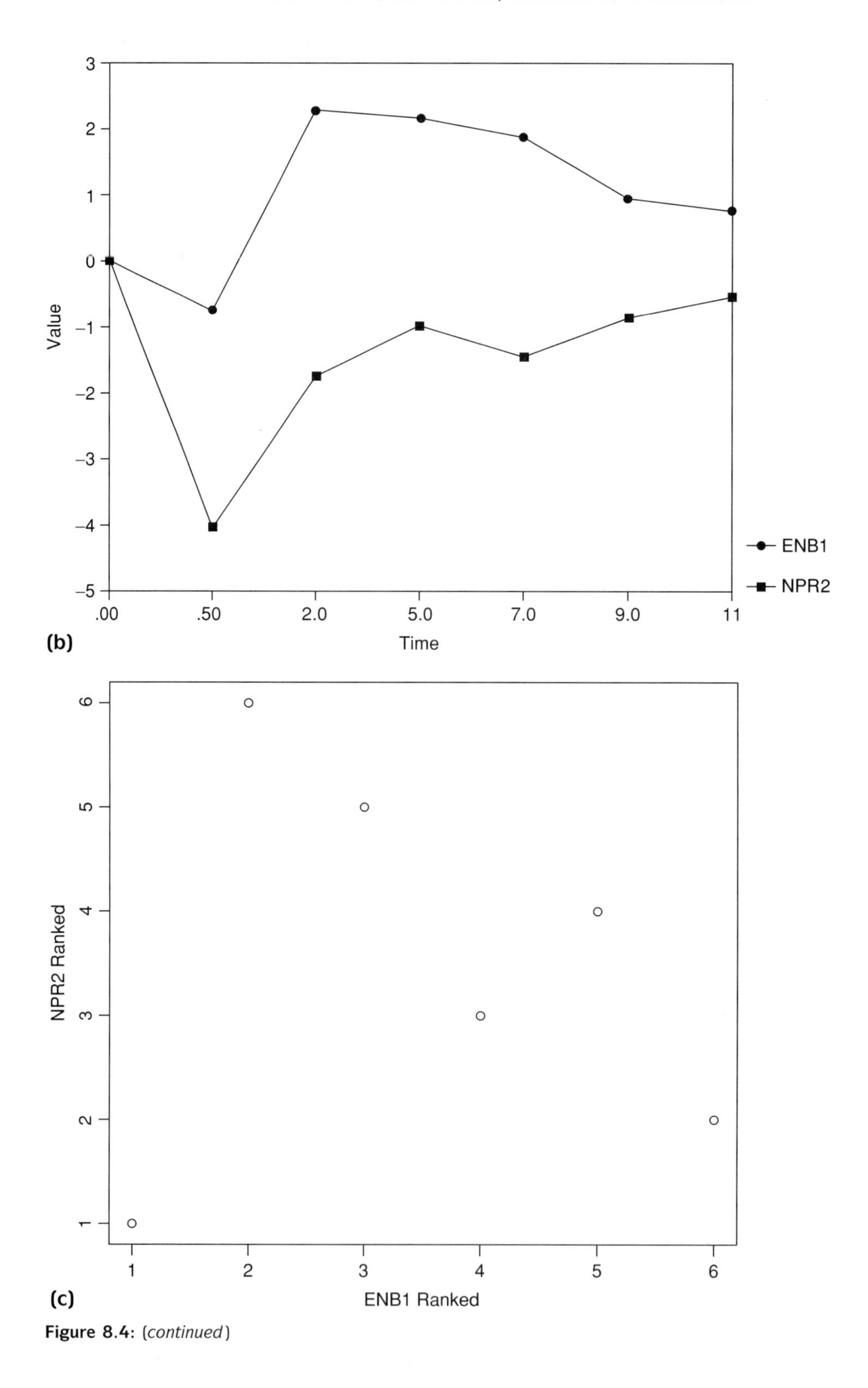

Figure 8.4: *(continued)*

TABLE 8.1: Spearman Correlation

Time	ENB1 Ratio	NPR2 Ratio	ENB1 Rank	NPR2 Rank
0.5	−0.76359	−4.05957	1	1
2	2.276659	−1.7788	6	2
5	2.137332	−0.97433	5	4
7	1.900334	−1.44114	4	3
9	0.932457	−0.87574	3	5
11	0.761866	−0.52328	2	6

Note: The uncentred ratios for each of the six time points for the genes ENB1 and NPR2 are shown in columns 2 and 3. The ranks are calculated by ordering the values for each gene and assigning the value 1 for the lowest measurement, up to 6 for the highest measurement. Equation 8.1 is then applied to calculate the correlation coefficient: in this case the correlation is −0.09 (marginally negatively correlated). However, the Pearson correlation is 0.63 (reasonably strongly correlated).

However, we can see from Figure 8.4a that the reason for this correlation is the outlier in the bottom-left corner; if this were removed, the points would not appear to be positively correlated. The situation is reflected in the time series charts (Figure 8.4b): the genes do not appear to be related.

Outliers are a serious and common problem with microarray data; this is partly because the data can be noisy, and partly because of the large number of genes being studied. Spearman correlation provides a measure of correlation that is robust to large outliers.

Spearman correlation works in a similar manner to the non-parametric tests described in Chapter 7. The true measurements or log ratios are replaced by ranks: 1 for the smallest value, 2 for the second smallest, and so on.[4] Equation 8.1 is then applied to the ranked data, producing a correlation coefficient that also lies between −1 and 1. To use Spearman correlation as a distance measure, we apply Equation 8.3 or 8.4 so that uncorrelated variables have distance 1 and correlated variables have distance 0.

EXAMPLE 8.4 SPEARMAN CORRELATION OF ENB1 AND NPR2

The procedure for calculating the Spearman correlation can be seen in Table 8.1. The Spearman correlation of these genes is −0.09, compared with the Pearson correlation of 0.63. This is a very different result: instead of being positively correlated, the time series are probably not correlated (Figure 8.4c).

In general, Spearman correlation is a more robust measure of correlation than Pearson correlation and can therefore be more appropriate for microarray data, particularly if the data are noisy. However, as with the centering of the data for Pearson correlation, the direction of regulation of the genes is lost during the ranking process.

[4] Tied data points are given the average of the tied ranks. For example, if the two smallest data points were tied, then both points would be given a rank of 1.5.

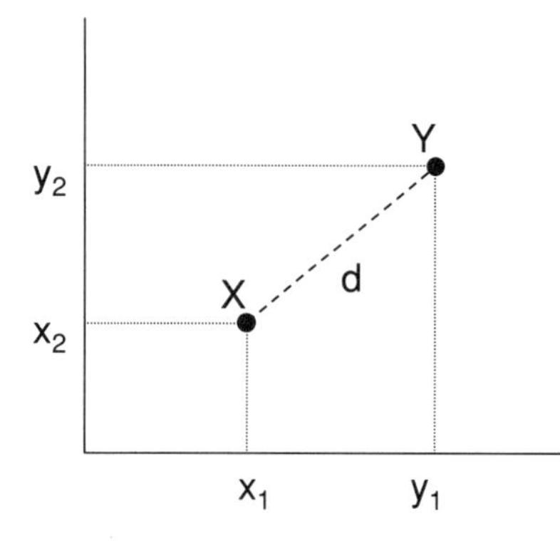

Figure 8.5: Euclidean distance. The distance between the two points X and Y in two-dimensional space, with coordinates (x_1, y_1) and (x_2, y_2), is given by the Pythagorean theorem:

$$d(X,Y) = \sqrt{(x_1 - y_1)^2 + (x_2 - y_2)^2}$$

This concept extends naturally to higher dimensions (Equation 8.5).

This is not a problem with patient data (e.g., data set 8B), but can give problems with time series data (e.g., data set 8A). We show an example of this later.

Euclidean Distance

Euclidean distance is very different from correlation as a measure of the relationship between gene expression profiles. Euclidean distance is an extension of distance that we are used to in real life: the straight-line distance between points in two- or three-dimensional space.

In two dimensions, the distance between two points is calculated using the Pythagorean theorem (Figure 8.5). When we use Euclidean distance with gene expression profiles, we extend the same idea to high dimensions. Using the same notation as before, the Euclidean distance between two profiles X and Y is given by the following equation:

$$d(X,Y) = \sqrt{\sum_{i=1}^{n} (x_i - y_i)^2} \qquad \text{(Eq. 8.5)}$$

EXAMPLE 8.5 EUCLIDEAN DISTANCE IS NOT SCALE INVARIANT

One of the key problems with Euclidean distance is that it is not scale invariant: two gene expression profiles with the same shape but different magnitudes will appear to be very distant. The genes BUR6 and IDH1 from data set 8A have similar profiles of up-regulation, reaching peak gene expression at 7 hours (Figure 8.6). However, BUR6

Figure 8.6: Euclidean distance is not scale invariant. The two genes BUR6 and IDH1 from data set 8B are shown. Both genes have similar profiles and are up-regulated, reaching maximum gene expression at 7 hours. **(a)** Log ratio data. Although the profiles have very similar shape, BUR6 is considerably more up-regulated, achieving a maximum of about 5 (approximately 30 times up-regulution) while IDH1 achieves of a maximum of about 2 (approximately 4 times up-regulation). The Euclidean distance is 5.8 (a large value). **(b)** The data have been scaled by dividing each profile by the standard deviation of the absolute values of the time points. This method preserves the direction of regulation of the genes relative to time zero. The Euclidean distance is 0.88 (a much smaller value).

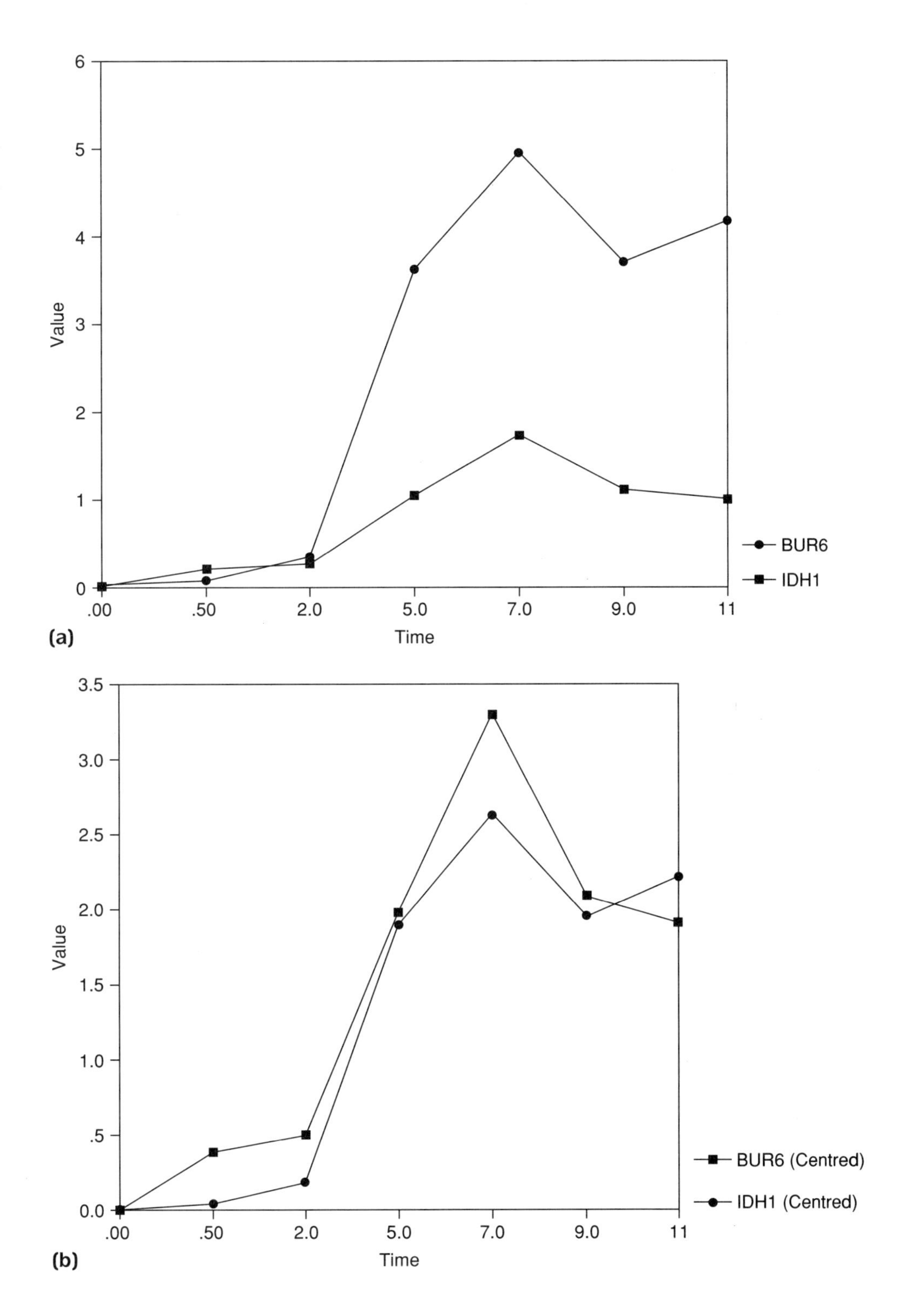
6
5
4
3
2
1
0
Value
.00
.50
2.0
5.0
7.0
9.0
11
Time
BUR6
IDH1
(a)
3.5
3.0
2.5
2.0
1.5
1.0
.5
0.0
Value
.00
.50
2.0
5.0
7.0
9.0
11
Time
BUR6 (Centred)
IDH1 (Centred)
(b)

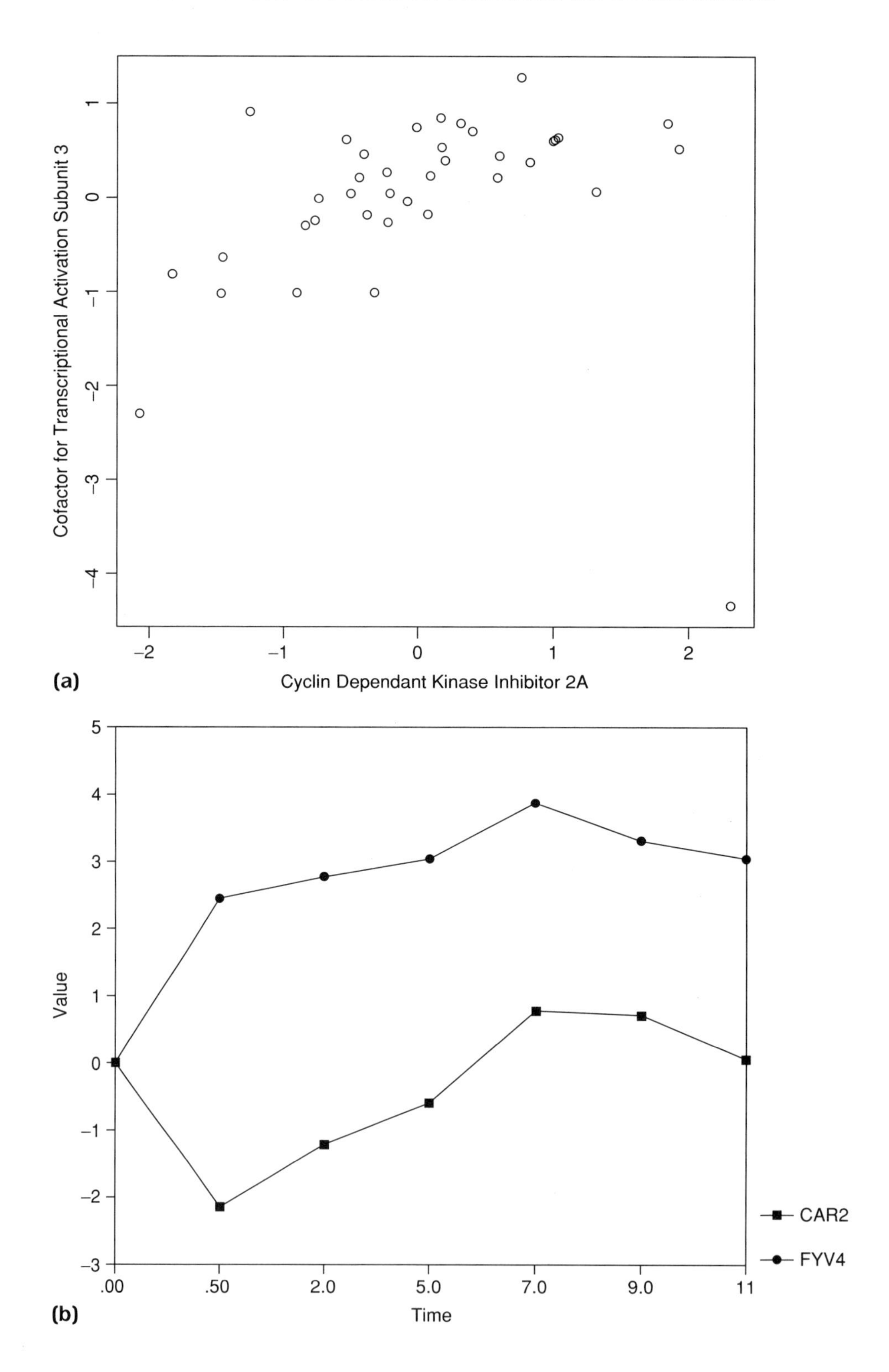

Cofactor for Transcriptional Activation Subunit 3
1
0
−1
−2
−3
−4
−2
−1
0
1
2
(a)
Cyclin Dependant Kinase Inhibitor 2A
Value
5
4
3
2
1
0
−1
−2
−3
.00
.50
2.0
5.0
7.0
9.0
11
CAR2
FYV4
(b)
Time

is more up-regulated than IDH1. The Euclidean distance between these genes is very large, even though the profiles are very similar in shape.

This problem can be resolved by centering the profiles. When the profiles have been centred, the Euclidean distance is very small (Figure 8.6b).

EXAMPLE 8.6 EUCLIDEAN DISTANCE AND CORRELATION CAN GIVE DIFFERENT RESULTS

In Figure 8.7, we show two examples of gene profiles that are similar by correlation but different by Euclidean distance.

The first example is from data set 8B, where we are looking to identify genes with similar behaviour in different patients. In this case, Spearman correlation gives a more robust result: the data are positively correlated, but a single outlier in the opposite direction from the correlation has resulted in a large Euclidean distance.

The second example is time series data from data set 8A: two genes have similar shape profiles, but one is up-regulated and the other is down-regulated. They appear to be correlated, although the expression is very different; the Euclidean distance is large, reflecting the difference between the patterns. In this case, Euclidean distance is probably a more realistic measure.

From these examples, we can see that when choosing a distance measure to use for further analysis, such as the cluster analyses we describe later in the chapter, there is no one answer as to what is the best measure. Different measures have different strengths and weaknesses and can be combined with different data scaling to produce different results (Table 8.2).

SECTION 8.3 DIMENSIONALITY REDUCTION

One of the central features of microarray data is that there is a lot of it. In mathematical terms, we talk about the data being high-dimensional: by this we simply mean that we are measuring a large number of genes, or a large number of samples, and frequently both (e.g., data set 8B).

Figure 8.7: Differences between Euclidean distance and Spearman correlation. (a) Data from data set 8B where the Euclidean distance is large (7.9) but the Spearman correlation is also strong ($r = 0.79$ or distance = 0.21). The plot shows centred data (to have mean 0 and standard deviation 1) for the two genes Cyclin Dependant Kinase Inhibitor 2A (CDKN2A) and Cofactor for Transcriptional Activation Subunit 3. The two genes are positively correlated, but there is a single outlier that is in the opposite direction of the main trend (bottom-right-hand corner of the figure). This single outlier results in a high Euclidean distance, while Spearman correlation detects the correlation. In this case, Spearman correlation is probably a better measure of similarity. **(b)** Data from data set 8A where Euclidean distance is large (8.5) but the Spearman correlation is very strong ($r = 0.91$ or distance = 0.09) for the two genes CAR2 and FYV4. The plot shows the log ratio data (to log 2); the Euclidean distance has been calculated on data that have been scaled by the standard deviation of the absolute values of the ratios. In this case, CAR2 is about 4-fold down-regulated after 30 minutes, while FYV4 is about 5-fold up-regulated at that time. After 30 minutes, the genes have similar shape, but FYV4 remains up-regulated, while CAR2 switches from being down-regulated to being up-regulated. In this case, Euclidean distance is probably a better measure of similarity.

TABLE 8.2: Strengths and Weaknesses of Different Distance Measures

Pearson Correlation	Spearman Correlation	Euclidean Distance
✓ Powerful	✓ Robust to outliers	✓ Geometric interpretation
✓ Spots positive and negative correlations	✓ Spots positive and negative correlations	✓ Can retain up- or down-regulation information with appropriate scaling
✓ Scale invariant on centred data	✓ Completely scale invariant: no scaling or centering required	✓ Can detect magnitude of changes if used without scaling
× Assumes linearity	× Less powerful	× Not scale invariant: results depend on scaling used
× Susceptible to outliers	× Ignores pattern of up- or down-regulation in time series	× Cannot detect negative correlations

Often, we want to visualise microarray data, either as an aid to visual analysis or as a precursor to the application of more sophisticated algorithms. The human brain has evolved to be able to visualise objects in two or three dimensions: we live in a three-dimensional world, and we see via the stereoscopic combination of two two-dimensional images from our eyes. Our principal tools for visualising microarray data are themselves two-dimensional: computer screens, projector images, research papers and books. This makes the visualisation of microarray data difficult: we are trying to represent very high-dimensional data in the two or three dimensions described.

This section describes two methods for visualising microarray data in two or three dimensions: **principal component analysis** and **multidimensional scaling**.

Principal Component Analysis

Imagine a cloud of points in three-dimensional space. Now imagine placing a piece of card behind the points and looking at the shadows of the points on the card: we have *projected* a three-dimensional group of points onto a two-dimensional space. Principal component analysis (PCA) is a method that projects a high-dimensional space onto a lower dimensional space. We choose the angle at which to look at the high-dimensional space so that we capture as much of the variability of the original data as we can in the lower dimensional space and then ignore the other dimensions.

EXAMPLE 8.7 PRINCIPAL COMPONENT ANALYSIS OF A PEN

Suppose we have a pen in three dimensions, and we want to construct a two-dimensional view of that pen. If we look at the pen from one end, all we see in two dimensions is a circle with a bump (Figure 8.8a). If we rotate the pen so that we are looking at it lengthwise, but hide the clip, we can see that we have a long object, which is fatter at the end with the cap (Figure 8.8b). In this rotation, we have resolved the first principal component by finding the axis with maximum variability in the shape of the pen: the long side.

If we rotate the pen further, so that we can see the clip, it is now recognisable as a pen (Figure 8.8c). In this second rotation, we have resolved the second principal component: we have found the axis that is perpendicular to the first axis that contains the most remaining variability.

Figure 8.8: Principal component analysis of a pen. (a) Pen looked at end-on. No principal components have been resolved. The pen is not recognisable. **(b)** Pen looked at sideways with clip obscured. The first principal component is resolved; we can see that the object is "long" and wider at one end, but that the clip is obscured. **(c)** Pen looked at sideways showing clip; it should be recognisable as a pen. The second principal is resolved; we can now see that the clip and the object is recognisable as a pen. Even in (c), some information is lost. We do not know that the pen is round; it could be a square pen.

How PCA Works

Suppose we want to reduce a microarray experiment with 10,000 genes into two or three dimensions. We first construct what is known as the *variance–covariance* matrix for these genes. This matrix captures the variability of each gene and the extent to which it co-varies (equivalent to correlation) with every other gene. So we would have a $10{,}000 \times 10{,}000$ element array. We use this array to identify a new variable that is a linear combination of the genes and that has the maximum amount of variance. This is the first principal component (Figure 8.9). We then find the variable that is orthogonal to this first variable and that maximises the remaining variance. This is the

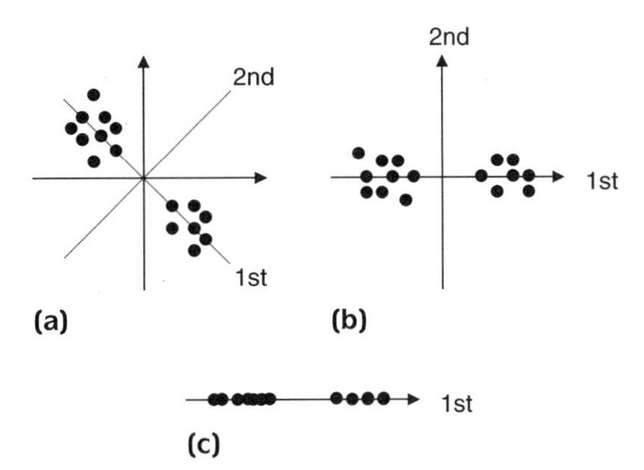

Figure 8.9: Principal component analysis on a simple, imaginary, two-dimensional data set. The data set is two-dimensional and consists of two elliptical "clusters." **(a)** The components are not resolved. **(b)** The first component is along the direction of the maximum variability, in which the clusters become separated. Because the data is only two-dimensional, we have no choice about the second component, which must be orthogonal to the first. **(c)** We can resolve these two-dimensional data into just one dimension and see the clusters in the data.

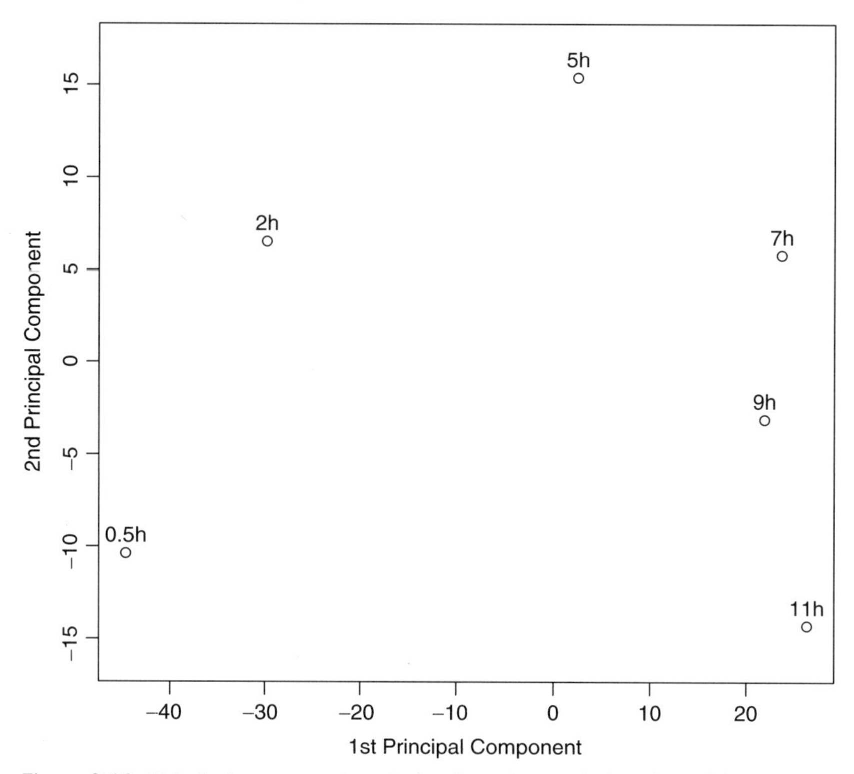

Figure 8.10: Principal component analysis of yeast sporulation data. Principal component analysis is performed on the 1,000 most varying genes from data set 8A in order to be able to visualise the similarity among the six samples. The plot shows a number of interesting features:

- The six samples follow a clear pattern in the principal component plot; the analysis makes biological sense.
- If we look at just the first principal component, then the samples taken at 7, 9 and 11 hours all cluster together, while the earlier samples resolve (in order). This implies that the three late samples are fairly similar to each other, and that the main transcription changes occur over the first 7 hours.
- The second principal component resolves the three later time points; interestingly, the latest time points have similar value to the earliest. This suggests that there may be some transient processes that are switched on over the first 5 hours and then switched off.

second principal component. We repeat the process until we have as many principal components as we are interested in.

EXAMPLE 8.8 PRINCIPAL COMPONENT ANALYSIS OF YEAST SPORULATION DATA

We want to see the relationship between the six time points of data set 8A. We cannot visualise all 6,000 genes, so we use PCA to visualise the relationship between the time points in a two-dimensional plot. This is a good illustrative example because we have an a priori expectation that neighbouring time points should be similar.

In this example, PCA reveals three interesting features about the data (Figure 8.10):

- Neighbouring time points are close to each other on the figure; we can even see the "path" taken by the time series through the principal component space. From this we conclude that the analysis makes biological sense.

TABLE 8.3: Principal Components of Data Set 8A

Principal Component	1st	2nd	3rd	4th	5th
Standard deviation	30.3	11.3	9.1	5.9	5.3
Proportion of variance	77%	11%	7%	3%	2%
Cumulative proportion	77%	88%	95%	98%	100%

Note: The first five principal components explain essentially all of the variability in the 1,000 genes used for this analysis. So although we are looking at a very large number of genes, there are not many different processes in this experiment: 88% of the variability is explained by the two components used for Figure 8.10. Therefore, this figure is a good representation of the similarities and differences among the six samples.

- If we just look at the first principal component (x axis), the samples at 7, 9 and 11 hours are clustered together, while the other three time points are resolved, in order, along the axis.
- The second principal component resolves the three later time points, but, interestingly, the trend in the second principal component is a rise in value to the 5-hour sample, and then a return, so that the latest samples are most similar to the earliest.

The behaviour of these two components suggests two sets of processes: one set of processes that changes over a period of 7 hours and persists in the altered state during sporulation (corresponding to the first principal component), and another set of processes that are activated over a period of 5 hours, but which then return to their original state (corresponding to the second principal component). So simply by looking at the principal component plot, we can gain an understanding of the underlying biology.

PCA also identifies the amount of variability captured in each of the components (Table 8.3). In this example, most of the variability (77%) is captured in the first component, and 88% in the first two; five components capture all of the variability of the data. We therefore expect that this particular example is biologically relatively simple, with a small number of pathways in action during sporulation. With more complex data, the variability can be spread over many principal components.

Multidimensional Scaling

Multidimensional scaling (MDS) is a different approach to dimensionality reduction and visualisation. Unlike PCA, it does not start from the data, but rather from the measurements of distance between the samples or profiles being compared.

We measure the distance between profiles using any of the measures described in Section 8.2. MDS then attempts to locate the profiles in two- or three-dimensional space in such a way that the distances in the two- or three-dimensional space are as close as possible to the distances measured between the profiles in the higher dimensional space.

TABLE 8.4: Multidimensional Scaling of Data Set 8A

	Measured Distance Between Profiles (Scaled Distance in Parentheses)				
	0.5 h	**2 h**	**5 h**	**7 h**	**9 h**
Euclidean distance					
2 h	31 (22)				
5 h	55 (54)	38 (33)			
7 h	71 (70)	58 (53)	31 (23)		
9 h	68 (67)	55 (53)	31 (27)	22 (9)	
11 h	73 (71)	61 (60)	39 (38)	30 (20)	21 (12)
Correlation distance					
2 h	0.23 (0.26)				
5 h	0.53 (0.53)	0.28 (0.33)			
7 h	0.53 (0.54)	0.36 (0.38)	0.07 (0.12)		
9 h	0.63 (0.63)	0.45 (0.46)	0.12 (0.15)	0.04 (0.10)	
11 h	0.67 (0.68)	0.53 (0.53)	0.18 (0.24)	0.09 (0.15)	0.06 (0.09)

Note: The MDS of the six profiles starts with the distance matrix of the measured distance between the profiles and finds points in two- (or three-) dimensional space so that the distances between those points are close to the distances in the matrix. Those points are shown in Figure 8.9. The actual distances are shown with the distances between the points in two-dimensional space shown in parentheses. Some of these are very close, for example, the distances from the 0.5 h sample. Where the distances are quite different (e.g., the distance between the 9 h and 11 h sample), it is implied that the mapping of the data into two dimensions is not accurate. Some information has been lost and would require an extra dimension to visualise. One way to think of it is that either the 9 h or 11 h sample wants to pop out of the page and be somewhere in the air above the book.

EXAMPLE 8.9 MULTIDIMENSIONAL SCALING ON YEAST SPORULATION DATA

We perform MDS on data set 8A; in this way, we can compare the results of MDS with the results of PCA on the same data. One of the advantages of MDS over PCA is that we can measure the distances between the samples in different ways. In this example, we have calculated the distance matrix using both Pearson correlation distance and Euclidean distance (Table 8.4).

Using these distance measures, we find points in two-dimensional space that have similar distances between them (Figure 8.11). The MDS plot using Euclidean distance is almost identical to the PCA plot. The MDS plot using correlation has some differences: the earlier time points are more spread out, whereas the later time points are closer together. In both cases, the inherent structure of the time series is preserved, with the points following a clearly recognisable "path."

Figure 8.11: Multidimensional scaling of yeast sporulation data. In both figures, we have mapped the six time points into two-dimensional space so that the distances between the points in two dimensions are as close as possible to the distances between the profiles using the 1,000 most varying genes. **(a)** Euclidean distance is used. The plot is almost identical to the PCA plot and shows the same clear progression along the time points. **(b)** Correlation distance is used. The plot is similar, but has some differences: the distances between the earlier samples are larger, while the three late samples are more closely clustered. The natural "path" through the two-dimensional space seen with PCA and with the Euclidean distance MDS is not present for the 7h, 9h and 11h samples.

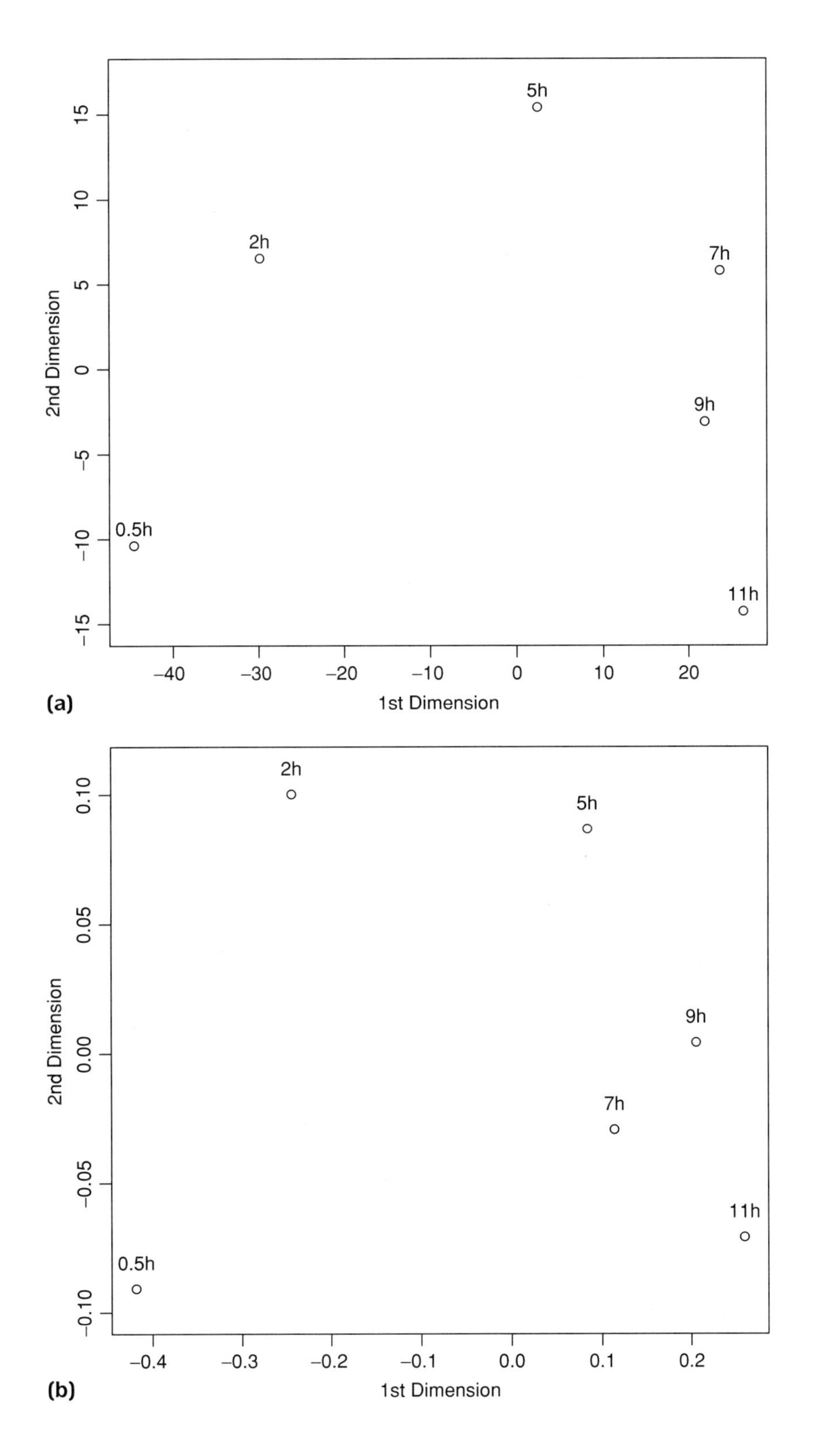
5h
2h
7h
9h
0.5h
11h
15
10
5
0
−5
−10
−15
2nd Dimension
−40
−30
−20
−10
0
10
20
1st Dimension
(a)
2h
5h
9h
7h
11h
0.5h
0.10
0.05
0.00
−0.05
−0.10
2nd Dimension
−0.4
−0.3
−0.2
−0.1
0.0
0.1
0.2
1st Dimension
(b)

TABLE 8.5: Comparison of Principal Component Analysis with Multidimensional Scaling

Principal Component Analysis	Multidimensional Scaling
✓ Can visualise high-dimensional data in two or three dimensions	✓ Can visualise high-dimensional data in two or three dimensions
✓ Do not impose any a priori structure on the relationships between genes and samples	✓ Do not impose any a priori structure on the relationships between genes and samples
✓ Can be used as inputs for classification techniques in Chapter 9	✓ Can resolve non-linear relationships if used with a non-linear distance measure
✓ Implemented in wide range of packages, including GeneSpring, J-Express, R and Matlab	✓ Allows visualisation of distance matrix to be used for cluster analysis and can be used to help select an appropriate distance measure
× Principal components are abstract concepts and have no concrete meaning	✓ Can be used as inputs for classification techniques in Chapter 9
× Can only resolve linear relationships between genes and samples	✓ Implemented in R and Matlab
× Susceptible to outliers – uses raw, scaled or centred data	× Dimensions have no meaning at all
× Difficult to visualise large numbers of genes or samples	× Different distance measures give different results
	× Difficult to visualise large numbers of genes or samples
	× Not currently implemented in GeneSpring, J-Express or other commonly used gene expression analysis packages

MDS is an excellent and very natural way to visualise the distance matrix between gene profiles or genes. The clustering algorithms described in the remaining sections of this chapter all work from a distance matrix, so the structure of the cluster analysis you perform should reflect the MDS plot. This can help you to determine distance measure to use for a cluster analysis. MDS can also be used as a guide to choosing the number of clusters for k-means clustering that we describe in Section 8.6.

The MDS plot has an advantage over a cluster plot in that it does not impose any structure on the data. It has the disadvantage of being difficult to use for large numbers of genes. We summarise the advantages and disadvantages of PCA and MDS in Table 8.5.

SECTION 8.4 HIERARCHICAL CLUSTERING

The next two sections discuss the most widely used analysis tool for gene expression data: hierarchical clustering. This is a methodology that arranges the gene or sample profiles into a tree so that similar profiles appear close together in the tree and dissimilar profiles are farther apart.

The technique has become popular for four reasons:

- It can simplify large volumes of data.
- The analysis reveals groups of similar genes that can then be studied in greater depth.

- It is possible to visualise the data in a hierarchical way using interactive computer programs.
- The results are visualised in the style of phylogenetic analyses, which are familiar to many geneticists.

In this section, we describe how hierarchical clustering works, by first applying it to a small number of genes and then to a larger number of genes. We then look at a number of methodological considerations of hierarchical clustering and at the effects of using different distance measures. Section 8.5 describes a method for determining the reliability and robustness of clusters in relation to the variability and noise in the microarray experiment.

EXAMPLE 8.10 THE BASIC METHOD OF HIERARCHICAL CLUSTERING

We demonstrate the method for hierarchical clustering through a simple example. We will cluster five genes from data set 8B: CREME9, ALOX5, HS2ST1, PELI1 and RDHL. In Table 8.6, we show the distance matrix for these five genes. This has been computed by Spearman correlation. The algorithm has four steps:

1. Look at the distance matrix and find the nearest entries (which may be genes or clusters of genes).
2. Join these entries together in the tree to form a new cluster.
3. Compute the distance between the newly formed cluster and the other genes and clusters.
4. Return to step 1 and repeat until all genes and clusters are linked.

If we look at Table 8.6a, we can see that the nearest two genes are CREME9 and RDHL. Therefore, these are the first two genes to be linked (Figure 8.12a), forming a new cluster that contains these two genes.

We proceed to step 3, in which we must calculate the distances between the remaining genes and the cluster containing CREME9 and RDHL (Table 8.6b). There are a number of different methods for doing this, and the trees formed by these methods frequently look different. Here, the distances have been calculated using **average linkage**: we discuss this and other methods later.

We now return to Step 1 of the algorithm: the nearest entries in the table are the genes ALOX5 and PELI1. These are joined to form a new cluster (Figure 8.12b), and we return again to step 1. This process continues until all the genes and clusters are combined (Table 8.6; Figure 8.12). The result is exactly the type of tree you will see in programs like TreeView, GeneSpring and J-Express. Traditionally, the height of the branches is proportional to the distance between the genes or clusters. Therefore, closer genes (e.g., CREME9 and RDHL) will be joined by shorter branches than more distant genes (e.g., ALOX5 and PELI1), which will be joined by taller branches.

TABLE 8.6A: Hierarchical Clustering

Gene	CREME9	ALOX5	HS2ST1	PELII	RDHL
CREME9	0.00				
ALOX5	0.57	0.00			
HS2ST1	0.79	0.46	0.00		
PELI1	0.39	0.39	0.51	0.00	
RDHL	0.03	0.62	0.79	0.43	0.00

Note: The distances between each of five genes from data set 8B are calculated. In the first step of the process, we identify the smallest distance: between CREME9 and RDHL. This is a triangular matrix; the distance between two genes is the same whichever way you look at them.

TABLE 8.6B: Hierarchical Clustering

Gene	CREME9-RDHL	ALOX5	HS2ST1	PELII
CREME9-RDHL	0.00			
ALOX5	0.59	0.00		
HS2ST1	0.79	0.46	0.00	
PELI1	0.41	0.39	0.51	0.00

Note: The genes CREME9 and RDHL have been combined to form a single cluster. We compute the distance between this cluster and each of the other genes. There are a number of methods for computing this which are discussed in Section 8.4; here, we use a method called *average linkage*. The closest entries in the table are now between ALOX5 and PELI1.

TABLE 8.6C: Hierarchical Clustering

Gene	CREME9-RDHL	ALOX5-PELI1	HS2ST1
CREME9-RDHL	0.00		
ALOX5-PELI1	0.50	0.00	
HS2ST1	0.79	0.46	0.00

Note: A new cluster has now been formed containing the genes ALOX5 and PELI1. We now calculate the distance between this cluster and the remaining genes. The closest entries are now between the cluster containing ALOX5 and PELI1, and the gene H2ST1.

TABLE 8.6D: Hierarchical Clustering

Gene	CREME9-RDHL	ALOX5-PELI1-HS2ST1
CREME9-RDHL	0.00	
ALOX5-PELI1-HS2ST1	0.60	0.00

Note: In the final step, the two remaining clusters will be joined.

EXAMPLE 8.11 HIERARCHICAL CLUSTERING ON A LARGER GENE SET

We apply hierarchical clustering to 15 genes from data set 8A. These genes have several functions: DNA repair, nucleotide excision repair, protein biosynthesis, stress response, transcription initiation and unknown function. From the time series profiles (Figure 8.13), you can see that some genes might cluster together; for example, CDC21 and DIN7 have very similar shapes, although slightly different scales.

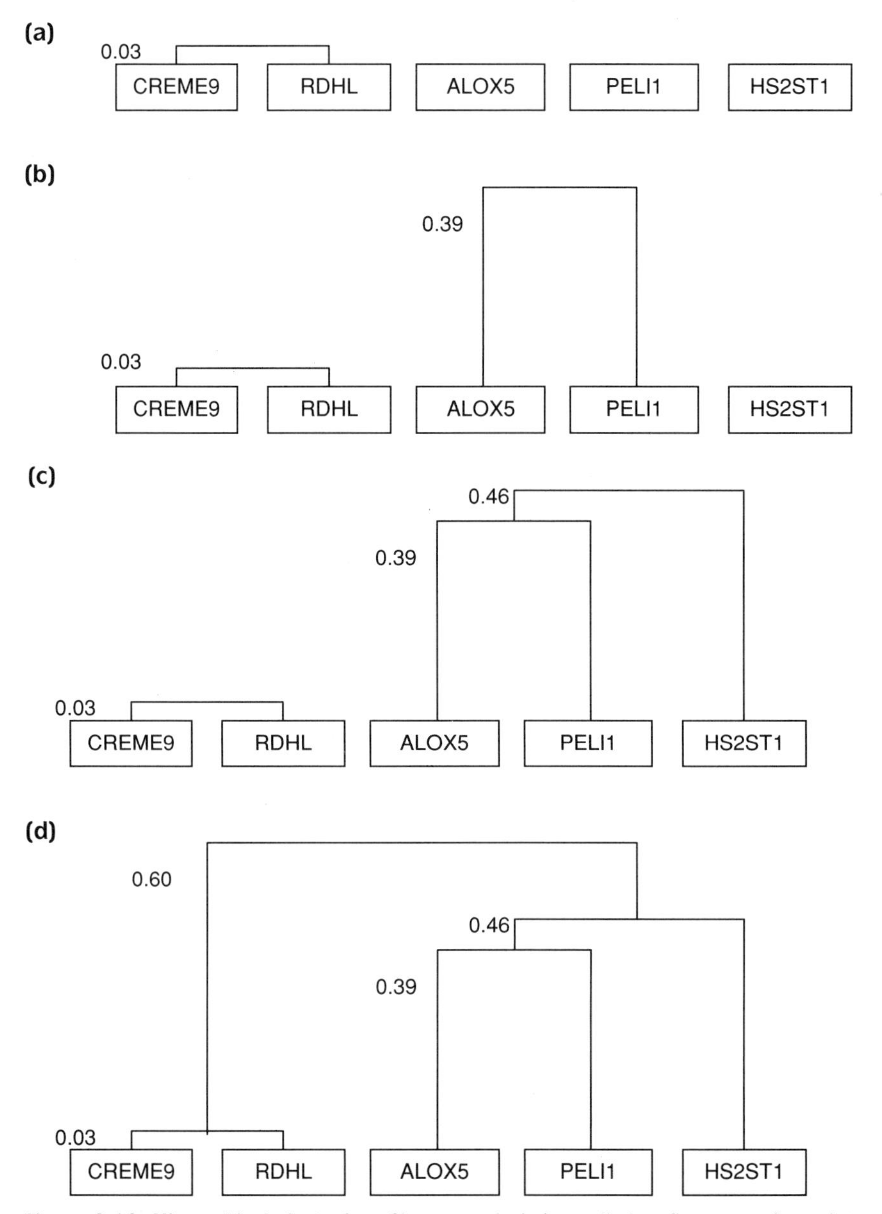

Figure 8.12: Hierarchical clustering. Cluster analysis is applied to five genes from data set 8B; the distance matrix for these genes is shown in Table 8.6. There are four steps: **(a)** The closest two genes, CREME9 and RDHL, are joined to form a cluster; the distance between them is 0.03. The algorithm continues with step 2, in which we calculate the distance between each of the remaining genes and the cluster containing CREME9 and RDHL. **(b)** The next smallest distance is between ALOX5 and PELI1; the distance is 0.39. These are joined to form a new cluster. **(c)** The next smallest distance is between the cluster containing ALOX5 and PELI1, and the gene HS2ST1, a distance of 0.46. These are joined to form a three-gene cluster containing ALOX5, PELI1 and HS2ST1. The cluster containing ALOX5 and PELI1 forms a subcluster of the three-gene cluster. **(d)** The two remaining clusters are joined to complete the tree. We have marked the distances on the tree; traditionally, the heights of the branches are proportional to the distance between the genes, so very close genes would have very short branches whereas distant genes would have long branches.

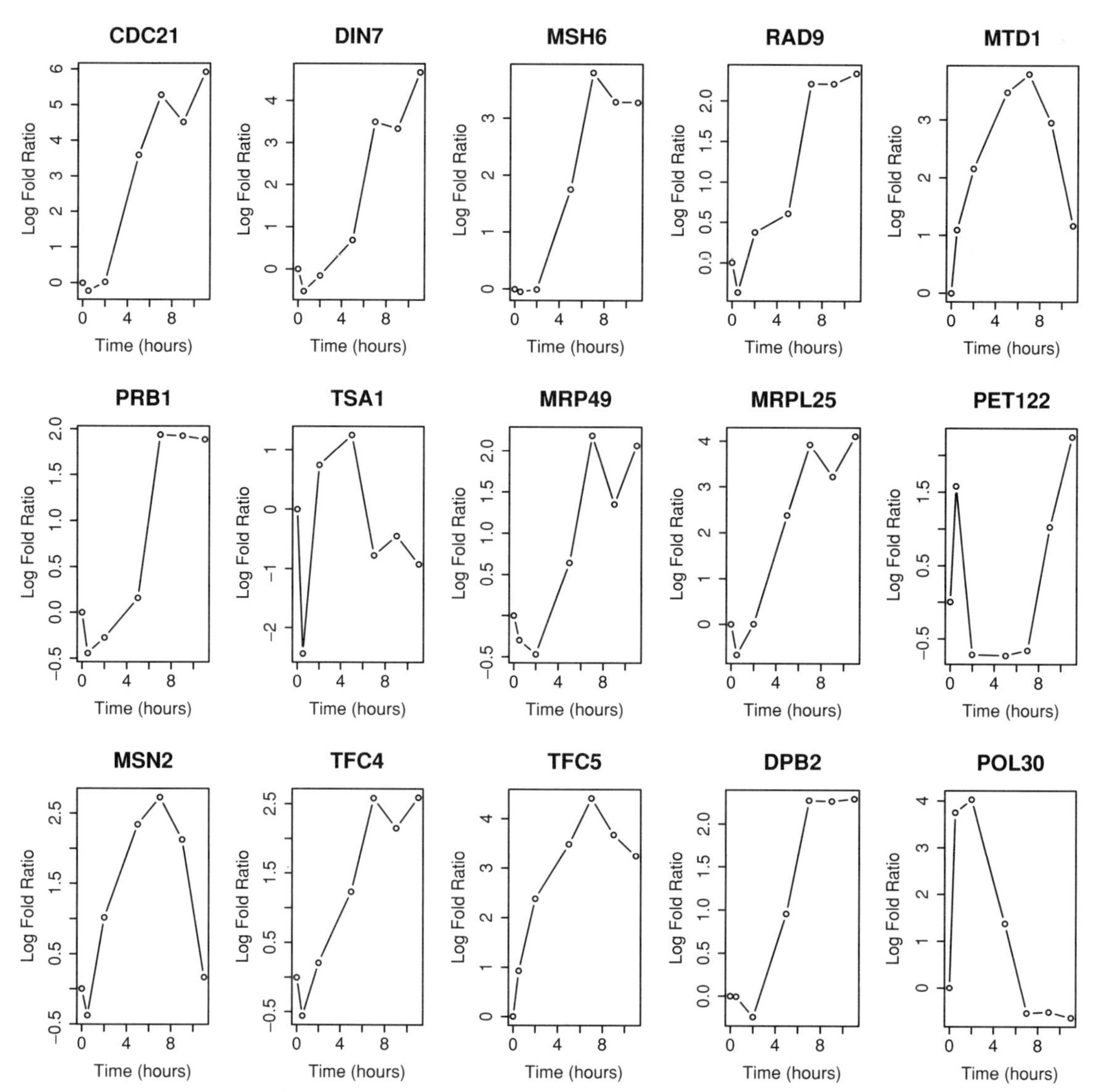

Figure 8.13: Gene profiles for clustering. We show the time series for the 15 genes from data set 8A that we cluster in the example cluster analyses. These genes have a number of different functions: DNA repair (DIN7, MSH6, RAD9), DNA replication (CDC21), nucleotide excision repair (DPB2, POL30), protein biosynthesis (MRP49, MRPL25, PET122), transcription initiation (TFC4, TFC5), stress response (TSA1, MSN2) and unknown function (MTD1 and PRB1). By looking at the profiles, you can see some of the "natural" clusters: genes whose transcription is persistently up- or down-regulated, and genes with transient up- or down-regulation. All the gene expression measurements are ratios relative to expression at time zero. We have included the point (0,0) in all of the plots, but have not included it in the analyses because it is not a measured value.

Figure 8.14c is a dendrogram that has been constructed using Pearson correlation and average linkage. The genes cluster into two broad groups: those that are persistently up-regulated during the time course, and those that show a transient response. Note that Pearson correlation detects negative correlations, so PET122 is reasonably close to MSN2 and MTD1 while POL30 is very close to DIN7 and MSH6.

Linkage Methods

In Step 3 of the algorithm, we join two genes or clusters together to form a new cluster and need to compute the distances between the new cluster and the remaining genes or clusters. There are a number of methods that can be used to calculate the new distances. Each method will produce a different clustering, so it is important to choose the method you use carefully.

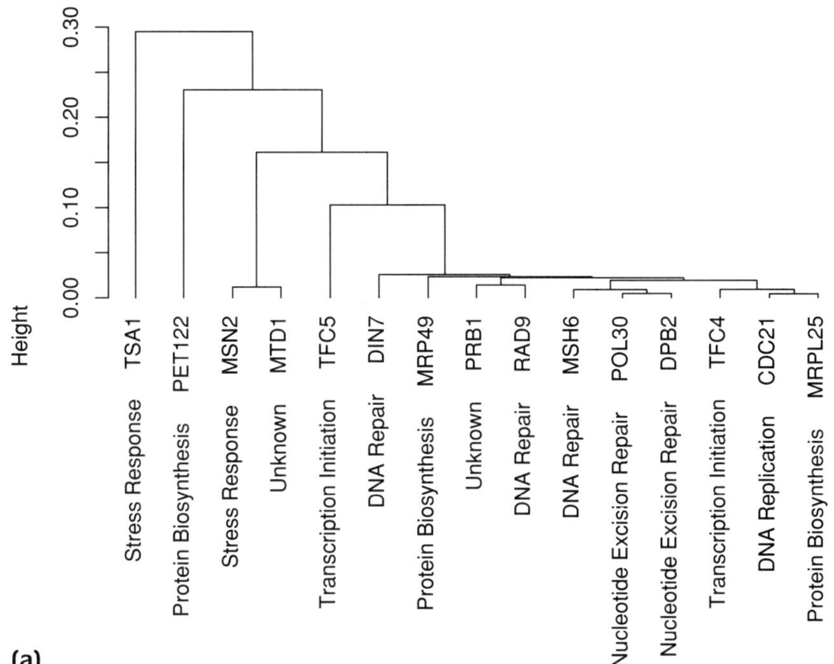

Figure 8.14: Hierarchical clustering. (a) Clustering with Pearson correlation and single linkage. Single linkage tends to produce *chaining*, which we see in this figure. The algorithm has found the large cluster of persistent genes, but the transient genes are then added one at a time to the large cluster rather than forming a cluster on their own. Single linkage is usually not a good method for microarray analysis. **(b)** Clustering with Pearson correlation and complete linkage. The two main clusters are well defined. The most obvious difference between this clustering and average linkage is the inclusion of TFC5 with the transient genes. **(c)** Clustering with Pearson correlation and average linkage. This is a very commonly used method for microarray data. There are two broad clusters corresponding to the persistent and transient responses. PET122 is clustered with the transient genes. TFC5 is clustered with the persistent genes, but lies outside the main group. Note that TFC5 could have been drawn in between the two main clusters, which would probably be a more intuitive place to put it. The large cluster has a great deal of fine structure. **(d)** Clustering with Spearman correlation and average linkage. Many genes have distance zero between them so there is no fine structure. The groups are quite different from the Pearson correlation groups: MSH6 and PRB1 are closer to the transient genes because the gene expression decreases very slightly. **(e)** Clustering with Euclidean distance and average linkage. This is very different from the other clusterings. Most importantly, the two genes that are negatively correlated with other profiles, PET122 and POL30, are outliers and not clustered with other genes. The persistent and transient groups are identified, and there is greater distance between the genes and less fine structure.

(*continued*)

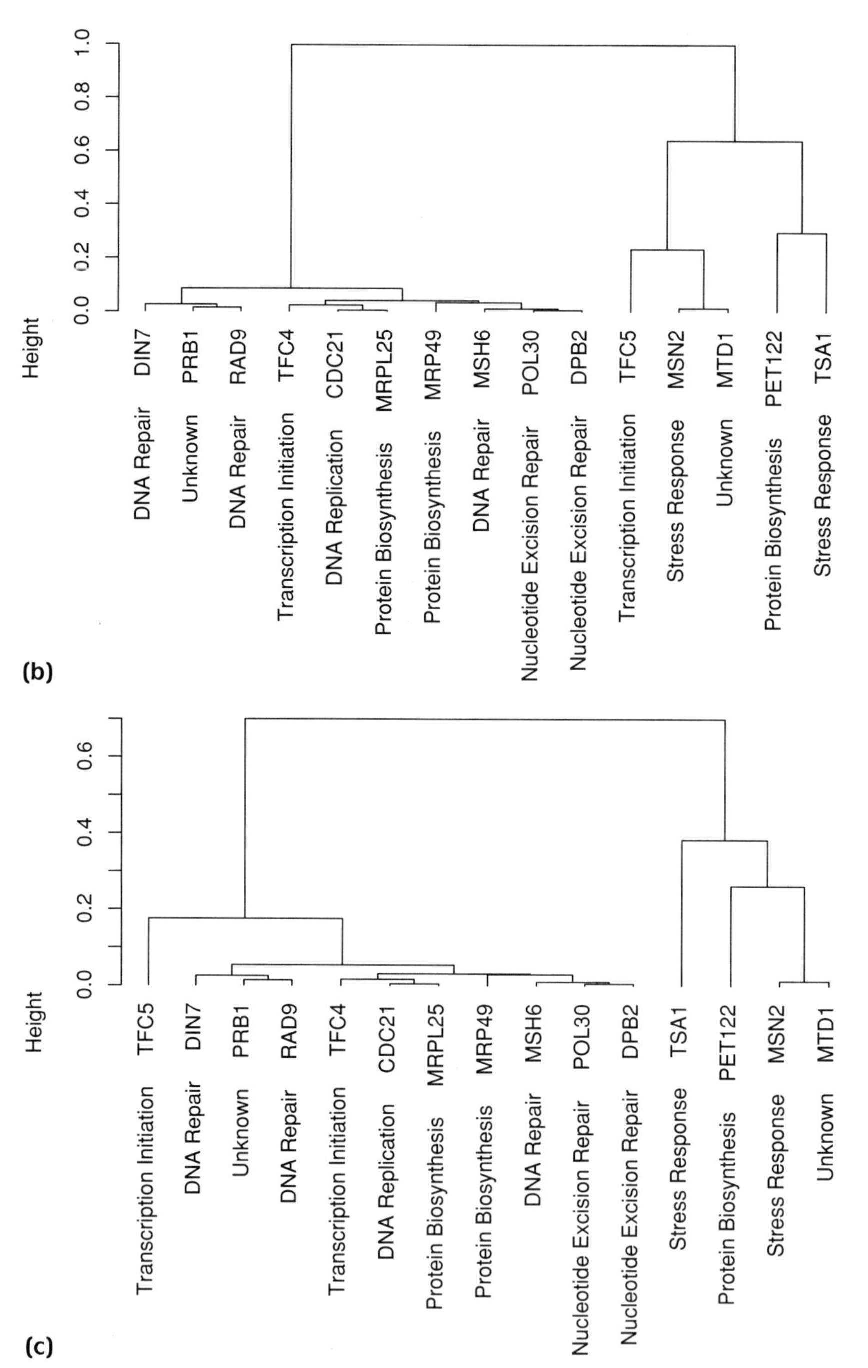

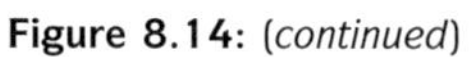

Figure 8.14: (*continued*)

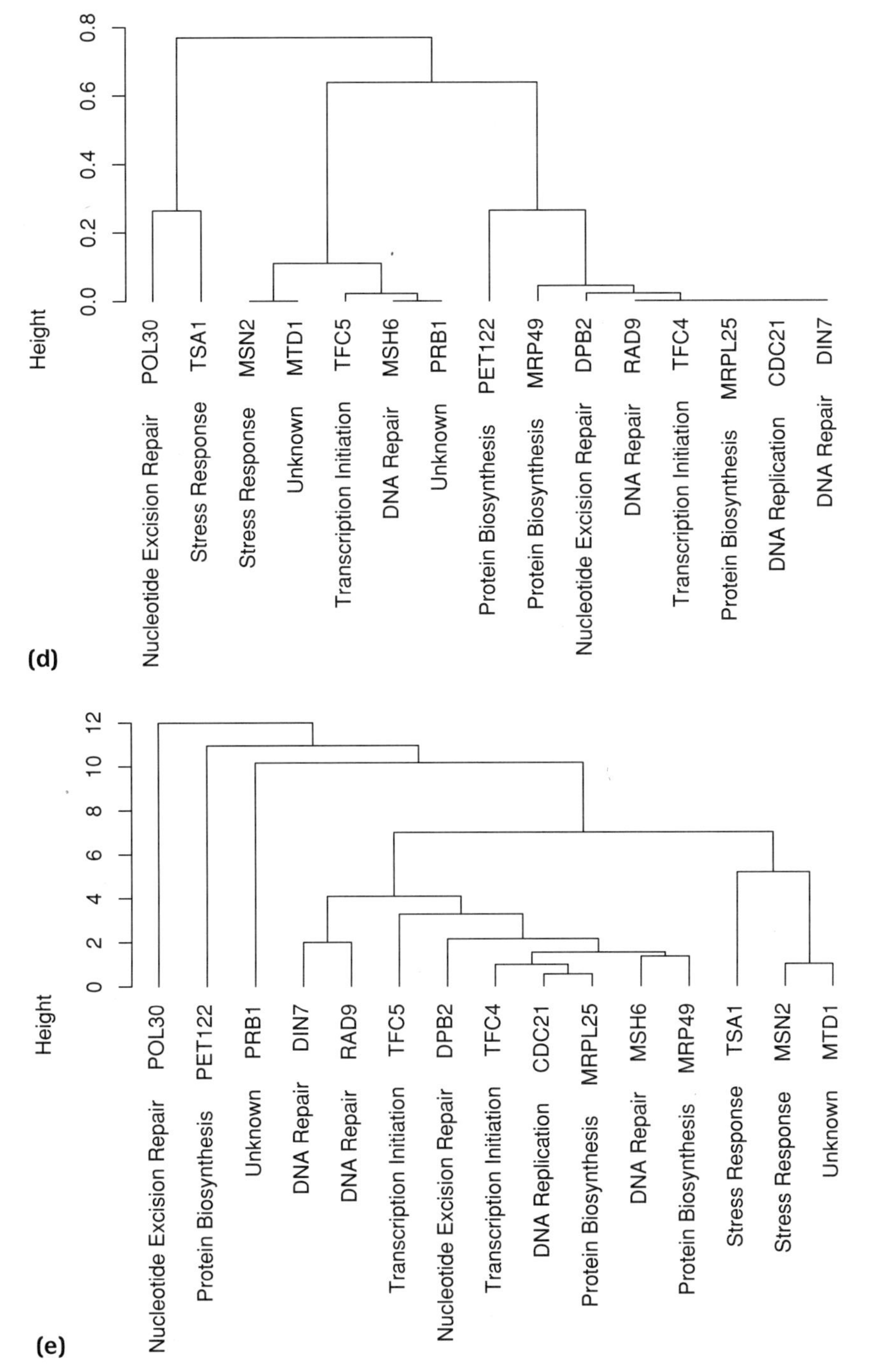

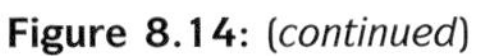
Figure 8.14: (*continued*)

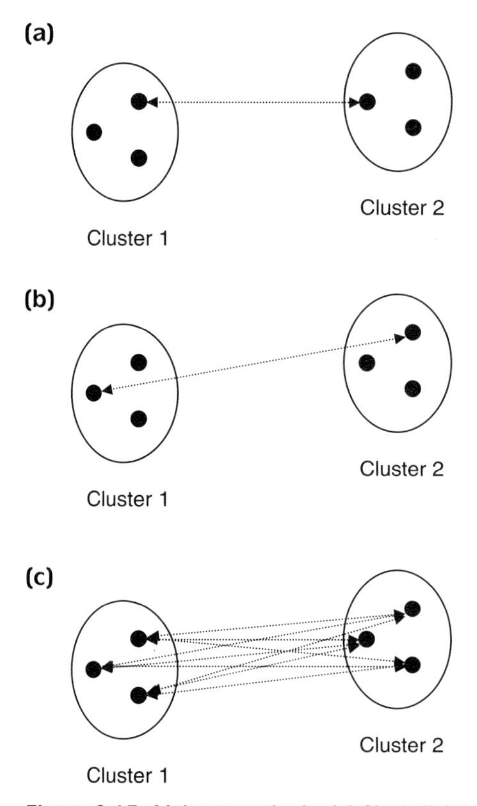

Figure 8.15: Linkage methods. (a) Single linkage. The distance between two clusters is defined as the distance between the two nearest points in the clusters. **(b)** Complete linkage. The distance between two clusters is defined as the distance between the two farthest points in the clusters. **(c)** Average linkage. The distance between two clusters is defined as the average of all of the distances between all of the points in the clusters.

In this section, we will discuss the three most commonly used methods: **single linkage**, **complete linkage** and **average linkage**. There are many other methods, and these have been implemented in statistics packages such as R and SPSS, as well as in dedicated microarray analysis packages such as GeneSpring and J-Express.

Single linkage defines the distance between two clusters as the distance between the nearest points in the clusters (Figure 8.15a). Clustering using single linkage (Figure 8.14a) tends to produce an effect called *chaining*: single genes are added to clusters one at a time. This can be seen in the left-hand portion of Figure 8.14a, where the transient genes are added to the main cluster one at a time. Single linkage can be useful when the data have natural clusters that are well defined but have irregular shapes, yet it is generally not recommended for microarray data.

Complete linkage defines the distance between two clusters as the distance between the farthest points in the clusters (Figure 8.15b). Complete linkage produces small, compact, well-defined clusters (Figure 8.14b). It performs well when there are well-defined clusters in the data and performs less well in fuzzier data.

Average linkage defines the distance between two clusters as the average of the distances between all pairs of points in the two clusters (Figure 8.15c). Average linkage is intermediate between single and complete linkage and tends to perform well in many microarray applications (Figure 8.14c).

The important feature to realise about these and other linkage methods is that they do produce different cluster diagrams (Figure 8.14). Individual genes may cluster differently according to the linkage method used; for example, the gene TFC5 clusters differently in the three dendrograms. Therefore, it is not wise to infer too much from any given cluster in any one dendrogram because these clusters may only be present as a result of the chosen methodology.

Distance Measures

We have already seen that the linkage method can produce different clusters. The distance metric you use can also result in different clusters. We demonstrate this by showing how the clustering differs when using Pearson correlation, Spearman correlation and Euclidean distance (Section 8.2).

Figure 8.14 shows dendrograms of the 15 genes of Example 8.11 that have been clustered using average linkage, but with the three distance measures described. There are three important features to note:

- **Negative correlations.** Pearson and Spearman correlation have the ability to spot negative correlations, that is, genes with opposite profiles. An example is the gene PET122. This has a profile that is similar in shape to MTD1 and MSN2, but which is opposite. When PET122 is up-regulated, MTD1 and MSN2 are down-regulated, and vice versa. In the clusterings produced by Pearson and Spearman correlation, these genes are close. But with Euclidean distance, the profiles are very distant, so they appear far away on the dendrogram.
- **Clusters with more than one gene.** Spearman correlation can produce a distance of zero if the gene profiles have exactly the same shape; the fine structure of the clusters disappears and is replaced with clusters containing a group of genes. An example is the cluster containing MRP49, RAD9, TCF4, MRPL25, CDC21 and DIN7. These all have exactly the same shape, and so have a distance of zero by Spearman correlation.
- **Larger distances.** Euclidean distance tends to produce larger distances than the correlations, so the clusters are generally "looser."

Which distance measure you use is up to you: there is no right answer. We recommend that you apply all distance measures to your cluster analysis and look at the results produced by all methods before drawing conclusions.

Isomorphisms

The final point about hierarchical clustering is the idea of **isomorphisms**. When we draw a cluster, each time we have a node, we have a choice: which gene (or cluster) do we place on which side of the node? Therefore, there are many ways[5] of drawing

[5] If there are n genes in the analysis, there are 2^{n-1} different ways of drawing the cluster diagram.

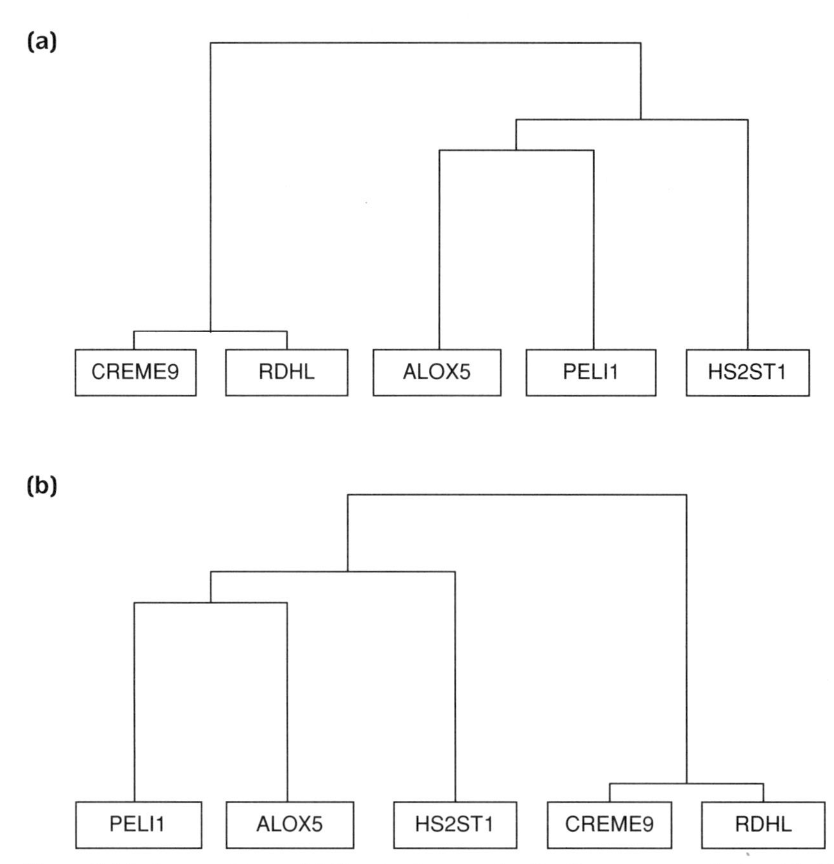

Figure 8.16: Isomorphisms. (a) The dendrogram for the five genes we clustered in Example 8.9. **(b)** An alternative dendrogram for the same clustering. The clustering is identical, but we have drawn the dendrograms differently.

the same clustering, and so it is important to remember that just because two genes or clusters are "near" each other on the dendrogram does not mean that they are near each other in the clustering. You must always look up the tree to see the lengths of the branches in order to establish proximity.

EXAMPLE 8.12 ISOMORPHIC CLUSTERS

We show two drawings of the same clustering of the five genes from data set 7A (Figure 8.16). Although the clustering is the same, the genes are in a different order.

SECTION 8.5 THE RELIABILITY AND ROBUSTNESS OF HIERARCHICAL CLUSTERING

We have described hierarchical clustering – the most commonly used clustering method for microarray analysis. There is a question that must always be asked of any analysis: how reliable are the results? Microarray data are noisy, often with large coefficients of variation. We need to be sure that the structures we see in the cluster diagrams represent truly related groups of genes and are not a representation of random noise or experimental error. Ideally, we want to be able to place a numerical confidence on a cluster, in a similar manner to placing a standard deviation or standard error on a straightforward numerical measurement.

There are three methods for assessing the reliability of a cluster analysis.

- **Visually.** We look at the gene expression profiles in the clusters and see whether they look similar. This is a useful exercise, but wholly subjective and unreliable, and it becomes difficult if the number of genes and/or samples is large.
- **By biological relevance.** If we assume that clusters of genes or samples should make biological sense, then we would expect biologically relevant genes or biologically similar samples to cluster together. For example, in Figure 8.12a, we see the DNA repair genes clustering together, and the stress response genes clustering together.

  This is also an important and valuable process, but it too is subjective, and it is often easy to come up with post hoc justifications as to why particular genes might cluster together without looking at the data set as a whole and considering the genes that are not clustered.
- **Using an appropriate statistical measure based on known and measured experimental variability.** By performing such an analysis, we relate the reliability of the clusters to the reliability of the experimental data from which they are built. From a statistical perspective, this is the best approach.

In this section we describe a statistical measure to quantitatively assess the reliability of clusters: the construction of a **consensus tree** using **parametric bootstrapping**. The method of consensus trees is well established in phylogenetic analysis and is an excellent method for microarray analysis.

Parametric Bootstrapping

We introduced the idea of bootstrapping in Section 7.4. The aim of the bootstrap is to create imaginary data sets that look very much like the original data set. For assessing the reliability of cluster analysis, we want to construct data sets that could represent a completely separate but identical experiment to the one we have performed. Any biological differences between samples or treatments would be the same. The only differences are that the bootstrap data sets would have different experimental errors as a result of being different experiments.

We apply the bootstrap via knowledge of the coefficient of variability of the experiment,[6] as measured using the methods of Chapter 6. We start with the real log ratio

[6] The procedure we describe is a parametric bootstrap because it constructs the bootstrap data by adding random deviates from a parametrised distribution. In an experiment with many replicates, it is also possible to apply a non-parametric bootstrap. Instead of adding random deviates to the data, we would, for each measurement of each gene in each sample, select one of the replicates at random and use that value as its measure of expression. Non-parametric bootstraps have the advantage of allowing different levels of error for different genes and non-normal error distributions. However, they require a well-replicated experiment to be successful. The parametric bootstrap has the advantage that it can be applied to any experiment, irrespective of the number of replicates used.

TABLE 8.7: Construction of Bootstrap Data for MSH6

Time	Real Data	Normal Random Variable	Bootstrap Data
30 minutes	−0.05	−0.10	−0.15
2 hours	−0.01	0.12	0.11
5 hours	1.75	−0.62	1.13
7 hours	3.80	−1.02	2.78
9 hours	3.28	0.43	3.71
11 hours	3.28	0.29	3.57

Note: To construct bootstrap data you start with the real data, compute normal random variables with standard deviation equal to the measured variability of the experimental data and add these to the real data. Note that in this example, the real data reach peak expression at 7 hours, but the bootstrap data reach peak expression at 9 hours. This is because the difference between the values at 7 hours and 9 hours is smaller than the level of experimental variability, so it cannot be relied upon as being a true difference in gene expression.

values for each gene. We then construct random numbers from a normal distribution with mean zero and standard deviation calculated using Equation 6.1, and we add these random numbers to the real log ratios. The resulting data set is a bootstrap data set: it looks like the original data set, and the errors we have added to it are exactly of the magnitude of the errors in the experiment.

EXAMPLE 8.13 BOOTSTRAP DATA FROM DATA SET 8A

We will construct a bootstrap data set for the gene MSH6 from data set 8A. This gene is initially not differentially expressed, becomes expressed at 5 hours, reaches a peak at 7 hours, and then appears to decrease in expression (Figure 8.13).

The coefficient of variability for this data set is about 40%. Using Equation 6.1, we calculate the standard deviation of the normal errors for the logged values to be 0.38. Because we are using log ratio data, we must multiply this by $\sqrt{2}$ to get a standard deviation of 0.54. This is essentially a standard deviation for each of the data points in Figure 8.13.

Now we construct the bootstrap data for this gene. We want the bootstrap data to look like the original data, with time points at 30 minutes, 2 hours, 5 hours, 7 hours, 9 hours and 11 hours. We construct six normal random variables[7] and add these to the real data (Table 8.7).

We repeat this procedure for every gene that we analyse with cluster analysis. Having constructed the bootstrap data for every gene, we perform the cluster analysis on the bootstrap data. This produces a new dendrogram that might be different from the

[7] It is straightforward to construct normal random numbers in all statistics packages, such as SPSS, R and SAS, and it is also possible to do this in Microsoft Excel using the Data Analysis add-in.

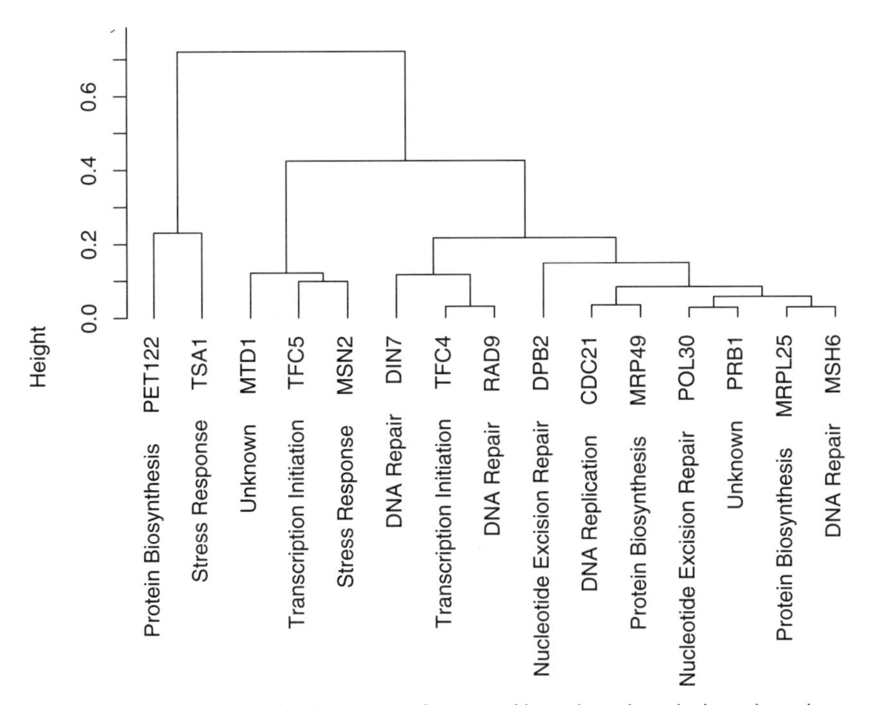

Figure 8.17: Clustering of a bootstrap data set. Normal random deviates have been added to the clustering data set and the data have been reclustered using Pearson correlation and average linkage. The clustering is similar to the clustering in Figure 8.14c: the data still splits into persistent and transient genes, and many of the genes that were similar continue to cluster together. But there are differences, particularly in the fine structure of the clusters. TCF5 has also moved from the persistent group into the transient group.

original dendrogram (Figure 8.17). In this example, the dendrogram is similar to the clustering of the real data: it identifies the two main clusters of persistent and transient response genes, and many of the genes that are close in Figure 8.14a are close in Figure 8.17. But there are also some differences; much of the fine structure of the clusters has changed, and the gene TFC5 has now clustered with the transient genes rather than the persistent ones.

Construction of a Consensus Tree

The bootstrap is applied not once, but many times. It is usual to create at least 1,000 bootstrap data sets, and the clustering algorithm is applied to each bootstrap data set. Next comes the important step. We are interested in clusters of genes that appear consistently in the bootstrap trees: these are the clusters that appear even when we add noise to the data that simulates the experimental errors, and so are the clusters that are robust to experimental error. Any cluster structure that does not appear consistently in the bootstrap trees is not robust to experimental errors, and it is difficult to draw scientific inference from such clusters.

Mathematicians have shown that the clusters that appear in more than 50% of the bootstrap trees form a consistent pattern, and it is possible to construct a tree from them, called a **consensus tree.** Not all genes may be resolved in a consensus tree: there may be groups of genes that cluster together but have no further structure. That is not a problem; it simply means that there is not enough experimental evidence to subdivide those genes into smaller clusters.

We also know the number of times each node on the consensus tree has been seen among the bootstrap trees. This can be used as a measure of confidence in each node. Therefore, we have achieved our objective – we have identified the cluster structure that is robust to experimental error and, on each node that remains, we have a quantification of the confidence in each node of the tree.[8]

EXAMPLE 8.14 CONSENSUS TREE FOR DATA SET 8A

We run the bootstrap 1,000 times for the 15 genes we selected from data set 8A, first with the measured experimental variability of 40%, and then with 30% variability to demonstrate the dependence of consensus trees on data quality.

When we use 40% variability, only two clusters appear in more than 50% of the trees: the cluster containing the 11 persistent genes appeared 590 times, and the cluster containing the 2 transient genes MSN2 and MTD1 appeared 535 times (Figure 8.18a). No other structure was observed. The genes TSA1 and PET122 were not clustered with any other genes. This implies that the fine structure seen in Figure 8.12a cannot be relied upon as biologically meaningful: it is all within the bounds of experimental error. Interestingly, the gene TFC5 has been clustered with the persistent genes; the decrease in gene expression after 7 hours is within the 40% experimental error and so is not significant.

When we use 30% variability, which is less than the variability in this particular experiment, but could be representative of a better quality experiment, a little more structure is seen. There is a separation between TFC5 and the other persistent genes, while PET122 and TSA1 cluster with MSN2 and MTD1 (Figure 8.18b). No other fine structure is observed.

[8] Consensus tree construction has been part of phylogenetic software such as Phylip for many years. At the time of this writing, these methods have not yet been included in any of the commonly used microarray analysis software packages.

Figure 8.18: Consensus tree construction. We construct consensus trees for the clustering shown in Figure 8.14c using 1,000 bootstrap replicates. **(a)** We use the experimental variability of 40%. Only two clusters are present in the consensus tree. The cluster containing the 11 persistent genes is observed in 590 of the 1,000 bootstrap trees, and the cluster containing MTD1 and MSN2 appears 535 times. The genes PET122 and TSA1 are not clustered. No other cluster structure is observed. Interestingly, the gene TFC5 clusters with the persistent genes: this is because the decrease in gene expression after 7 hours is within the bounds of experimental error. It is difficult to infer any further cluster structure from this data, such as the structure seen in Figure 8.14c, because this structure is too sensitive to experimental error. **(b)** The consensus tree is constructed with experimental variability of 30%. The clusters from (a) are present with greater certainty; there is also more structure present, with PET122 and TSA1 clustering with MSN2 and MTD1, and TFC5 being slightly different from the other 10 persistent genes. We conclude that we can get more cluster information from a better experiment.

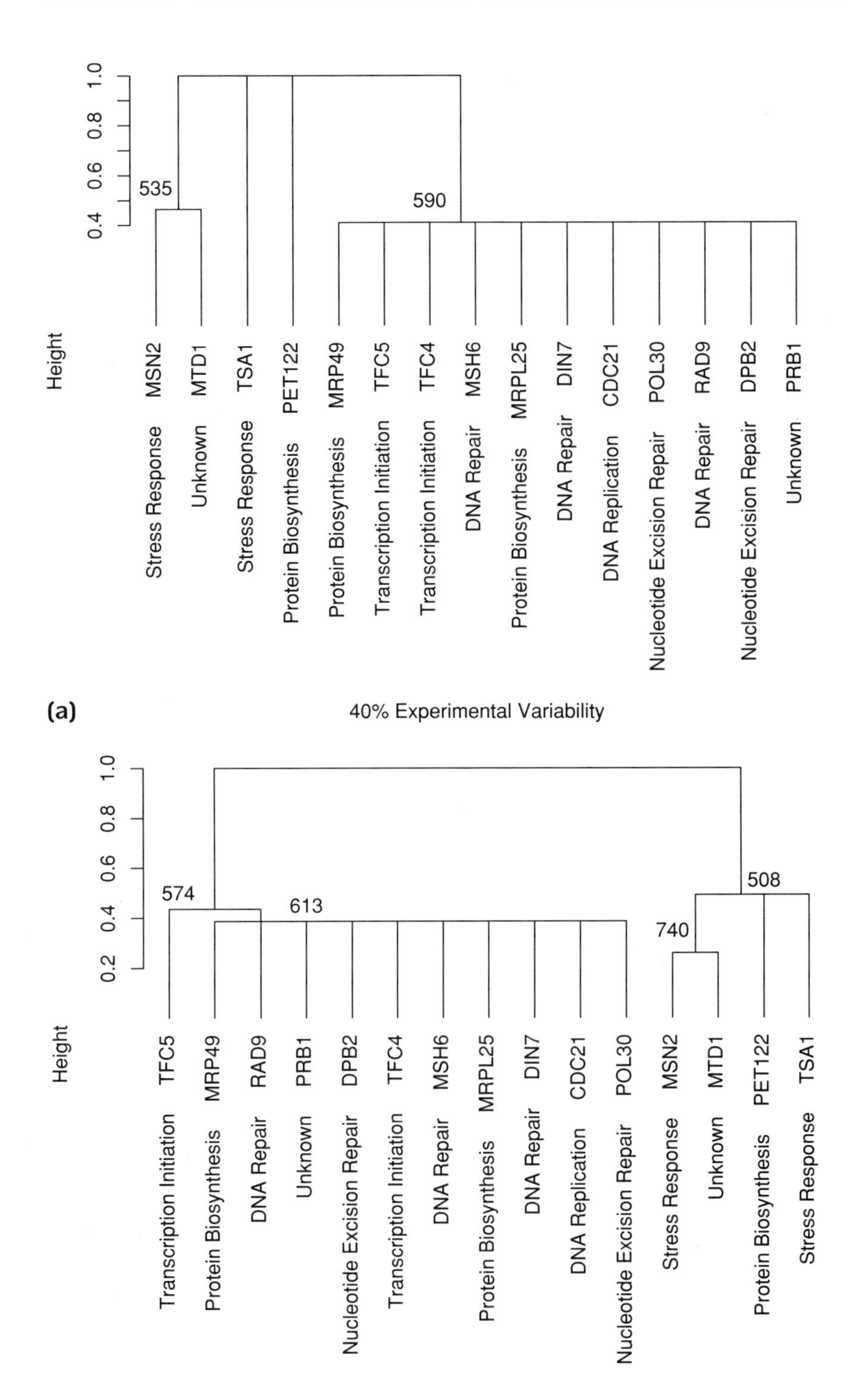
Height
0.4
0.6
0.8
1.0
535
590
Stress Response MSN2
Unknown MTD1
Stress Response TSA1
Protein Biosynthesis PET122
Protein Biosynthesis MRP49
Transcription Initiation TFC5
Transcription Initiation TFC4
DNA Repair MSH6
Protein Biosynthesis MRPL25
DNA Repair DIN7
DNA Replication CDC21
Nucleotide Excision Repair POL30
DNA Repair RAD9
Nucleotide Excision Repair DPB2
Unknown PRB1
(a)
40% Experimental Variability
Height
0.2
0.4
0.6
0.8
1.0
574
613
740
508
Transcription Initiation TFC5
Protein Biosynthesis MRP49
DNA Repair RAD9
Unknown PRB1
Nucleotide Excision Repair DPB2
Transcription Initiation TFC4
DNA Repair MSH6
Protein Biosynthesis MRPL25
DNA Repair DIN7
DNA Replication CDC21
Nucleotide Excision Repair POL30
Stress Response MSN2
Unknown MTD1
Protein Biosynthesis PET122
Stress Response TSA1
(b)
30% Experimental Variability

In summary, we learn two lessons from consensus trees:

- Much of the fine structure seen in cluster analyses may not represent when we take into account experimental variability. The consensus tree shows the structure that is reliable.
- The better you run your experiment, in terms of reducing experimental errors and variabilities, the more information you can get from your cluster analysis.

SECTION 8.6 MACHINE-LEARNING METHODS FOR CLUSTER ANALYSIS

There are two methods that have come from the machine-learning community that are implemented in many of the gene expression analysis software packages: **k-means clustering** and **self-organised maps**. This section gives a brief description of these methods and shows them applied to the data set we have used for hierarchical clustering.

K-Means Clustering

K-means is a clustering algorithm that differs from hierarchical clustering in three important ways:

- The number of clusters has to be specified in advance.
- There is no hierarchy or relationship between the clusters, nor is there any hierarchy or relationship between the genes or samples within the clusters; the clusters are just groups of similar gene expression profiles.
- K-means starts by randomly allocating genes or samples into clusters. Therefore, different runs of k-means can give slightly different results.

The K-Means Algorithm

The k-means algorithm has six steps:

1. Choose the number of clusters, denoted k.
2. Randomly assign each gene expression profile to one of the k clusters.
3. Calculate the centroids of each of the k clusters.
4. For each profile in turn, calculate the distance between it and the centroids of each of the k clusters.
5. If that profile is closest to a cluster different from the one in which it currently belongs, move the profile to the new cluster and update the centroids of both clusters.
6. Go back to step 4 and repeat until no profiles change cluster membership.

EXAMPLE 8.15 K-MEANS CLUSTERING OF DATA SET 8A

We apply k-means clustering to the 15 genes to which we have applied the hierarchical clustering using correlation as the distance measure. The results for 2, 3, 4 and 5 clusters are shown in Table 8.8. This clustering also finds the groups of persistent and

TABLE 8.8: K-Means Clustering of Genes from Data Set 8A

Number of Clusters (k)			
$k=2$	$k=3$	$k=4$	$k=5$
Cluster 1	**Cluster 1**	**Cluster 1**	**Cluster 1**
MTD1	MTD1	MTD1	MTD1
MSN2	MSN2	MSN2	MSN2
TSA1	TSA1	TSA1	TSA1
POL30	**Cluster 2**	TFC5	TFC5
Cluster 2	POL30	**Cluster 2**	**Cluster 2**
PET122	**Cluster 3**	PET122	PET122
TFC5	PET122	**Cluster 3**	**Cluster 3**
DIN7	TFC5	POL30	POL30
RAD9	DIN7	**Cluster 4**	**Cluster 4**
PRB1	RAD9	DIN7	DIN7
CDC21	PRB1	RAD9	RAD9
MSH6	CDC21	PRB1	PRB1
MRP49	MSH6	CDC21	**Cluster 5**
MRPL25	MRP49	MSH6	CDC21
TFC4	MRPL25	MRP49	MSH6
DPB2	TFC4	MRPL25	MRP49
	DPB2	TFC4	MRPL25
		DPB2	TFC4
			DPB2

Note: We apply k-means clustering to the 15 genes studied, varying the number of clusters k from 2 to 5. The clustering algorithm finds the persistent and transient groups. When $k = 2$, POL30 is clustered with the transient genes, probably because there are too few clusters. When $k = 4$, the cluster structure seems to fit the data well. When $k = 5$, the persistent genes split into two groups: the data has probably been overclustered.

transient response genes. When $k = 2$, the gene POL30 is grouped with the transient genes. This is probably an anomalous result because there are too few clusters. The value for $k = 4$ looks like a good result. When $k = 5$, the persistent genes split into two groups and the data are probably overclustered.

How to choose a good value of k?

The number of clusters has to be chosen in advance. This means that the user must make some attempt to estimate the number of clusters before running the algorithm. There are two approaches to estimating the number of clusters to use: via MDS and empirically. We recommend both.

MDS (Section 8.3) allows you to visualise the distances between the genes or samples in a two-dimensional space. By using the distance measure you intend to use for the clustering you can gain an indication of whether there is a natural number of clusters in the data.

EXAMPLE 8.16 MDS TO HELP CHOOSE k

We apply MDS to the 15 genes (Figure 8.19). There is a main cluster of genes on the left-hand edge of the space; this might be two clusters with MSH6, MRPL25 and TCF4

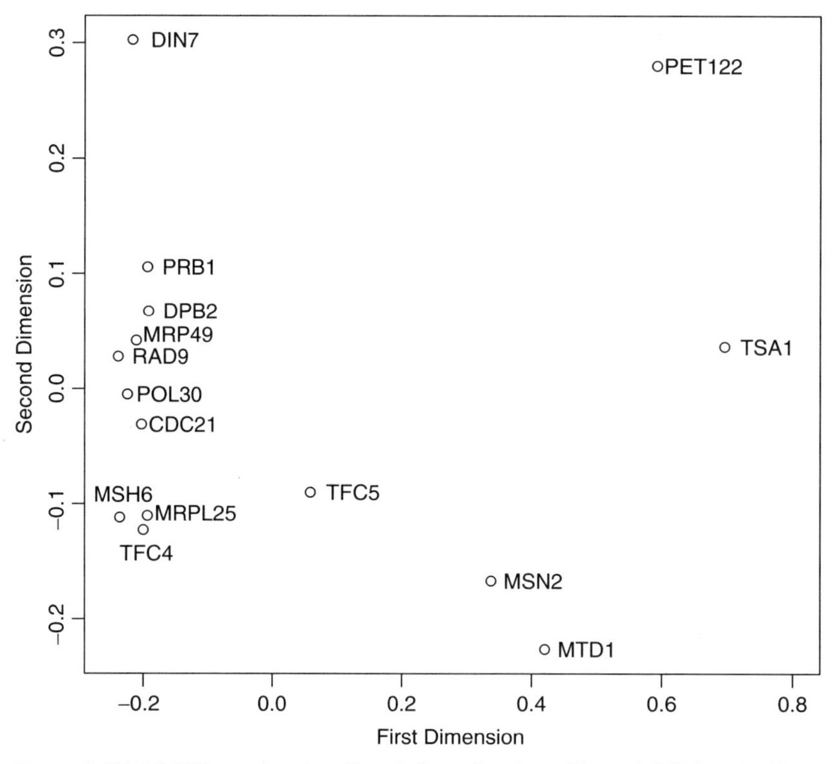

Figure 8.19: Multidimensional scaling of clustering data. We apply MDS to the 15 genes we are using for clustering, using Pearson correlation as a measure of distance. There is a large cluster of genes at the left-hand edge of the figure, which has a subcluster of MSH6, MRPL25 and TFC4. The genes MSN2 and MTD1 are close together, and TFC5 is in between these and the main cluster. DIN7, PET122 and TSA1 appear to be outliers. We would estimate from this figure that there might be between four and six natural clusters in the data.

slightly away from the main group. MTD1 and MSN2 are close together and might form another cluster; TFC5 appears to be between these and the big cluster. DIN7, PET122 and TSA1 are away from the other genes. So, we would certainly need at least three clusters, and probably at most six clusters.

Empirical choice of *k*means running the k-means algorithm with different numbers of clusters and different distance metrics, and testing the reliability of the clusters using one (or all) of the following methods described. We would then select a number of clusters that gives robust, reliable and meaningful results.

Validating K-Means Clustering

As with hierarchical clustering, it is essential to validate the results of k-means clustering. There are four ways to validate your results; these are similar to the methods described in Section 7.4.

- **Visually.** Look to see if genes or samples in the same cluster have similar profiles.

- **By biological validity.** Look to see if the genes in the same clusters have similar or complementary biological functions, or if the samples in the same clusters are derived from similar biological sources.
- **By reclustering the data with the same value of *k*.** There is a random element in k-means clustering. If the same clusters are coming up, then the clustering is performing well. If each clustering gives different clusters, then you may be using a bad value for *k*, a bad distance measure or this may not be a good method to analyse the data.
- **By statistical analysis.** Parametric bootstrapping (Section 8.5) can also be applied to k-means clustering. One should only believe clusters that appear in the majority of bootstrap clusterings, and the bootstrap will allow you to place a measure of confidence on those clusters.

Self-Organised Maps

The final clustering algorithm we look at in this chapter is self-organised maps. These are clustering methods that are similar to k-means in that the user specifies a predefined number of clusters. However, unlike k-means, the clusters relate to one another via a spatial topology. Usually, the clusters are arranged on a square grid. The algorithm is quite detailed and we refer the interested reader to the reference at the end of the chapter. There are three important properties of self-organised maps:

- The clusters relate to each other in a spatial topology, usually in a grid.
- The size of the grid (number of clusters) must be chosen in advance: it is usually wise to try different sizes and see what works best.
- The genes are at first allocated to the clusters at random, so different runs of a self-organised map may give different results.

EXAMPLE 8.17 SELF-ORGANISED MAPS ON DATA SET 8A

We run a self-organised map with a 2×2 grid on the same 15 genes that we have been using for all of the clustering methods. We show the average profiles of the four clusters in Figure 8.20. The four clusters have a spatial relationship: the two clusters on the top represent the persistent responses, while the two clusters on the bottom are the transient responses. In this case, the two top clusters are very similar, probably because even a 2×2 grid contains too many clusters for this data set.

In Table 8.9 we show the allocations of genes from two separate runs of a self-organised map. The allocation of the genes to clusters is different. This is a serious problem when using self-organised maps for analysing microarray data and must be considered whenever using them.

Choosing the Size of a Self-Organised Map

Exactly the same principles should be applied to choosing the size of a self-organised map as applied for choosing the number of clusters *k*. You can use MDS to look to

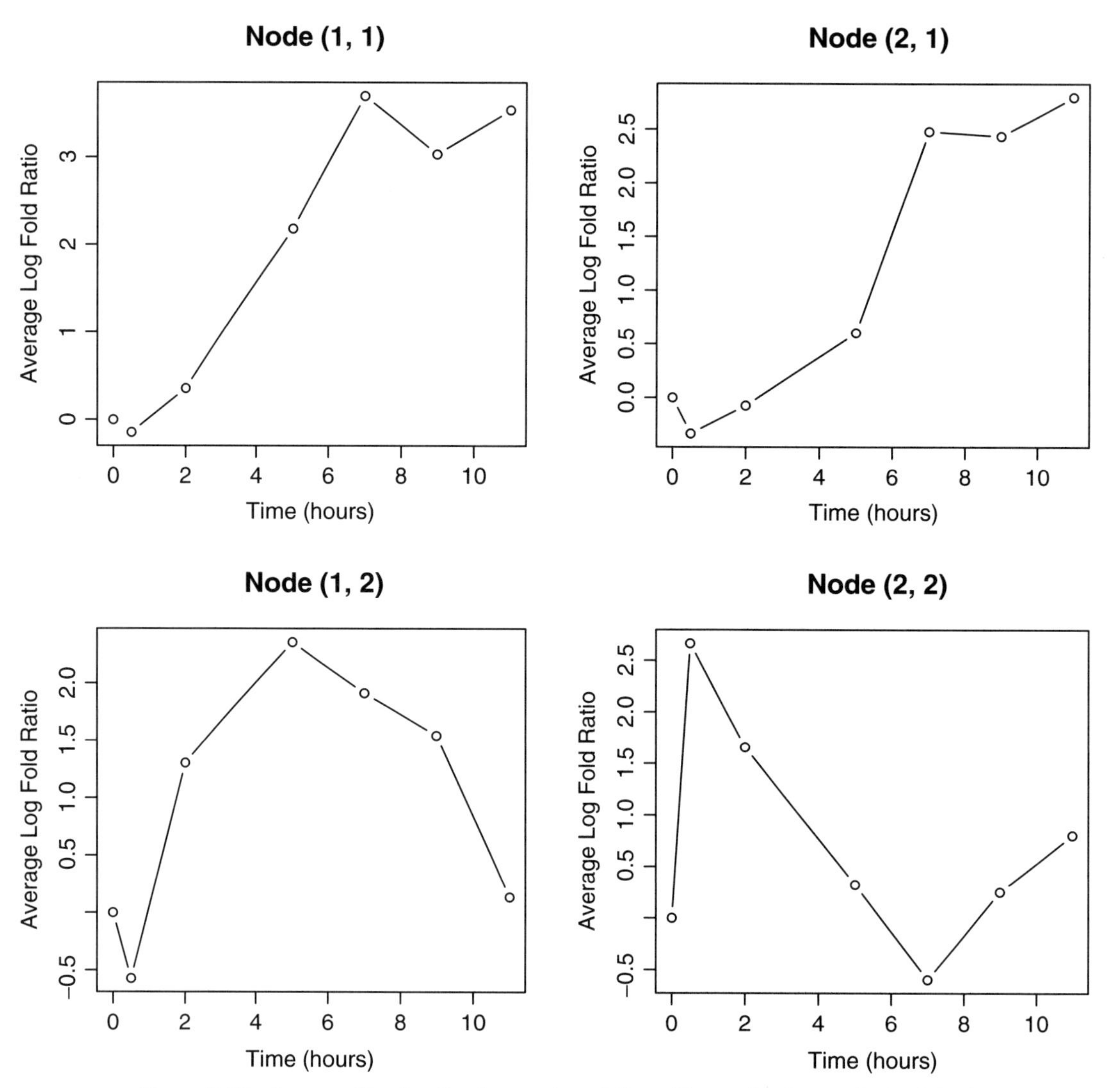

Figure 8.20: Gene profiles for the four nodes of the self-organised map. Each of the four nodes of the self-organised map has average profiles shown in the four panes of the figure. The top two nodes are fairly similar; both contain persistently regulated genes. The bottom two nodes contain transiently regulated genes.

see if there is a natural number of clusters, and you can choose the size of the map empirically.

EXAMPLE 8.18 CHOOSING THE SIZE OF A SELF-ORGANISED MAP

Looking again at the MDS plot (Figure 8.19), we can see that four clusters would be a reasonable number to use, but the nine clusters given by a 3×3 grid would be too many clusters for these data. Hence, in Example 8.17 we used a 2×2 grid to give four clusters.

TABLE 8.9A: Self-Organised Maps Applied to Data Set 8A

CDC21	DIN7
MSH6	RAD9
MRP49	PRB1
MRPL25	DPB2
TFC4	
TFC5	
MTD1	POL30
MSN2	PET122
TSA1	

Note: Allocations of genes to clusters from the run of a 2 × 2 self-organised map that was used to create Figure 8.16.

TABLE 8.9B: Self-Organised Maps Applied to Data Set 8A

CDC21	PET122
MSH6	
MRP49	
MRPL25	
TFC4	
TFC5	
DIN7	
RAD9	
PRB1	
DPB2	
MTD1	POL30
MSN2	TSA1

Note: Allocations of genes to clusters from a separate run of a 2 × 2 self-organised map. Note that the allocations are different in the two runs. There is a random element to self-organised maps, which means that two runs can be very different.

Validation of Self-Organised Maps

As with k-means clustering, self-organised maps can be validated four ways:

- **Visually.** Looking to see that the profiles are similar in the clusters and are different between clusters.
- **By biological relevance.** Check that genes with similar biological function or samples from similar biological sources are present in the same or nearby clusters.
- **By reclustering.** Does the self-organised map change dramatically when it is re-run? If so, the clusters that change may be difficult to interpret.
- **By statistical analysis.** Although it is possible to apply the bootstrap methods described in Section 8.5 to self-organised maps, the consensus clusters may

TABLE 8.10: Advantages and Disadvantages of Different Clustering Methods

Hierarchical	Hierarchical Consensus Tree	K-Means	Self-Organised Maps
✓ Easy to understand algorithm ✓ Intuitive interpretation of results ✓ Same data give the same results every time ✓ Implemented in many gene expression packages × No a priori reason why genes should relate on a binary tree × Noise and errors can adversely influence a tree × No measure of confidence in results	✓ Intuitive interpretation of results ✓ Robust to noise and errors ✓ Hierarchy is representative of statistically significant differences in gene expression profiles ✓ Provides a measure of confidence in clusters × Not yet implemented in gene expression analysis software × Can give slightly different answers on different runs	✓ Intuitive interpretation of results ✓ No preimposed hierarchy ✓ Implemented in many gene expression packages × User has to specify number of clusters in advance × Can give different results when run different times × No measure of confidence in results although can construct consensus clusters	✓ Implemented in many gene expression packages × User has to specify number of clusters in advance × No a priori reason why gene expression clusters should fit a two-dimensional topology × Can get very different results when run different times × Cannot easily construct consensus clusters without losing topology

not relate to each other according to any spatial topology, and so the original topology of the self-organised maps may be lost.

KEY POINTS SUMMARY

In Table 8.10, we summarise the key advantages and disadvantages of the different clustering methods. The key points to take from this chapter are as follows:

- There are a number of ways to measure the similarity between gene expression profiles and the measure you use will affect your results. Thus, it is recommended to run your analyses with a number of different distance measures.
- PCA and MDS provide a good way to visualise your data without imposing any hierarchy on them.
- Hierarchical clustering can be used to identify related genes or samples and portray them using a dendrogram.
- There are many variants of hierarchical clustering, each of which can produce different results. You should always try different linkage methods and distance metrics.
- Machine-learning methods can also be used to define relationships between genes or samples but can produce different results each time you run them.
- You should always perform a statistical validation of your results (e.g., using a bootstrap algorithm).

SUPPLEMENTARY INFORMATION AND REFERENCES

Books

Everitt, B.S. 1993. *Cluster Analysis* (3rd edition). Edward Arnold: London.

Covers all cluster analysis methods.

Chatfield, C. and Collins, A.J. 1980. *Introduction to Multivariate Analysis.* Chapman and Hall: London.

Covers many multivariate techniques, including principal component analysis, multidimensional scaling and cluster analysis.

Kohonen, T. 2000. *Self Organised Maps.* Springer: Berlin.

Covers self-organised maps and related techniques.

Papers

Data Set 8A

Chu, S., DeRisi, J., Eisen, M., Mulholland, J., Botstein, D., Brown, P.O., and Herskowitz, I. 1998. The transcriptional program of sporulation in budding yeast. *Science* 282: 699–705.

Data Set 8B

Alizadeh, A.A., Eisen, M.B., Davis, R.E., Ma, C., Lossos, I.S., Rosenwald, A., Boldrick, J.C., Sabet, H., Tran, C., Powell, J.I., Yang, L., Marti, G.E., Moore, T., Hudson, J. Jr., Lu, L., Lewis, D.B., Tibshirani, R., Sherlock, G., Chan, W.C., Greiner, T.C., Weisenberger, D.D., Armitage, J.O., Warnke, R., Levy, R., Wilson, W., Grever, M.R., Byrd, J.C., Botstein, D., Brown, P.O., and Staudt, L.M. 2000. Distinct types of diffuse large B-cell lymphoma identified by gene expression profiling. *Nature* 403: 503–11.

Eisen, M.B., Spellman, P.T., Brown, P.O., and Botstein, D. 1998. Cluster analysis and display of genome-wide expression patterns. *Proceedings of the National Academy of Sciences* 95: 14863–68.

The original paper applying cluster analysis to microarray data.

Zhang, K. and Zhao, H. 2000. Assessing reliability of gene clusters from gene expression data. *Functional and Integrative Genomics* 1: 156–73.

An excellent description of the use of bootstrapping and consensus trees to assess the reliability of hierarchical clusters.

Internet Resources

http://genome-www5.stanford.edu/MicroArray/SMD/

The Stanford Microarray Database. All the data sets we refer to in this chapter are available for download from this database. There are also links to the papers from which the data derive.

http://www.r-project.org/

R package is freely available statistical software that can run on Windows, MacIntosh and Unix computers. There is a great deal of microarray analysis software written for R.

http://www.molmine.com

Homepage of Molmine, the distributors of J-Express. This is a gene expression analysis package that is available to academics for free and which contains many of the algorithms described in this chapter.

http://www.sigenetics.com

Homepage of Silicon Genetics, the company that produces GeneSpring. This is the most commonly used commercially available microarray analysis software. It implements many of the methods described in this chapter.

CHAPTER NINE

Classification of Tissues and Samples

SECTION 9.1 INTRODUCTION

One of the most exciting areas of microarray research is the use of microarrays to find groups of genes that can be used diagnostically to determine the disease that an individual is suffering from, or prognostically to predict the success of a course of therapy or results of an experiment.

In these studies, samples are taken from several groups of individuals with known pathologies, outcomes or phenotypes and hybridised to microarrays. The aim is to find a small number of genes that can predict to which group each individual belongs. These genes can then be used in the future as part of a molecular test on further individuals, either using a focussed microarray, or a simpler method such as quantitative polymerase chain reaction (PCR).

EXAMPLE 9.1 DATA SET 9A

Bone marrow samples are taken from 27 patients suffering from acute lymphoblastic leukemia (ALL) and 11 patients suffering from acute myeloid leukaemia (AML) and hybridised to Affymetrix arrays.[1] We want to be able to diagnose the leukemia in future patients using either Affymetrix technology or using more focussed arrays with a small number of genes. How do we choose a set of rules to classify these samples?

The development of such predictive models depends on statistical and computational techniques, many of which are still the subject of active research. There are essentially three parts to developing a predictive model, and so the chapter is arranged into three further sections:

Section 9.2: Methods of Classification, looks at a number of commonly used methods for distinguishing between groups of individuals based on a given set of measurements. There are several well-established methods for doing this, many of which have been shown to work well with microarray data.

Section 9.3: Validation, looks at methods to verify the results of a classification analysis. It discusses the two most commonly used approaches: training and test sets, and cross-validation.

[1] The data are from the paper of Golub et al. (1999). The reference is given at the end of the chapter. The data are available from the Stanford Microarray Database.

Section 9.4: Dimensionality Reduction, looks at a number of methods that can be used to find appropriate small numbers of genes from the large numbers of genes on a microarray that can be used to distinguish between the groups of individuals in a robust and reliable manner. This is a topic of open research; we describe some of the methods that are used, but there are no well-established standards.

SECTION 9.2 METHODS OF CLASSIFICATION

In this section we describe methods that allow you to predict the class to which an individual belongs, based on gene expression measurements. To do this, we build predictive models using the gene expression measurements of a number of individuals with known class membership. In the machine-learning community, this is known as **supervised learning**. Throughout this section, we assume that we have already selected a small number of genes whose expression measurements we use, and are not using all the genes on the microarray. In Section 9.4, we discuss methods we would use to select genes that allow us to build good predictive models.

We start by describing two concepts that are central to classification: **separability** and **linearity**. We then describe five different classification methods: **k-nearest neighbours**, **nearest centroid**, **linear discriminant analysis**, **neural networks** and **support vector machines**. These are all methods that have been applied to microarray data analysis.

Separability

Suppose we have gene expression measurements for a number of samples. In classification analyses, it is helpful to think of each sample as occupying a location in a high-dimensional space. Each axis of the space is the measurement of expression of one of the genes. If we are using the measurements of two genes, then the samples can be thought of as having locations in two-dimensional space; if we are using measurements of three genes, then we would think of the samples as having locations in three-dimensional space; and if we are using the measurements of ten genes, we would think of the samples as having locations in ten-dimensional space.

EXAMPLE 9.2 CONCEPTUALISING SAMPLE MEASUREMENTS IN GENE EXPRESSION SPACE

In data set 8A, there are 27 ALL patients and 11 AML patients. If we consider only the genes ubiquinol cytochrome c reductase core protein II and defender against cell death I then we can think of each sample as having a position in a two-dimensional space, with each axis being the level of gene expression of each of the two genes (Figure 9.1a).

Within this conceptual framework, there are two extreme scenarios:

- **Separable:** The different groups to which the samples belong occupy different regions of the gene expression space.

- **Non-separable:** The different groups to which the samples belong are mixed together in the same region of gene expression space.

In many cases, the data may only be partially separable, with the different groups broadly occupying different regions of space but with some overlap at the boundaries. The aim of the classification methods described in this section is to find a way to partition the space so that each group is in a different region and to describe the partitions, so that given a new sample, we can determine to which group it belongs. As part of this process, we can quantify the extent to which the data is separable.

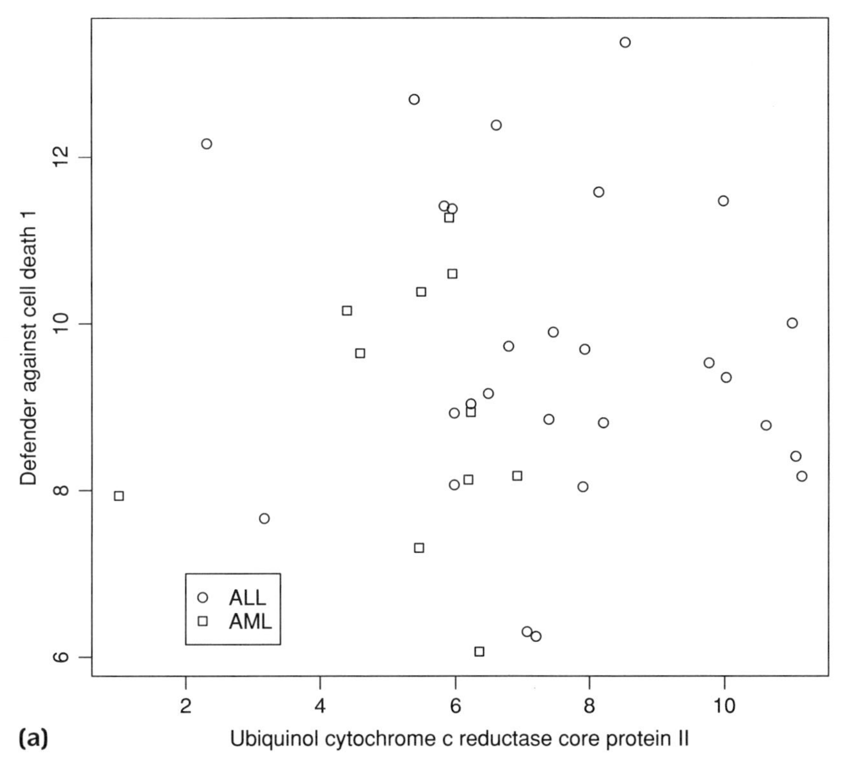

Figure 9.1: Separability of classes. Twenty-seven samples from patients suffering from ALL and 11 samples from patients suffering from AML have been hybridised to microarrays (data set 9A). We would like to find groups of genes that can distinguish between the two groups. For illustrative purposes, we show examples with just two genes. In general, it would be common to use more than two genes for classification analyses. **(a)** The genes ubiquinol cytochrome c reductase core protein II and defender against cell death 1 cannot distinguish between the two groups. We call data that behaves in this way *inseparable*. **(b)** The genes C-myb gene and leptin receptor divide the data clearly into two groups. In this case, it would be possible to draw a straight line between the ALL and AML classes to separate the groups; we call the data *linearly separable*. Methods such as linear discriminant analysis would work well with such data. **(c)** The genes immunoglobulin alpha heavy chain and DNA G/T mismatch binding protein also separate the two classes. But in this case, there is no straight line that could be used to divide the two groups. We would need either two straight lines or a curved line to distinguish the classes. We call this data *non-linearly separable*.

(*continued*)

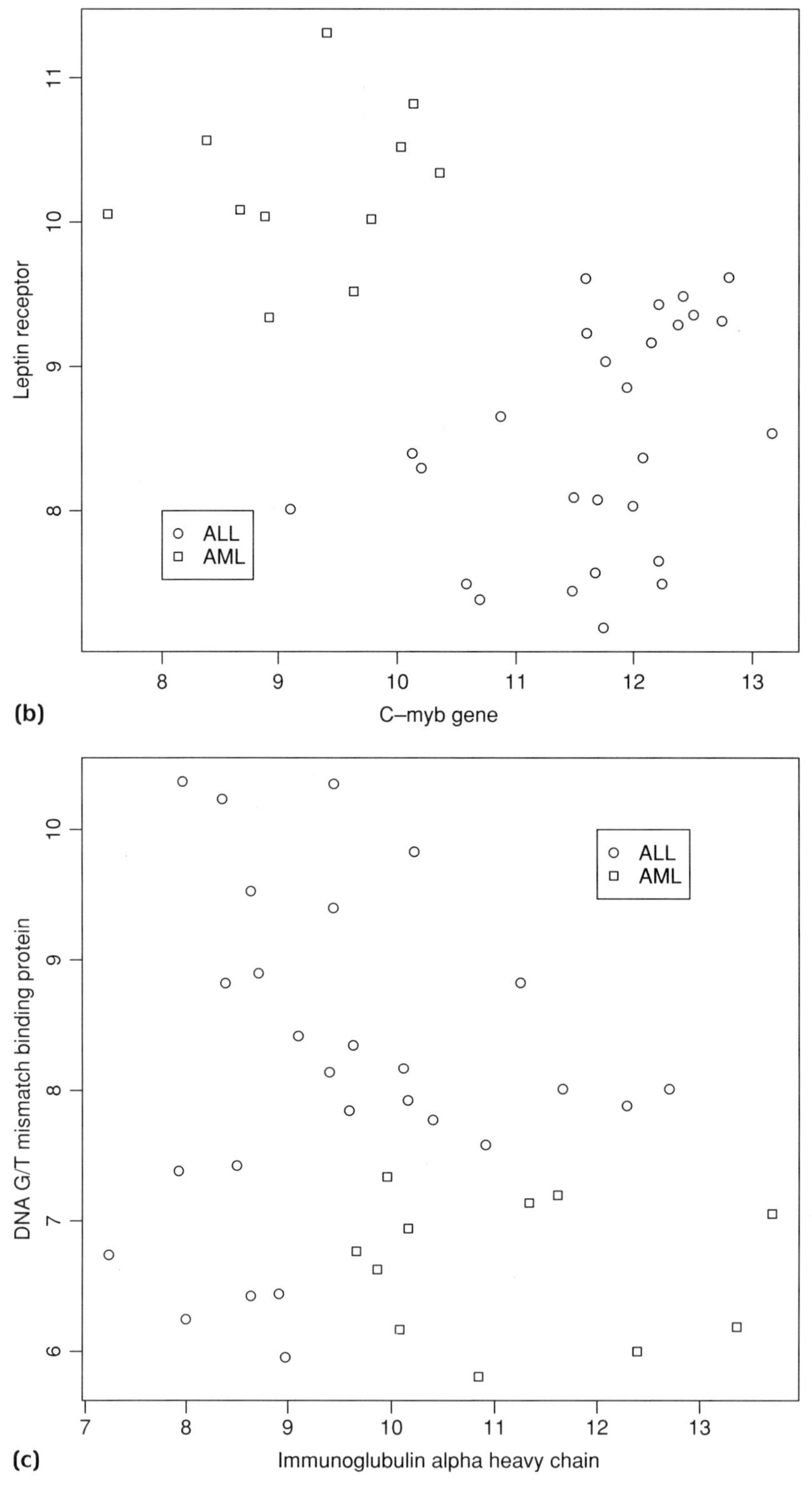

Figure 9.1: (*continued*)

In a microarray experiment, the level of separability of the data will be determined by the group of genes whose expression measurements are being considered. Part of the remit of the dimensionality reduction methods described in Section 9.4 is to find groups of genes that maximise the separability of the data.

EXAMPLE 9.3 SEPARABLE AND NON-SEPARABLE DATA

If we consider the measurement of the sample in data set 9A in two-dimensional space with the genes ubiquinol cytochrome c reductase core protein II and defender against cell death I, then the two groups of patients (ALL and AML) do not occupy different regions of space (Figure 9.1a). The data are not separable.

If we consider the same patients but using the measurements of C-myb gene and leptin receptor, then the two groups are in distinct regions of space (Figure 9.1b). The data are separable, and it is possible to build a classifier using these genes.

Linearity

When we consider separable data, there are two possibilities for how the samples can be separated in space:

- **Linearly separable.** Data are linearly separable if it is possible to partition the space between the two (or more) groups using straight lines.
- **Non-linearly separable.** Data are non-linearly separable if the groups are separable, but it is not possible to partition the groups using straight lines.

We will describe some methods that only apply linear separation techniques, and other methods that are able to classify non-linearly separable data. Microarray data are frequently non-linear, so it is often recommended to try both linear and non-linear methodologies.

EXAMPLE 9.4 LINEARLY AND NON-LINEARLY SEPARABLE DATA

Data set 9A can be both linearly and non-linearly separable, depending on which gene expression measurements are considered. With the genes C-myb gene and leptin receptor, the data are linearly separable (Figure 9.1b). The genes immunoglublin alpha heavy chain and DNA G/T mismatch binding protein also separate the groups, but the separation is non-linear (Figure 9.1c).

Number of Classes

Throughout this chapter we use data set 9A, which consists of two classes, AML and ALL. The reason for this is that it is easiest to understand classification methods from the perspective of distinguishing between two groups. However, many applications of classification using microarrays may have more than two groups of individuals. Three of the five methods we will describe extend very naturally to such data.

TABLE 9.1: Advantages and Disadvantages of Five Classification Algorithms

K-Nearest Neighbours	Centroid Classification	Linear Discriminant Analysis	Neural Networks	Support Vector Machines
✓ Intuitive and easy to understand ✓ Extends naturally to more than two classes ✓ Separates non-linearly separable data ✓ It is very fast: there is no training time × Not robust to outliers	✓ Intuitive and easy to understand ✓ Extends naturally to more than two classes ✓ It is very fast: there is no training time × Gives incorrect results on non-linear data	✓ Strong statistical theory ✓ Generally outperforms centroid classification × Does not extend naturally to more than two classes × Gives incorrect results on non-linear data	✓ Able to discriminate non-linearly separable data ✓ Extends naturally to more than two classes × Slow to train × Have to optimise architecture	✓ Able to discriminate non-linearly separable data ✓ Faster to train than neural networks × Does not extend naturally to more than two classes × Have to optimise kernel function

EXAMPLE 9.5 MICROARRAY DATA WITH FOUR CLASSES: DATA SET 9B

There are four types of small round blue cell tumours of childhood: neuroblastoma (NB), non-Hodgkin lymphoma (NHL), rhabdomyosarcoma (RMS) and Ewing tumours (EWS). Sixty-three samples from these tumours have been hybridised to microarrays.[2] We wish to be able to diagnose these tumours in children using a focussed microarray and a set of rules to distinguish the tumour types.

Classification Methods

We now describe five different methods for partitioning space and predicting the group of a new sample:

- K-nearest neighbours
- Centroid classification
- Linear discriminant analysis
- Neural networks
- Support vector machines

These are all well-established and commonly used methods; they are implemented in most specialist data analysis software packages, such as R or Matlab. The advantages and disadvantages of these methods are summarised in Table 9.1.

K-Nearest Neighbours

K-nearest neighbours (KNN) is the simplest method for deciding the class to which a sample belongs (Figure 9.2). We have a number of samples with known class membership. We want to classify a new sample with unknown class membership. There are three steps:

1. We look at the gene expression measurements for the sample we are trying to classify.
2. We find the nearest of the known samples as measured by an appropriate distance measure (typically Euclidean distance; see Section 8.2).
3. The class of the sample is the class of the nearest samples.

There are two parameters in using a KNN algorithm. The first parameter is k: this is the number of nearest samples to look at; typically, we use $k = 3$, so that we look at the nearest 3 samples, but some people use $k = 5$ or $k = 1$. The second parameter, l, is the smallest margin of victory for a definite decision to be made; otherwise, the individual will be unclassified. So if $k = 3$ and $l = 3$, then all three nearest neighbours must be in the same class for a classification to be made: if one of the three nearest neighbours is in a different class, no classification will be made. If $k = 3$ and $l = 1$, then a simple majority vote will always result in a classification.

[2] The data are from the paper of Khan et al. (2001). The reference and URL for the data are given at the end of the chapter.

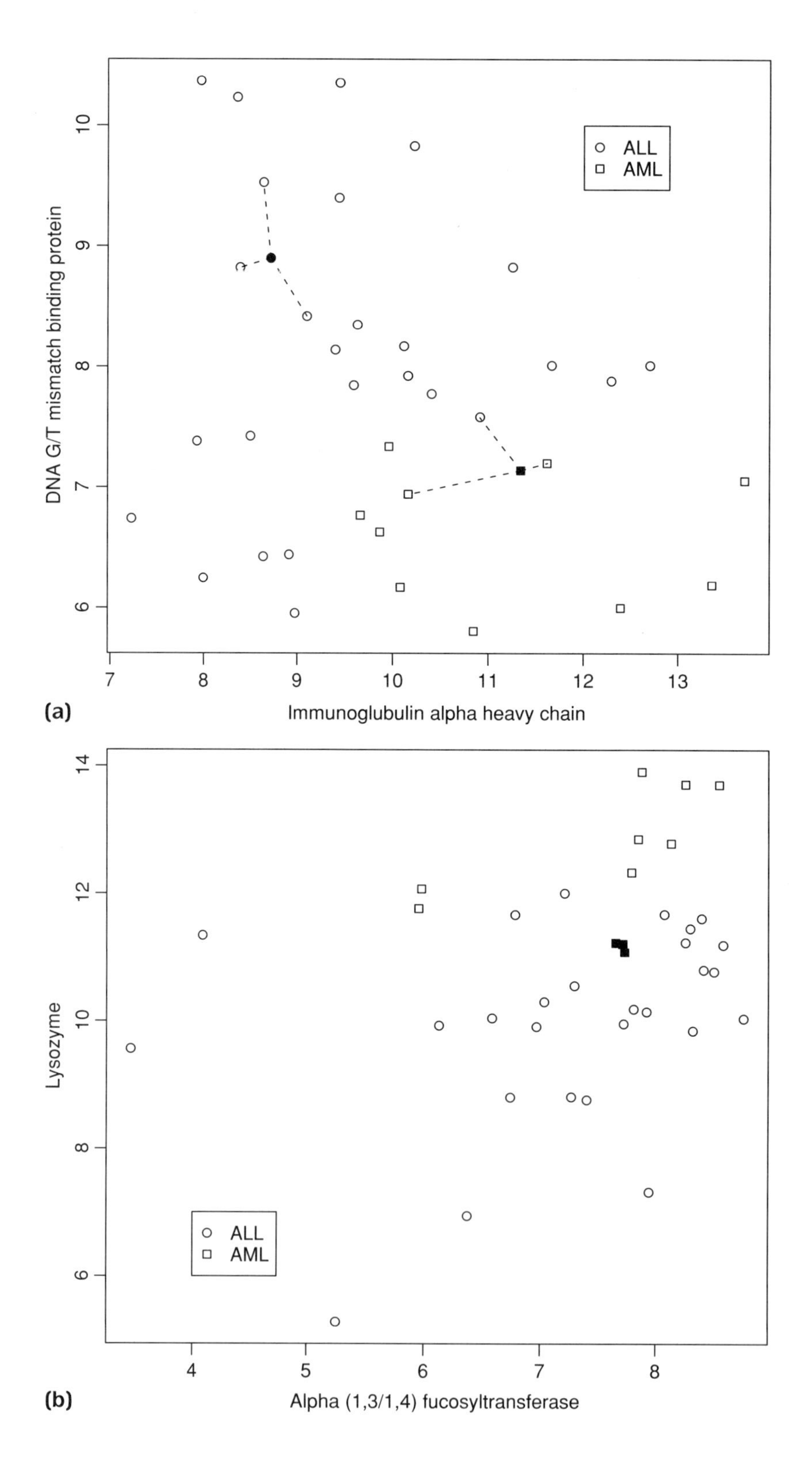

ALL
AML
DNA G/T mismatch binding protein
Immunoglubulin alpha heavy chain
(a)
ALL
AML
Lysozyme
Alpha (1,3/1,4) fucosyltransferase
(b)

KNN has the advantages of making intuitive sense and being easy to understand; it extends easily to more than two samples and can separate classes that occupy non-linear regions of space (Figure 9.1c). However, it is not robust to outliers: a small number of outliers can result in a false or non-classification (Figure 9.2b).

Centroid Classification

This is also a very simple method for classification. There are three steps to classifying a sample with unknown class membership:

1. For each class, we calculate the centre of mass of the points of the representative samples (Figure 9.3a).
2. Calculate the distance between the position of the sample to be classified and each of the centres of mass of the classes using an appropriate distance measure.
3. Assign the sample to the class whose centre of mass is nearest to it.

Centroid classification is also simple to understand and implement, and extends very naturally to more than two samples. It is fast and uses all the data. However, it can give completely incorrect results if the data is not linearly separable (Figure 9.3b).

Linear Discriminant Analysis

This is a very different type of analysis from the previous analyses. Unlike KNN or centroid classification, linear discriminant analysis (LDA) is what is known as a parametric method: we construct a statistical model from the data and base the classification on that model. There are two steps to classifying a sample:

1. Calculate a straight line (in two dimensions) or hyperplane (in more than two dimensions) that separates two known classes so as to minimise the within-class variance on either side of the line and maximise the between-class variance (Figure 9.4).
2. The class of the unknown sample is determined by the side of the hyperplane on which the sample lies.

LDA has strong statistical theory behind it. Because it takes into account the variability of the data, it tends to perform better than centroid classification. However, it does not extend naturally to more than two classes and only works if the data are linearly separable.

Figure 9.2: K-nearest neighbours algorithm. (a) The KNN algorithm applied to data set 9A, with $k = 3$ and $l = 3$. The three nearest neighbours of the highlighted ALL sample are all ALL samples; therefore, this sample would be classified as ALL. However, the highlighted AML sample has two nearest neighbours that are also AML samples, but one neighbour that is an ALL sample. With $l = 3$, this sample would not be classified, because it would be uncertain to which class it belongs. With $l = 1$, a single dissenter would be allowed, and so this sample would be classified as AML. **(b)** The potential problem with the KNN algorithm is that, in this case, the data are not very well separated. There are three AML samples very close together in an area where it is difficult to discriminate between the two groups. If we were classifying a new sample, and it was close to the cluster of AML samples, it would be classed as AML, when we might prefer if it were unclassified.

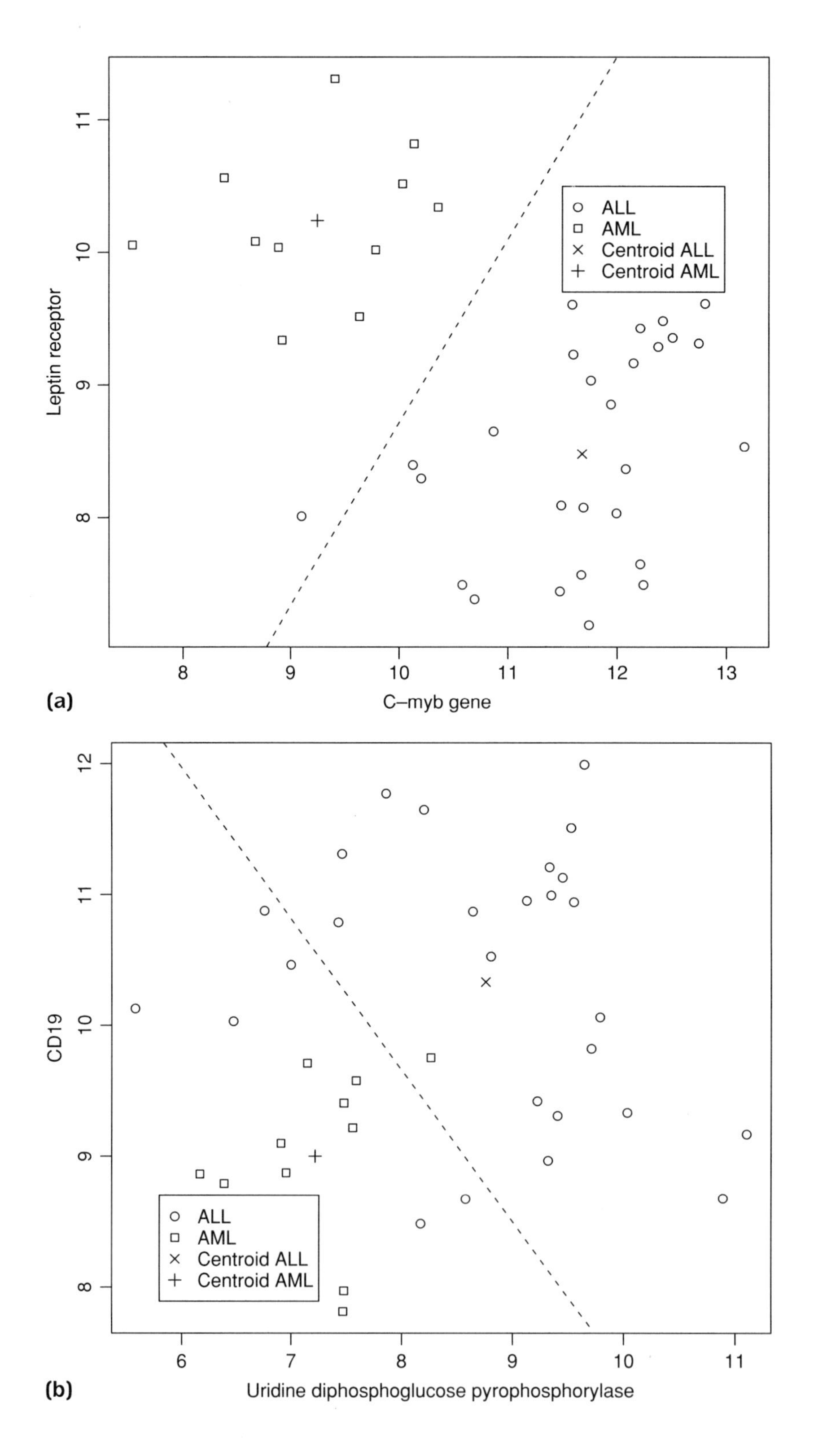

Leptin receptor
C–myb gene
ALL
AML
Centroid ALL
Centroid AML
8
9
10
11
12
13
(a)
CD19
Uridine diphosphoglucose pyrophosphorylase
ALL
AML
Centroid ALL
Centroid AML
6
7
8
9
10
11
12
(b)

Neural Networks

Neural networks are a method of separating space into classes in a way that can include non-linear separation. Neural networks are based on a model of the working of the brain: the network is organised as a series of nodes (simulating neurons), which have inputs and outputs (Figure 9.5a). The output of the nodes depends on the input into the nodes; the different inputs all have relative importance, which are determined by a set of parameters known as *weights*.

The neural network has the ability to learn by adjusting the weights. It is trained by giving it examples of samples to be classified; the network adjusts the weights on the input of the nodes so that it produces the correct output. The network is trained until it shows no further improvements in predicting the classes of the training set.

There are two steps for using a neural network to predict the class of an individual:

1. Train the neural network using the samples with known classes.
2. Apply the neural network to the new individual to determine its class.

Neural networks have the key advantage of being able to discriminate non-linearly separable data (Figure 9.5b) and extend naturally to the analysis of more than two classes. For these reasons, neural networks have become widely used tools in many fields. However, neural networks require training and optimisation, which makes them slower techniques to use. There is also no generic architecture for neural networks: the number of hidden nodes (Figure 9.5a) has to be identified empirically.

Support Vector Machines

Support vector machines (SVMs) are the most modern method applied to classification. SVMs are similar to LDA: they work by separating space into two regions by a straight line or hyperplane in higher dimensions. The hyperplane is chosen so as to minimise the misclassification error of the SVM.

The power of SVMs is that the data are first projected into a higher dimensional space and then separated using a linear method. This allows non-linear separation of the data. There are a number of different ways to project the data into a higher dimension (the method chosen is known as a *kernel function*). We do not go into the details of the mathematics here, but refer the reader to the reference at the end of the

Figure 9.3: Nearest centroid algorithm. (a) Calculate the centroids of the two groups; these are shown by a cross for the ALL samples and a plus for the AML samples. New samples are classified by whichever centroid is the nearest. Two classes are separated by a perpendicular bisector of the line joining the two centroids. In this case, all but one of the training samples are classified correctly by this method. **(b)** Centroid algorithms can go wrong with non-linearly separable data. Here, the genes uridine diphosphoglugose pyrophosphorylase and CD19 separate the data into two regions, but the ALL samples "wrap around" the AML samples. Because of this, many of the ALL training samples are closest to the AML centroid, and would be misclassified. Nearest centroid algorithms have problems with non-linearly separable data and should be avoided unless it is clear that the separation of the classes is linear.

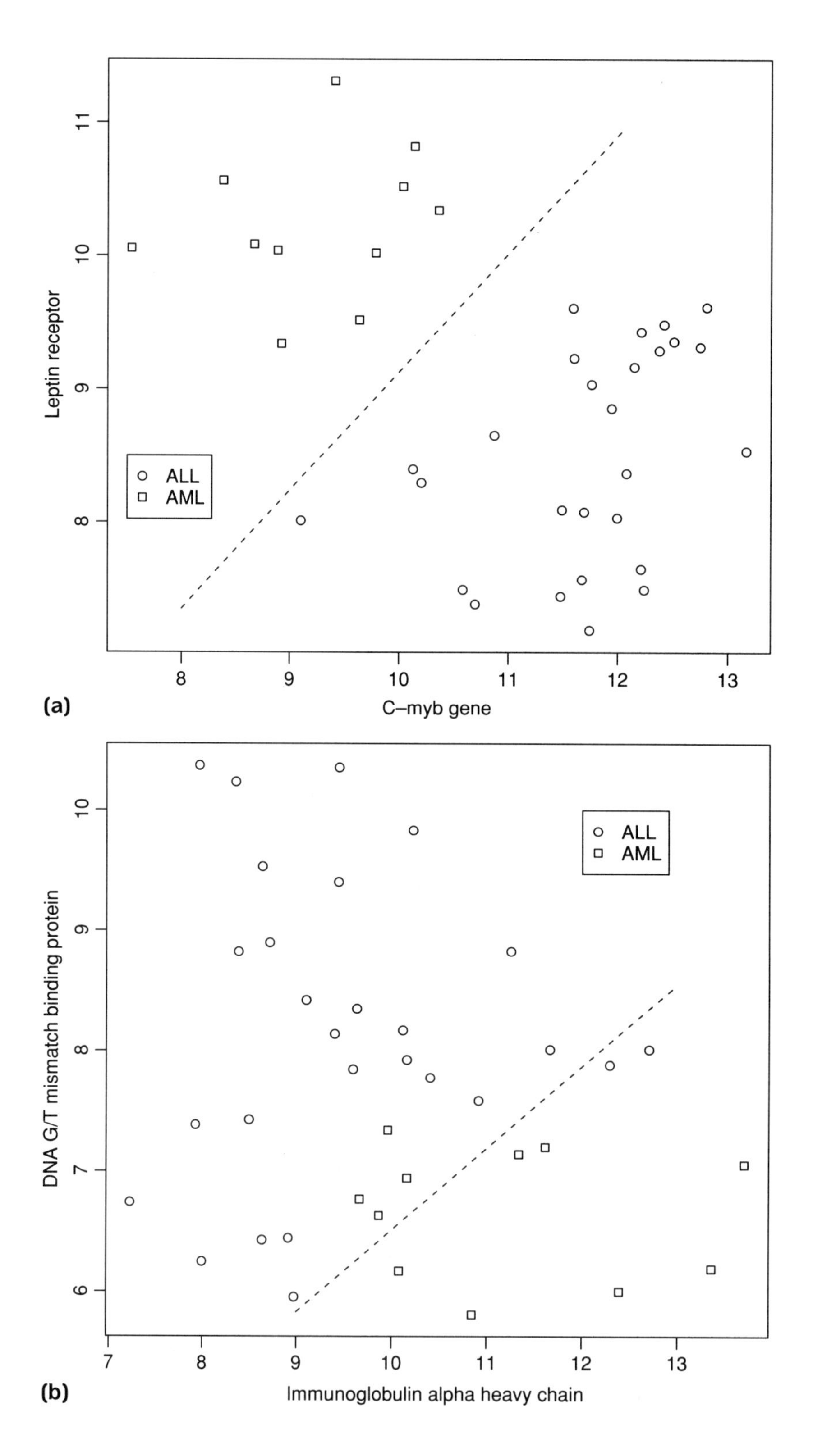
Leptin receptor
C–myb gene
ALL
AML
8
9
10
11
12
13
(a)
DNA G/T mismatch binding protein
Immunoglobulin alpha heavy chain
ALL
AML
6
7
8
9
10
11
12
13
(b)

chapter. There are three steps in applying a support vector machine:

1. Project the data from the known classes into a suitable high-dimensional space.
2. Identify a hyperplane that separates the two classes.
3. The class of the new individual is determined by the side of the hyperplane on which the sample lines.

SVMs can discriminate non-linear regions of space and are faster to train than neural networks. However, they do not naturally extend to more than two classes, and there is no natural kernel function to use: the user must optimise this empirically.

SECTION 9.3 VALIDATION

There is a core problem in the classification methods we have just described: given a classification algorithm that has been trained on two or more groups of individuals with known classes, how do we know that this method is going to be generally applicable to new individuals? It is possible that the individuals we have used to train the algorithm are in some way not representative of the groups to which they belong and, because of this, the algorithm may fail to classify subsequent individuals correctly.

There are two methods that are commonly used to resolve this problem: the use of **training and test sets** and **cross-validation**. If you are going to perform a classification analysis, we recommend that you use one or both of these validation methods. Training and test sets are more effective, but less powerful with small data sets. Cross-validation can be used with smaller data sets and is typically used as part of the training stage to help optimise the parameters of the algorithm.

Training and Test Sets

This is the most widely used and accepted method for validating the results of a classification algorithm. Approximately two-thirds of the data are used to *train* the algorithm: the algorithm is optimised to classify this training data as best as possible. After training, the algorithm is tested on the remaining third of the data to provide independent verification and quantification of the success of the algorithm.

EXAMPLE 9.6 TRAINING AND TEST DATA SETS FROM DATA SET 9A

In data set 9A, the authors had 62 patients available in their study, 41 suffering from ALL and 21 suffering from AML. The authors chose to use a training set of 38 patients,

Figure 9.4: Linear discriminant analysis (LDA). (a) LDA finds the straight line between the two groups that best separates them. When the groups are linearly separable, LDA does better than the nearest centroid algorithm, because it takes the variability within and between the groups into consideration (c.f. Figure 9.3a). **(b)** LDA goes wrong with non-linearly separable data. Here, the genes separate the data, but no straight line can be drawn that separates the classes. If we were to use LDA on the original data, many of the samples in both classes would be misclassified. LDA should be avoided unless it is clear that the separation of the classes is linear.

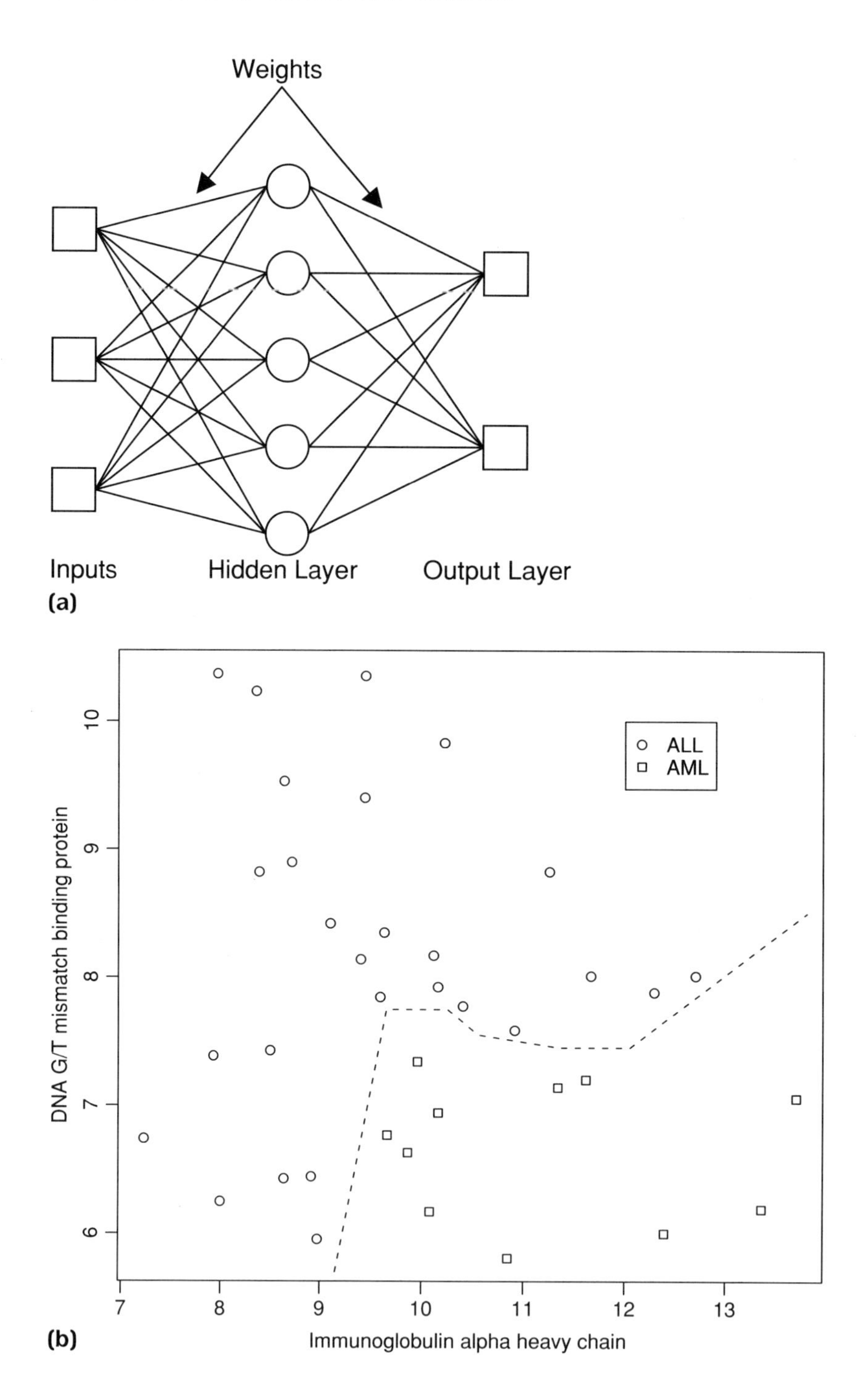
Weights
Inputs
Hidden Layer
Output Layer
(a)
ALL
AML
DNA G/T mismatch binding protein
Immunoglobulin alpha heavy chain
(b)

this being just over 60% of the data set – 27 ALL patients and 11 AML patients. The remaining 24 patients (14 ALL and 10 AML) were used by the authors as a test set to test the algorithms they developed. The success of any algorithm can be described by the number of correct classifications in the test set.

Cross-Validation

The alternative to using training and test sets is to use cross-validation to measure the success of an algorithm. Cross-validation has a *fold* associated with it that determines how the algorithm is implemented. k-fold cross-validation divides the data randomly into k equal (or almost equal) parts. The algorithm is then run k times, using $k-1$ of the parts as a training set and the other part as a test set. Each time the algorithm is run, a different test set is used, so that over the k runs of the algorithm, all data points are used as a test set. The success of the algorithm is the sum of the correct classifications over each of the runs.

EXAMPLE 9.7 CROSS-VALIDATION OF A CLASSIFICATION ALGORITHM

Data set 9A has 62 patients, 41 suffering from ALL and 21 from AML. A 3-fold classification would divide the data into three groups:

- Group A: 14 ALL patients and 7 AML patients
- Group B: 14 different ALL patients and 7 different AML patients
- Group C: the remaining 13 ALL patients and the remaining 7 AML patients

The cross-validation is run in three steps:

1. Groups A and B are used for training and group C is used for testing.
2. Groups A and C are used for training and group B is used for testing.
3. Groups B and C are used for training and group A is used for testing.

Figure 9.5: Neural networks. (a) A neural network comprises a series of ordered nodes that are modelled on neurons in the brain. Each of the inputs would be genes (or principal components) and would connect to the nodes in the hidden layer. Each node in the hidden layer thus receives inputs from all of the inputs; each input into each node is weighted, and the node responds to the inputs according to their weighted sum. The hidden layer nodes will either "fire" or "not fire," depending on whether the sum crosses a set threshold. The hidden-layer nodes send output to the output layer. These nodes behave in exactly the same way as the hidden-layer nodes and will fire according to the weighted sum of their inputs from the hidden layer. The number of classes the neural network can discriminate depends on the number of output nodes: in this case, the neural network has two output nodes and so can discriminate four classes. The number of nodes in the hidden layer determines the extent to which the neural network can separate non-linear classes. With no hidden nodes, the neural network can only separate linearly separable classes. The neural network is *trained* by showing it examples where the desired output is known in advance. The weights on all of the connections – both from the inputs to the hidden layer and from the hidden layer to the outputs – are adjusted so that the neural network provides correct output for each of the known examples. The trained network is then used to classify samples with unknown group membership. **(b)** Separation of the AML and ALL patients from data set 9A using a neural network. The neural network has been able to distinguish the two groups which are non-linearly separable. The line is the approximate boundary of the classification regions of the neural network. For this example, we used six hidden nodes and needed only one output node (there are only two classes). The number of hidden nodes needs to be determined empirically; this is a major disadvantage of neural networks.

Leave-One-Out Cross-Validation

A particularly important special case of cross-validation is called **n-fold cross-validation** or, frequently, **leave-one-out cross-validation**. In this method, all but one of the samples are used to create a classifier, and the algorithm is tested on the left-out sample. This is repeated, leaving out each of the samples in turn, and the number (or proportion) of samples correctly classified is reported as the success of the algorithm.

Cross-validation is particularly useful during the training stage of a classification algorithm, in which parameters may need to be tuned so that the algorithm fits the training set. However, there is a disadvantage to cross-validation, which is that the results generated are not independent and so are not as reliable as using pure training and test sets.

SECTION 9.4 DIMENSIONALITY REDUCTION

The microarray experiment generates very high dimensional data; for example, data set 9A has 6,817 genes. Modern microarrays can contain up to 30,000 genes, and each sample has a measurement for each gene. The examples we have seen in Section 9.2 have looked at just two genes at a time – a two-dimensional measurement space – but the microarray experiments from which these examples derive have measurements in a much higher dimensional space – several thousand dimensions in each of the two data sets we have described.

One of the tasks involved in building a good classifier is to reduce the dimensionality of the data: instead of looking at all 6,000 gene expression measurements of a sample, we look at a small number of measurements. In the examples we showed earlier, we showed just 2 genes; in reality, we may want to use more genes, perhaps 5 to 20 genes. There are a number of reasons to seek to reduce the dimensionality of the system:

- **Removal of noise and irrelevant information.** Many genes do not contain information that is useful for determining the differences between the samples. These genes should not be used for classification; indeed, sometimes they may even contain noise that can lead to incorrect classification.
- **Speed of training of methods.** A number of methods we described, such as neural networks, work better with less input information. We need to reduce the dimensionality of the data before we can use these methods effectively.
- **Identical information.** Some genes are highly correlated and contain exactly the same information. Inclusion of all of such genes can cause some methods to be unreliable.
- **Multiplicity.** When we are looking at many thousands of genes in parallel, it is possible that some of these genes may appear to be differentially expressed between the different samples, but in fact these differences may be due to random variation.

TABLE 9.2: Advantages and Disadvantages of Four Dimensionality Reduction Methods

Principal Component Analysis	Individual Gene Selection	Pairwise Gene Selection	Genetic Algorithms
✓ Quick and easy to use × Does not provide a subset of classifying genes × Principal components may not separate the classes	✓ Quick and easy to implement ✓ Generates a subset of classifying genes × Best individual genes may not make the best classifiers × Need to combine selected genes to produce classifier	✓ Can generally find good classifiers ✓ Generates a subset of classifying genes × Need to combine selected genes to produce classifier × Slower than individual gene selection	✓ Finds the best classifiers ✓ Generates a subset of classifying genes × Slowest algorithm × Need programming skills to implement

- **Diagnostic tool.** Frequently, the aim is to produce a prognostic or diagnostic tool for the diseases or treatments being studied. While it might be feasible to use a microarray as such a tool, in many cases it will be cheaper and more efficient to produce a more focussed tool, such as quantitative PCR, that uses fewer, more relevant, genes.
- **Hypothesis generation.** A classification based on a small number of genes can be the basis of scientific hypotheses about the role of the relevant genes in the different diseases or treatments being studied. To do this, we need to find those genes.

The selection of an appropriate subset of genes is a difficult problem and is the subject of ongoing research. In the theory of computer science, a problem is classed as *hard* if the number of steps to evaluate the solution increases exponentially with the size of the problem. In this case, the number of possible subsets of N genes is 2^N, so the evaluation of all possible groups of genes increases exponentially with the number of genes in the study.

The four methods we describe are all commonly used and are seen frequently in the literature. All methods have their advantages and disadvantages. The methods we describe are

- Principal component analysis
- Individual gene selection
- Pairwise gene selection
- Genetic algorithms

To illustrate these methods, we will apply them to data set 9A using a number of the classification methods described in Section 9.2, and the training and test sets of Example 9.6. In Table 9.2, we summarise the advantages and disadvantages of each of these methods.

Principal Component Analysis

The first method we describe does not actually find a subset of relevant genes, but uses principal component analysis (PCA) to reduce the dimensionality of the data. PCA is described in full in Section 8.3; it is straightforward to use because it is implemented in all gene expression packages, as well as more advanced data analysis packages such as R and Matlab.

EXAMPLE 9.8 PCA APPLIED TO DATA SET 9A

When PCA is applied to data set 9A, many the classification methods (with the exception of KNN) have been able to separate the data (Figure 9.6; Table 9.3). However,

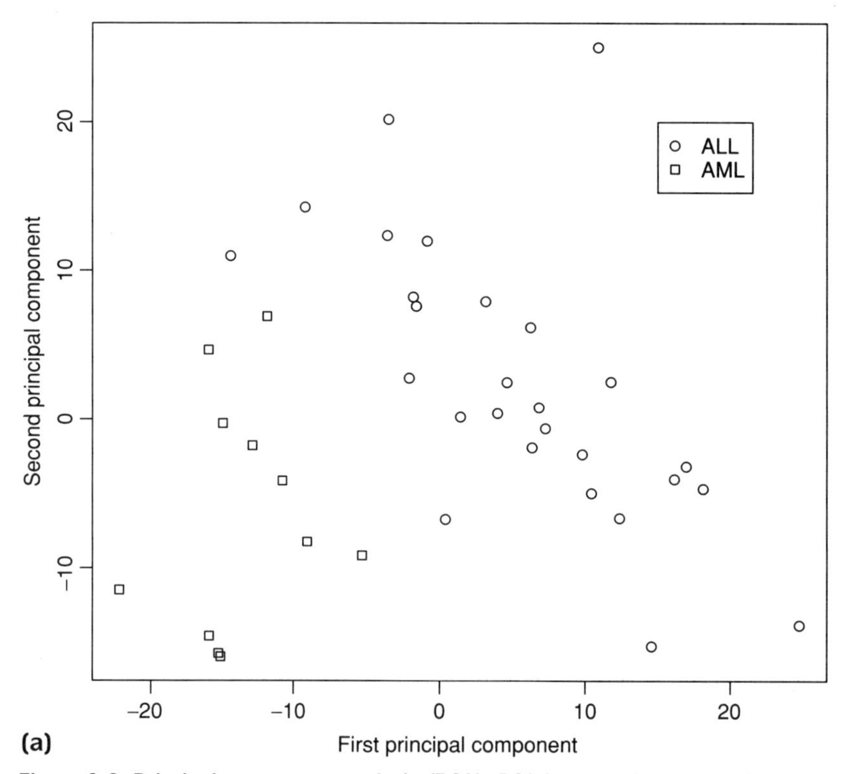

Figure 9.6: Principal component analysis (PCA). PCA is a popular method for reducing the dimensionality of the gene expression matrix before applying a classification algorithm. Here, it has been applied to data set 9A. **(a)** The two groups of patients (ALL and AML) plotted in the first two principal components. In this case, the data are separable by principal components. This is not necessarily always true with microarray data and for this reason principal components are not necessarily the best dimensionality reduction method. **(b)** The first 10 principal components all contribute to the overall variability of the data. It would make sense to include all of these components in the classification analysis. Other data sets (e.g., data set 8A) have fewer principal components contributing to the overall variability (Table 8.3), so in that case one would use a smaller number of principal components in the classification analysis. **(c)** The data points from the independent test set are mapped onto the first two principal components of the training set. In general, the separation is good; there are a small number of samples in the boundary region that would be misclassified by the first two principal components.

(continued)

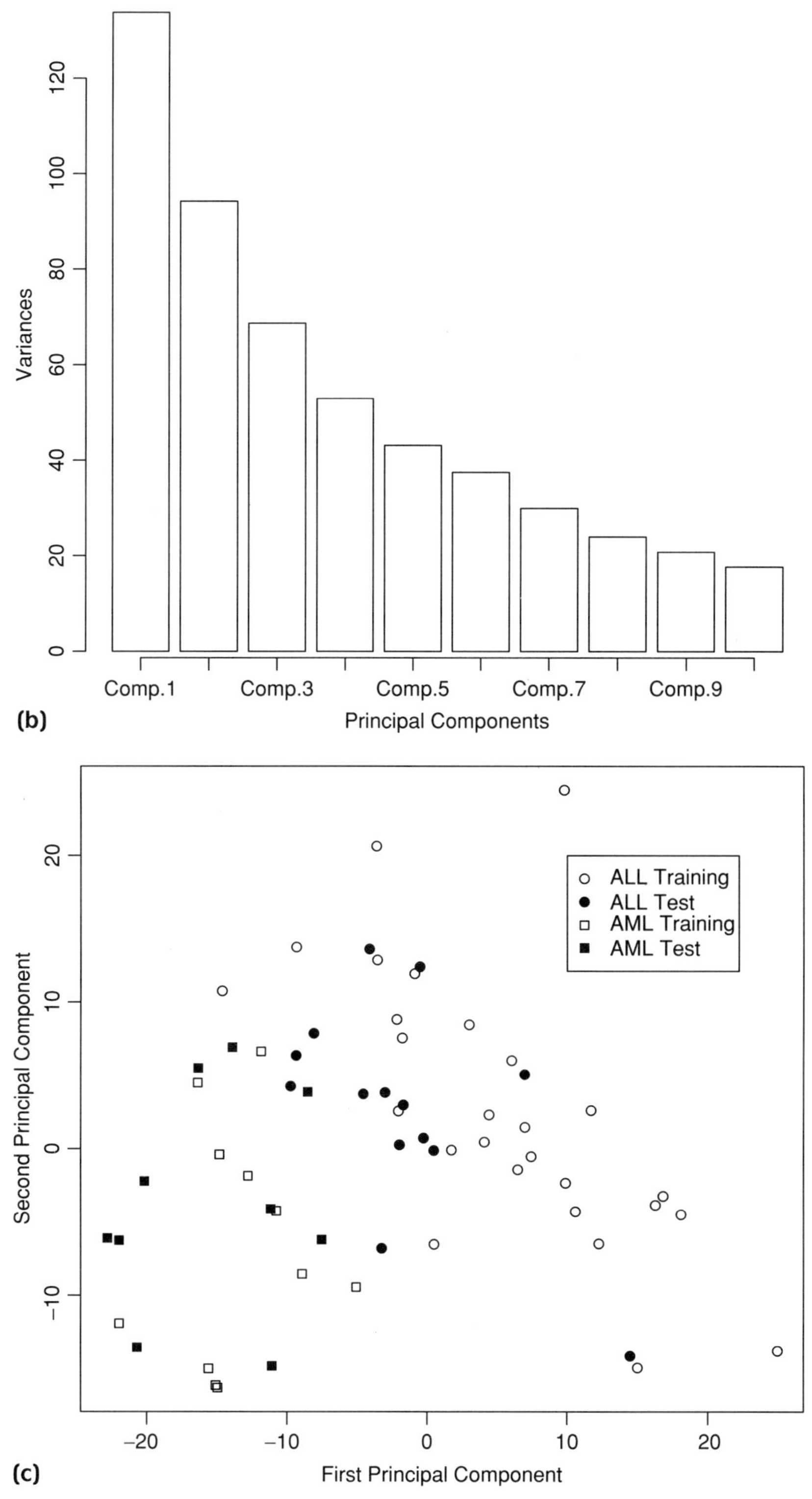

Figure 9.6: (*continued*)

TABLE 9.3

Predictions of different algorithms applied to data set 9A using the 10 most significant principal components and using 38 patients in a training set and 24 patients in a test set. The number of samples in each of the classes is given in parentheses. With most algorithms, PCA worked very well. This is not always the case.

Method	Training ALL (27)	Training AML (11)	Test ALL (14)	Test AML (10)
Nearest centroid	26	11	14	10
KNN ($k = 3, l = 3$)	26	11	12	3
LDA	26	11	13	10
Neural network (10 HNs)	27	11	14	8
Neural network (20 HNs)	27	11	14	10

although this is true of this particular data set, it is not necessarily true of other data sets.

There are several factors that have to be considered when using PCA for classification:

- PCA still requires the measurements of large numbers of genes. If the aim is to have a small number of genes that can be measured for future diagnostic use, PCA is useless.
- PCA finds axes that capture the variability in the overall data. These may not necessarily be the axes that separate the classes. If the classes are separated by the principal components, then PCA is an excellent method; if the principal components do not separate the classes, then PCA must be abandoned and a different method must be used.
- PCA is based on linear combinations of genes; if non-linear combinations of genes are needed to separate the data, PCA will not work.

Individual Gene Selection

The simplest method that actually chooses genes is to rank the genes that individually discriminate best between the two classes and then use the best genes as classifiers. A good measure of discrimination that is most commonly used is the t-statistic (Section 7.3). This captures the difference between the means of the classes as a ratio of the standard deviation of the two groups.

EXAMPLE 9.9 HIGHEST RANKED INDIVIDUAL GENES FOR DATA SET 9A

The top genes that can discriminate between the ALL and AML patients of data set 9A all have very good p-values associated with their t-statistics (Table 9.4). However, even the best gene does not separate the classes very well compared with the separation that can be achieved by two or more genes (Figure 9.7). Two of the genes in this table, C-myb and leptin receptor, form a good classifier as a pair of genes (Table 9.5). However, other genes that form good pairwise classifiers are not necessarily good classifiers individually.

TABLE 9.4: Top 10 Genes That Individually Classify the AML and ALL Patients

Gene	t-test p-value
CD33 antigen	1.9E-09
C-myb gene	6.32E-09
Leptin receptor	8.93E-08
Cathepsin D	1.99E-07
Transcription factor 3	2.02E-07
Connective tissue activation peptide III	3.48E-07
Myosin light chain	4.19E-07
Granulin	4.31E-07
Retinoblastoma binding protein P48	5.32E-07
NADPH-flavin reductase	6.85E-07

Note: The top 10 genes according to p-value of a t-test applied to the two groups on a gene-by-gene basis (Section 7.2). The genes have been prefiltered to include only genes that have been expressed in all patients. Note that in this particular case, C-myb gene and leptin receptor together are a good predictor of the two classes (Table 9.6). However, most of the other genes that are good predictors in pairs are not necessarily good at predicting the classes in isolation.

The classification methods described in Section 9.2 can be used with the top discriminating genes (Table 9.4) to build a classification model for data set 9A, either directly (Table 9.5) or using a voting algorithm (see later discussion). Although this method has worked well with the training set, it has not performed as well with the

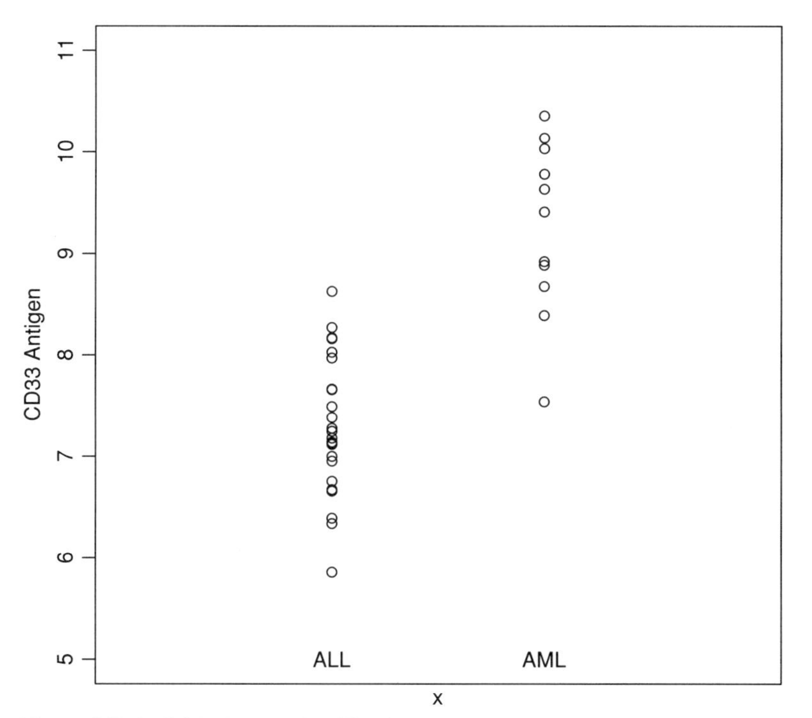

Figure 9.7: Individual gene classification. The gene that best distinguishes between the ALL and AML patients as measured by the t-statistic is CD33 antigen (Table 9.4). It does not separate the classes very well: there is substantial overlap in the gene expression values. Methods for gene selection based on choosing the best individual genes are not generally very good.

TABLE 9.5

Predictions of different algorithms on data set 9A, using the training and test sets of Example 9.6 and using the 10 best genes as determined by single-gene classification (Table 9.4). In this case, the training data are linearly separable using the 10 best genes; we have already seen this in Figure 9.1b, where the genes C-myb and leptin receptor separate the classes. Using these genes, all methods have performed comparably, but no methods have been able to predict all of the test AML cases correctly. It would appear likely that the test data are not separable with these 10 genes.

Method	Training ALL (27)	Training AML (11)	Test ALL (14)	Test AML (10)
Nearest centroid	27	11	14	8
KNN ($k = 3, l = 3$)	27	11	14	7
LDA	27	11	14	8
Neural network (10 hidden nodes)	27	11	14	5
Neural network (20 hidden nodes)	27	11	14	7

test data, with only between 5 and 8 of the 10 AML patients being classified correctly. It is likely that the training data are not separable using these 10 genes.

In general, individual gene selection is the weakest method for selecting genes to use for discrimination. The reason for this is that genes that are good individual classifiers may not work well together to classify samples. Conversely, genes that do work well together to classify the samples may not be good individually. For this reason, it is recommended that you use a method that looks at pairs or groups of genes to classify the data.

Pairwise Gene Selection

This is a more sophisticated method than relying on single genes: it looks at pairs of genes that are best able to discriminate the samples using the chosen method and then combines these genes to generate the overall predictor.

For example, with a KNN algorithm, we would apply the algorithm to all pairs of genes, and then select (say) the five best pairs of genes for a 10-gene predictor. A similar approach could be used for each of the methods.

EXAMPLE 9.10 PAIRWISE GENE SELECTION USING KNN

Using data set 9A, the KNN algorithm has been applied to the 500 most varying genes to identify 10 pairs of genes that each correctly classifies 37 out of the 38 patients in a cross-validation of the training set (Table 9.6). These correspond to 14 unique genes, which can be used to build a classifier. If KNN is applied using all 14 genes, then every sample in the training set is classified correctly. However, the algorithm performs less well on the test set, with only 15 out of 24 correct (Table 9.7).

Pairwise gene selection can be used to reduce dimensionality for all of the methods described in Section 9.2. In general, pairwise gene selection performs better than individual gene selection (Table 9.7). However, it is slower than individual gene selection,

TABLE 9.6

Pairwise gene selection has been applied to the training set of data set 9A using the KNN algorithm. There are 10 pairs of genes that correctly predict 37 out of 38 patients in a cross-validation of the training set. There are 14 unique genes in this list, which can be used as genes for classification of these tissues.

Genes		Training ALL (27)	Training AML (11)
Ferritin heavy chain	CD33 antigen	27	10
RLIP76 protein mRNA	CD33 antigen	27	10
Casein kinase 1 delta	CD33 antigen	27	10
DNA-damage-inducible transcript 1	CD33 antigen	26	11
NADPH-flavin reductase	CD33 antigen	27	10
LEPR leptin receptor	C-myb gene	26	11
Cholinergic receptor, nicotinic, alpha polypeptide 7	C-myb gene	26	11
Cholinergic receptor, nicotinic, alpha polypeptide 7	Topoisomerase (DNA) II beta	27	10
Catalase	Cytoplasmic dynein light chain 1	26	11
Nucleoside-diphosphate kinase	Retinoblastoma binding protein P48	27	10

because the number of pairs of genes is approximately half the square of the number of genes.

Voting Algorithms

There are two ways to combine classifying genes that are selected using single gene or pairwise gene selection algorithms. The first method is to use all the genes together in the algorithm. In Example 9.10, there are 10 pairs of genes selected for KNN, with 14 unique genes among them. These 14 genes can be used to build a single KNN predictor.

TABLE 9.7

Predictions of different algorithms using the 10 best genes selected by pairwise gene selection on data set 9A. Most of the algorithms performed better with pairwise gene selection than with individual gene selection.

Method	Training ALL (27)	Training AML (11)	Test ALL (14)	Test AML (10)
Nearest centroid	26	11	14	10
KNN ($k = 3, l = 3$)	27	11	11	4
KNN ($k = 3, l = 0$)	27	11	13	8
LDA	26	11	13	10
Neural network (10 HNs)	27	11	14	8
Neural network (20 HNs)	27	11	14	10

TABLE 9.8

The 14 test samples are classified using voting from the 10 pairs selected by KNN (Table 9.6). In each row, we count the number of times the individual was classified as ALL, unclassified or classified as AML by the 10 pairs. All ALL samples are correctly classified; there is more difficulty with the AML samples – 6 out of 10 are correctly classified.

Sample	ALL	Unclassified	AML
ALL 1	8	1	1
ALL 2	7	3	0
ALL 3	8	0	2
ALL 4	9	0	1
ALL 5	10	0	0
ALL 6	10	0	0
ALL 7	10	0	0
ALL 8	9	1	0
ALL 9	10	0	0
ALL 10	9	1	0
ALL 11	10	0	0
ALL 12	9	1	0
ALL 13	10	0	0
ALL 14	10	0	0
AML 1	0	1	9
AML 2	0	2	8
AML 3	0	0	10
AML 4	4	5	1
AML 5	3	3	4
AML 6	2	3	5
AML 7	6	4	0
AML 8	4	4	2
AML 9	1	2	7
AML 10	7	2	1

The second method is to use a *voting* algorithm. In a voting algorithm, each sample is classified by each gene or pair of genes selected, and the majority classification is used as the class of that gene.

EXAMPLE 9.11 USING A VOTING ALGORITHM WITH PAIRWISE KNN

The 10 pairs of genes are used to classify each of the 24 test samples in turn. The first ALL sample is classified 8 times as ALL, is unclassified once, and classified once as AML. All 14 ALL samples have been correctly classified, and 6 of the 10 AML samples have been correctly classified (Table 9.8).

Genetic Algorithms

The most powerful method we describe to choose gene subsets is a method known as a *genetic algorithm*. Genetic algorithms are computational methods for solving difficult problems that have their inspiration in evolutionary biology.

In biology, organisms with different genotypes have different phenotypes, which are more or less fit, and so pass on more or fewer offspring to the next generation. In genetic algorithms, there is a population of solutions to a given problem; each solution has a "genotype," which describes the parameters of the solution. The fitness of the solution is its ability to solve the problem. The most fit individuals are selected in each generation to produce offspring for the next generation.

As with real biology, the individual solutions can produce offspring both asexually and sexually. With asexual reproduction, the offspring are identical to the parent, with the possibility of changes via random mutation. With sexual reproduction, the "genomes" (parameters) of two individuals are recombined to produce a new individual.

We will describe a simple genetic algorithm that can be used for choosing gene subsets for classification analyses. As you will see, there are many ways in which genetic algorithms could be implemented, and they can be used with any of the classification methods. The implementation we describe is not necessarily the best of these: it is included for illustrative purposes, and we will also describe some possible modifications.

The Algorithm

There is a population of N individuals, each of whom has a genome consisting of a list of n genes that will be used together as a classifier. There are five steps:

1. Each individual starts with a random choice of n genes.
2. Construct a new, larger population from the old population; this will typically be of size $3N$, using three methods of reproduction:
   (a) **Cloning.** N individuals in the new population are identical to the N individuals in the old population.
   (b) **Mutation.** N individuals are created from each of the N individuals in the old population, which are identical to each of their parents, but with one gene randomly changed to a different gene.
   (c) **Recombination.** N individuals are created by randomly selecting two parents from the previous generation, combining their genes into a single pool, and then selecting n genes from the combined pool.
3. Calculate the *fitness* of each of the $3N$ individuals. In this case, the fitness will be the number of samples that are correctly classified in a leave-one-out cross-validation of the classification method applied to the training set using the n genes of that individual.
4. Select the best N individuals to form the next generation.
5. Return to step 2 and continue until the population contains sufficiently good solutions.

EXAMPLE 9.12 GENETIC ALGORITHM WITH KNN

As an example, we apply this genetic algorithm to select a group of 8 genes that will classify the ALL and AML samples of data set 9A. In this case, we will use a population size of 50, and have 8 genes in each classifier. In the seventh generation of the

TABLE 9.9: Genes Selected by Genetic Algorithm with KNN

A simulation of the genetic algorithm described in the text produced an 8-gene solution that was able to correctly classify all 38 patients in a leave-one-out cross-validation of the training set of data set 9A using a KNN algorithm with $k = 3$.

Tyrosine Phosphatase Epsilon
Basigin
Transcription elongation factor B
TCF3 transcription factor 3
Cholinergic receptor nicotinic alpha polypeptide 7
Transmembrane protein
Mitochondrial 60s ribosomal protein L
Cathepsin D

simulation I ran, there was a classifier that reported 38 out of 38 correct classifications on a cross-validation of the training set (Table 9.9). When this classifier is applied to the test set, all 14 ALL patients were correctly classified, and 7 out of 10 AML patients were correctly classified.

This algorithm is just one example of how a genetic algorithm could be implemented to solve this problem. There are many modifications that could be applied to the algorithm, including:

- Varying the population size;
- Varying the number or proportion of offspring created at each generation via cloning, mutation or recombination;
- Allowing different individuals to have different numbers of genes, possibly including increased fitness for classification using fewer genes.

Genetic algorithms are widely applicable to many difficult computation problems. However, there is a cost, too: genetic algorithms are notoriously slow, because they use random events to generate the next generation.

KEY POINTS SUMMARY

- There are several methods for classifying samples, each with advantages and disadvantages, including:
  - K-nearest neighbours
  - Centroid classification
  - Linear discriminant analysis
  - Neural networks
  - Support vector machines

- With microarray data, we need to reduce the dimensionality of the data to find groups of genes that are able to separate the data. There are several methods for doing so, including
  - Principal component analysis
  - Individual gene selection
  - Pairwise gene selection
  - Genetic algorithms
- Classification analyses should be verified using training and test sets and/or cross-validation.

FURTHER READING AND RESOURCES

Data Set 9A

Golub, T.R., Slonim, D.K., Tamayo, P., Huard, C., Gaasenbeek, M., Mesirov, J.P., Coller, H., Loh, M.L., Downing, J.R., Caligiuri, M.A., Bloomfield, C.D., and Lander, E.S. 1999. Molecular classification of cancer: Class discovery and class prediction by gene expression monitoring. *Science* 286: 531–36.

Classification has been performed using individual gene selection and a voting algorithm.

Data Set 9B

Khan, J., Wei, J.S., Ringner, M., Saal, L.H., Ladanyi, M., Westermann, F., Berthold, F., Schwab, M., Antonescu, C.R., Peterson, C., and Meltzer, P.S. 2001. Classification and diagnostic prediction of cancers using genes expression profiling and artificial neural networks. *Nature Medicine* 7: 673–79.

Classification has been performed with neural networks.

Tibshirani, R., Hastie, T., Narasimhan, B., and Chu, G. 2002. Diagnosis of multiple cancer types by shrunken centroids of gene expression. *Proceedings of the National Academy of Sciences* 99: 6567–72.

Classification using a version of centroid classification.

Bo, T.H. and Jonassen, I. 2002. New feature subset selection procedures for classification of expression profiles. *Genome Biology* 3: research0017.1–0017.11.

Description of pairwise gene selection applied to microarray data.

Li, L., Weinberg, C.R., Darden, T.A., and Pedersen, L.G. 2001. Gene selection for sample classification based on gene expression data: Study of sensitivity to choice of parameters of the GA/KNN method. *Bioinformatics* 17: 1131–42.

Classification using K-nearest neighbours and a genetic algorithm.

Furey, T.S., Christianini, N., Duffy, N., Bedarski, D.W., Schummer, M., and Haussler, D. 2000. Support vector machine classification and validation of cancer tissue samples using microarray expression data. *Bioinformatics* 16: 906–14.

Classification using support vector machines.

Software

http://www.r-project.org/

The R statistical package is available for free and contains implementations of all the methods we describe in this chapter.

Books

Webb, A. 2002. *Statistical Pattern Recognition.* John Wiley & Sons. 2nd Ed. New York.

Hastie, T., Tibshiramni, R., and Friedman, J. 2001. *The Elements of Statistical Learning, Data Mining, Inference and Prediction.* Springer: New York.

CHAPTER TEN

Experimental Design

SECTION 10.1 INTRODUCTION

The design of experiments is one of the most important areas of microarray bioinformatics and is a long-standing topic in classical statistics. The reason for good experimental design is that it allows you to obtain maximum information from an experiment for minimum effort – which translates into time and money. The alternative to good experimental design is to perform microarray experiments which produce data that cannot be analysed.

You might ask why it is that this topic appears at this point in the book, after data analysis rather than earlier in the book, alongside the material on the design of microarrays themselves. There are two reasons for this. The first is that the topics in this section use concepts from some of the earlier chapters, most importantly the ideas of hypothesis tests and *p*-values introduced in Chapter 7. But there is also a more philosophical reason why I have chosen to place the material on experimental design after the material on data analysis. In my view, it is absolutely critical to understand the scientific questions you are trying to answer, or even the scientific hypotheses you are seeking to generate, before you design your experiment. To this end, you should have a clear idea of the structure of the data you are seeking to produce and the types of data analysis you intend to employ before you design an experiment.

This chapter considers three areas of experimental design:

Section 10.2: Blocking, Randomisation and Blinding, looks at the statistical problems of confounding and bias, and the methods that are used to resolve these issues.

Section 10.3: Choice of Technology and Arrangement of Samples, discusses the relative benefits of Affymetrix and two-colour array platforms; and the arrangement of samples on arrays in a number of types of microarray experiment.

Section 10.4: How Many Replicates?, describes statistical methods to determine the number of replicates you would need to use in microarray experiments to obtain data that can detect the effects you are looking for.

SECTION 10.2 BLOCKING, RANDOMISATION AND BLINDING

We introduce the topic with a simple example, for which we describe three experimental designs: the first two are seriously flawed, and the third is better.

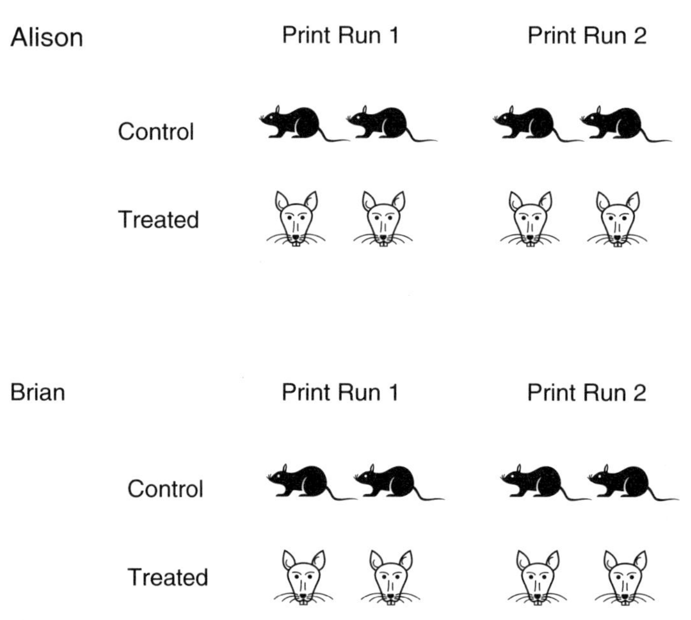

Figure 10.1: Blocked balanced design. There are 16 rats being used in the experiment, 8 treated with the toxin benzo(a)pyrene, and 8 treated with a control substance. Samples will be hybridised to two batches of 8 arrays from two print runs, and there are two researchers, Alison and Brian. Alison and Brian each treat 4 rats with toxin and 4 rats with the control substance; they prepare samples and hybridise 2 samples from each of the treated and control rats to 2 arrays from each of print runs 1 and 2. This is an example of a balanced design, which maximises the power of the experiment.

EXAMPLE 10.1 TOXIC RESPONSE TO BENZO(A)PYRENE

In an experiment to investigate the effects of benzo(a)pyrene, a known hepatotoxin, 8 rats will be treated with benzo(a)pyrene, and 8 rats will be treated with a control substance. Liver samples will be prepared from all 16 rats and hybridised to 16 arrays. The sample preparations and hybridisations will be performed by two researchers, Alison and Brian, and the 16 arrays have come in two batches of 8 arrays from two separate print runs.

Experimental Design 1

Alison chooses 8 rats and treats them with benzo(a)pyrene. She prepares liver samples from the rats and hybridises them to the 8 arrays from the first print run. Brian takes the remaining rats and treats them with the control substance; he prepares samples and hybridises them to the 8 arrays from the second print run.

Experimental Design 2

Alison chooses eight rats, and treats four with benzo(a)pyrene and four with the control substance. She chooses four arrays from each of the print runs, and hybridises samples from two treated rats and two control rats to each of the batches of four arrays. Brian does likewise with the remaining eight rats and eight arrays (Figure 10.1).

Experimental Design 3

Eight rats are randomly allocated to Alison; similarly, four arrays from each of the two print runs are also randomly allocated to Alison. Four preparations of benzo(a)pyrene and four preparations of the control compound are given to Alison in a way that she does not know the identity of any of the preparations. The arrays are prearranged for Alison so that she will hybridise two treated and two control rats to four arrays from each batch, with random allocation. Brian does likewise.

Although the faults of the first two designs may appear obvious to many readers, it is remarkable how many stories I have heard of microarray experiments being run in the manner of design 1. The problem with design 1 that is resolved in design 2 is **confounding**; the problem with design 2 that is resolved in design 3 is **bias**.

Confounding and Blocking

Suppose we use design 1 for the experiment, and a statistical analysis (such as a *t*-test or bootstrap test described in Chapter 7) is applied to identify genes that are up- or down-regulated in the treatment group relative to the control group. We would like to be able to say that these genes are up- or down-regulated as an effect of the hepatotoxin benzo(a)pyrene. However, the observed differences in gene expression could be because Alison and Brian handle the samples in a different way, and may not be related to the toxin. Alternatively, the observed differences in gene expression could have resulted from differences in the two batches of arrays and may not be related to the toxin. With this experimental design, we cannot know which of the three factors – treatment, researcher or batch – is responsible for the differences in gene expression. We say that these factors are confounded.

The problem of confounded variables is resolved via a technique called **blocking**. Experimental design 2 is an example of a blocked experiment. In this example, there are two blocking factors: the experimenter and the print run. Each of the eight arrays is allocated evenly between the two blocking factors (Figure 10.1). Therefore, if there is a significant difference in gene expression between the treated and control rats, it is possible to attribute that difference to the treatment, and it cannot be because of researcher or print run.

This is also an example of a **balanced design**. An imbalanced design might have uneven numbers of rats in the control and treatment groups, or uneven numbers of the two groups allocated to Alison and Brian. Balanced designs are more powerful than imbalanced designs; we discuss the meaning of power in Section 10.3.

It is important to notice that running the experiment in a balanced and blocked fashion adds no extra cost and no extra time to the experiment but makes a very important difference in how you can interpret the results.

Bias, Randomisation and Blinding

Experimental design 2 suffers from a problem known as bias. When Alison chooses the rats, she might choose rats that are in some way similar: they might be the healthiest looking, or the most docile, or anything else. There is no suggestion of impropriety on Alison's part: all people make subconscious choices without realising it. Therefore, by having Alison choose the rats she uses, we introduce a potential variability between the two groups of rats used by the two researchers.

Experimental design 3 uses **randomisation** to remove this bias by randomly allocating rats to the two researchers.

There is a second source of bias in experimental design 2. If Alison and Brian know to which rats they are giving the toxic benzo(a)pyrene, and to which rats they are giving the control substance, one or both of them might behave differently in the way they treat the two groups of rats. Again, there is no suggestion of impropriety: it may be purely subconscious factors that result, for example, in Brian treating the poisoned rats with greater care than the control rats.

Experimental design 3 uses **blinding** to avoid this bias: Alison and Brian are not aware which rats are treated with toxin and which rats are treated with control compound. As a result, they will treat both groups of rats in the same way.

SECTION 10.3 CHOICE OF TECHNOLOGY AND ARRANGEMENT OF SAMPLES

The problems of confounded variables and bias are ubiquitous to all experimental design and not just microarray experiments. This section discusses three problems specifically associated with microarray experiments:

- Is it better to use Affymetrix arrays or a two-colour array system?
- If using a two-colour array system, is it better to use a reference sample?
- If using a two-colour array system, what is the best arrangement of samples on the slides?

There is no universally correct answer to these questions. We will look at three types of experiment and show how the considerations of each case lead to different conclusions. There are also many factors that are not going to be determined by statistics; for example, whether to use Affymetrix arrays or two-colour microarrays may be determined by the facilities available to your laboratory. Similarly, whether to use a design that requires 20 or 40 arrays may be determined by financial constraints. However, there are statistical reasons why some of these designs are better than others. These are the areas we focus on in this section.

EXAMPLE 10.2 HEPATOCELLULAR CARCINOMAS

Samples are taken from disease and healthy tissue from patients suffering from hepatocellular carcinomas and hybridised to microarrays. We would like to identify genes that are up- or down-regulated in hepatocellular carcinomas relative to healthy tissue.

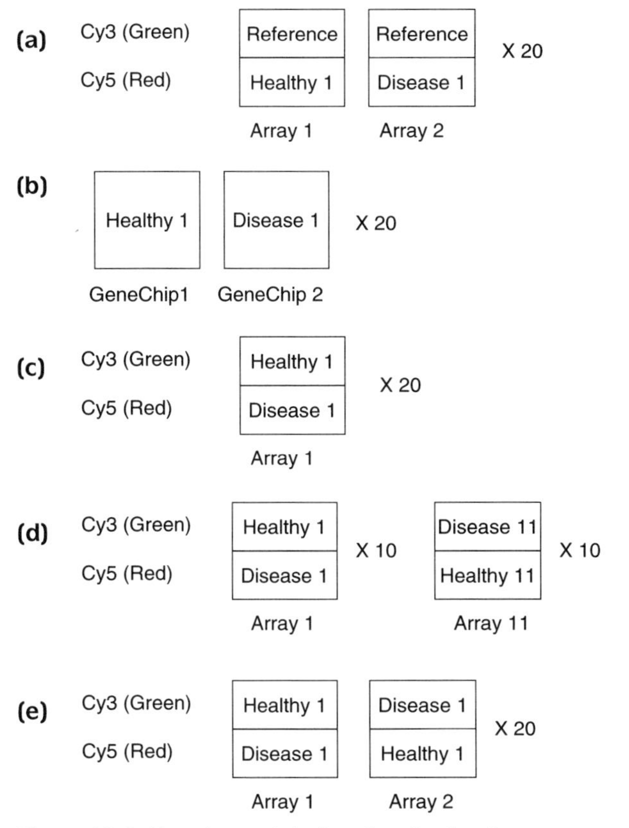

Figure 10.2: Experimental designs for simple microarray experiment. In an experiment to find differentially expressed genes in 20 patients suffering from hepatocellular carcinomas, healthy and disease tissue are taken from each patient and the prepared samples are hybridized to microarrays. Even this simple experiment lends itself to five different experimental designs. **(a)** Each of the 40 samples is hybridised to separate microarrays with a reference sample hybridised to the second channel. **(b)** Each of the 40 samples is hybridised to a separate Affymetrix GeneChip. **(c)** The two samples from each patient are hybridised to the same array, with the healthy sample in Cy3 and the disease sample in Cy5; only 20 arrays are needed.**(d)** As in (c), except that 10 patients have the healthy tissue in Cy3 and the disease tissue in Cy5, and the other 10 patients have healthy tissue in Cy5 and disease tissue in Cy3. **(e)** The samples from each patient are labelled twice: once with Cy3 and once with Cy5. These samples are then hybridised to two arrays in a dye-reversal experiment. As with (a) and (b), 40 arrays are needed.

Experimental Design 1

Forty microarrays are used. Every healthy and tumour sample is prepared with Cy5 (red). A reference sample of relevant cell lines is labelled with Cy3 (green). Each array is hybridised with a liver sample in Cy5 and the reference sample Cy3 (Figure 10.2a).

Experimental Design 2

Forty Affymetrix arrays are used. Each array is hybridised with a different sample (Figure 10.2b).

Experimental Design 3

Twenty microarrays are used; on each array, the healthy sample is hybridised with Cy3 and the tumour sample from the same patient is hybridised with Cy5 (Figure 10.2c).

Experimental Design 4

Twenty microarrays are used. On 10 arrays, the healthy sample is hybridised with Cy3 and the tumour sample is hybridised with Cy5. On the other 10 arrays, the healthy sample is hybridised with Cy5 and the tumour sample with Cy 3 (Figure 10.2d).

Experimental Design 5

Forty microarrays are used. Every healthy and tumour sample is labelled twice, once with Cy3 and once with Cy5. The healthy and tumour samples from each patient are hybridised to two arrays, once with the healthy Cy3 and the tumour Cy5, and once with the healthy Cy5 and the tumour Cy3 (Figure 10.2e).

The first point to observe is that we have described five different experimental designs for what is a very simple microarray experiment. There is a lot of choice in how to run the experiment, and the choices you make will have an impact on how you interpret the data. There are many obvious differences between the designs: designs 1, 3, 4 and 5 use two-colour arrays, whereas design 2 uses Affymetrix arrays; designs 1, 2 and 5 use 40 arrays, whereas designs 3 and 4 use 20 arrays. Design 1 includes a reference sample. What is the best way to run this experiment?

This example is a paired experiment and will require a paired analysis (Section 7.1). We are comparing two samples from the same patient in order to identify genes that are up- or down-regulated, and the two samples from each patient bear an obvious relationship to each other. Because of this, there is a clear case for using a two-colour array and to hybridise the two samples from the same patient to the same arrays.

Estimating Variability

We can estimate the variability of the different experimental designs with simple calculations based on the log-normal model introduced in Chapter 6. If the coefficient of variability of the hybridisation signal on an array is v, then the variance of the log of the signal, σ^2, is given by the following equation:

$$\sigma^2 = \ln(v^2 + 1) \qquad \text{(Eq. 10.1)}$$

Equation 10.1 is identical to Equation 6.1. If, for example, the coefficient of variability is 30%, then the variance is 0.086, and the standard deviation in the log signal is 0.29. Note that this is a natural logarithm, and to compute the standard deviation to base 2, divide this standard deviation by ln(2). In this case, the standard deviation in log to base 2 would be 0.42.

With experimental design 1 (reference sample), the log ratio of gene expression between the two samples is calculated indirectly by computing the log ratios of the healthy and disease samples to the reference samples and then subtracting these two log ratios. Because there are four samples involved in this calculation, there are four contributions of the variance σ^2 to the total variance, which is $4 \times 0.086 = 0.344$. So the standard deviation of the log ratio is $\sqrt{0.344} = 0.59$.

With experimental designs 2 (Affymetrix) and 3 (same array hybridisation), the log ratio is calculated directly. In design 2, we calculate the log ratio of the samples as the log of the ratio of the two Affymetrix signals, and in design 3, we calculate the log ratio of the red and green signals on the array. There are only two contributions of the variance σ^2 to the total variance. So the total variance is 0.172 and the standard deviation of the log ratio is 0.41. Therefore, experimental designs 2 and 3 have smaller errors and so both are better than experimental design 1 for this type of experiment.

This calculation is not exact; we have assumed that the variability of hybridisations between arrays is equal to the variability of hybridisations to the same array. In Chapter 6 we saw that this is not the case. The variability between arrays is generally higher than the variability of signals on the same array, so this experiment is better performed on two-colour arrays where the two samples from each patient can by hybridised to the same array than on Affymetrix arrays where the two samples from each patient have to be on different arrays.

Confounding and Colour Swaps

Experimental design 3 is not the best design for this experiment for other reasons. There is a problem: all of the healthy samples are labelled with Cy3, and all of the disease samples are labelled with Cy5. If, in the analysis, we see a gene that appears to be differentially expressed, this could result from the disease state, or it could result from differential incorporation of the Cy dyes. With experimental design 3, the labelling is confounded with the factor of interest (disease/healthy tissue), and we cannot tell which of these factors would be responsible for an observed differentially expressed gene.

In experimental designs 4 and 5 this problem has been resolved. Experimental design 4 is a balanced blocked design. The red and green dyes are a blocking variable, and so it is possible to determine the genes that are differentially expressed in diseased tissue.

Experimental design 5 is what is known as a full-factorial design, because each patient has had each sample hybridised twice, once with each Cy dye. There are two advantages of experimental design 5 over experimental design 4, and one disadvantage.

The first advantage of design 5 is that there are two measurements of log ratio for each patient, using the same number of arrays as experimental design 1 (reference sample). This is a form of technical replication and reduces the standard deviation of the measured log ratio by a factor of $\sqrt{2}$; for example, with a coefficient of variability of 30%, the standard deviation of the log ratio in experimental design 5 would be 0.29, compared with 0.41 in experimental design 4, and 0.59 in experimental design 1.

The second advantage is that the data can be analysed with a straightforward t-test or bootstrap t-test, whereas experimental design 4 will require a more sophisticated ANOVA analysis, and a more complex bootstrap to obtain p-values (Section 7.6).

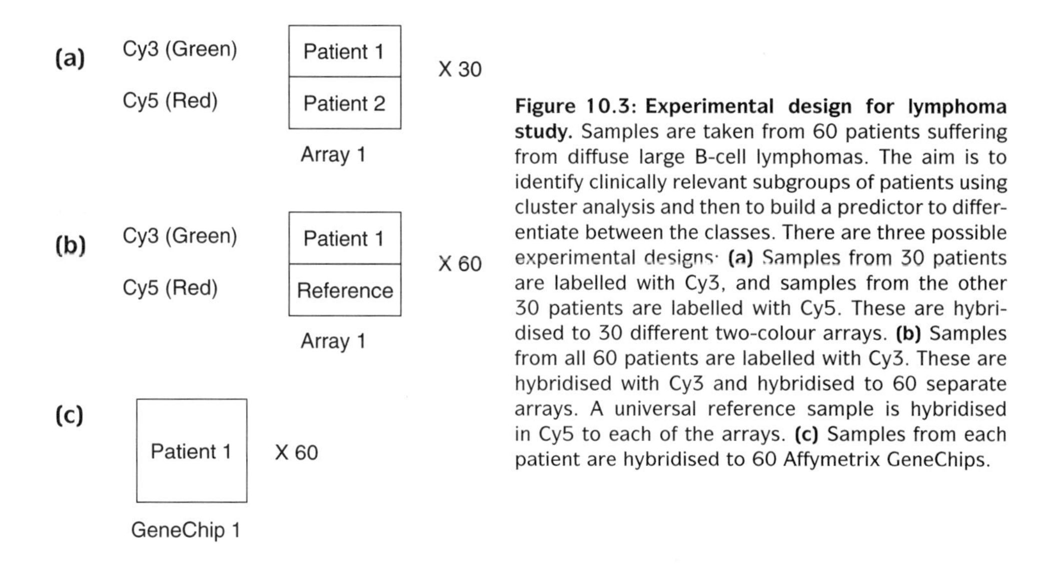

Figure 10.3: Experimental design for lymphoma study. Samples are taken from 60 patients suffering from diffuse large B-cell lymphomas. The aim is to identify clinically relevant subgroups of patients using cluster analysis and then to build a predictor to differentiate between the classes. There are three possible experimental designs: **(a)** Samples from 30 patients are labelled with Cy3, and samples from the other 30 patients are labelled with Cy5. These are hybridised to 30 different two-colour arrays. **(b)** Samples from all 60 patients are labelled with Cy3. These are hybridised with Cy3 and hybridised to 60 separate arrays. A universal reference sample is hybridised in Cy5 to each of the arrays. **(c)** Samples from each patient are hybridised to 60 Affymetrix GeneChips.

The obvious disadvantage of experimental design 5 is that it requires twice as many arrays as design 4 and twice as many labelling reactions as any of the other designs. The design you choose would therefore depend on the financial resources available to your laboratory.

EXAMPLE 10.3 B-CELL LYMPHOMAS

Samples are taken from 60 patients suffering from B-cell lymphomas and are hybridised to microarrays. The aim of the experiment is to identify clinically relevant subgroups of patients using a cluster analysis and then to build a classification model to differentiate between the subgroups.

Experimental Design 1

Thirty patient samples are prepared and labelled with Cy3, and 30 patient samples are prepared and labelled with Cy5. These are hybridised to 30 different two-colour arrays (Figure 10.3a).

Experimental Design 2

The samples from each patient are prepared and labelled with Cy3 and hybridised to 60 different two-colour arrays; a universal reference sample is hybridised in Cy5 to each array (Figure 10.3b).

Experimental Design 3

The samples from each patient are prepared and hybridised to 60 different Affymetrix arrays (Figure 10.3c).

Experimental design 1 is not a good design. In order to be able to apply the clustering methods of Chapter 7 or the classification methods of Chapter 8, we need to be able to compare each of the samples with the other on an equal footing. However, this design does not allow us to do this. Although the pairs of unrelated samples that have been hybridised on the same array can be compared easily, it is very difficult to make a comparison between two samples hybridised on different arrays, particularly if they are also labelled with different dyes. So although it might be tempting to use a design that requires half the number of arrays as designs 2 or 3, the data derived from this experiment would not naturally lend themselves to analysis.

Experimental design 2, on the other hand, is much better suited for the analyses that will be performed. Each sample can be normalised relative to the reference sample, so each sample can be meaningfully compared with the others for the purposes of cluster analysis or classification analysis.

Experimental design 3 is also a good design. The uniformity of the Affymetrix platform makes comparisons between the samples meaningful; any "dark" arrays can be normalised using between-array normalisation (Section 5.4).

EXAMPLE 10.4 TIME SERIES

Budding yeast can reproduce sexually by producing haploid cells through a process called sporulation. Yeast was placed in a sporulating medium and samples were taken at seven time points from the start of sporulation. We are interested in identifying genes that show similar profiles in the timecourse.

Experimental Design 1

The samples from the seven time points are hybridised to seven Affymetrix arrays (Figure 10.4a).

Experimental Design 2

The samples from the six time points after time zero are prepared and labelled with Cy3. A larger sample from the time zero time point is prepared and labelled with Cy5 as a reference sample.[1] The samples are hybridised to six arrays, with each time point in the Cy3 channel and the time zero sample in the Cy5 channel (Figure 10.4b).

Experimental Design 3

The samples from the seven time points are each labelled twice: once with Cy3 and once with Cy5. The arrays are hybridized to seven arrays as shown in Figure 10.4c. This is known as a **loop** design.

Experimental design 1 has a serious problem, which would also be true of performing this type of experiment using one colour on a microarray, or using radioactively

[1] Early time-course experiments used the sample at time zero as a reference sample. More recently, researchers are employing the better practice of using a mixture of sample from all time points as a common reference sample. This has the advantage of ensuring that there is signal in the reference sample from all genes that are expressed at some point during the time course.

(a)

Time 0	Time 1	Time 2	Time 3	Time 4	Time 5	Time 6
GeneChip 1	GeneChip 2	GeneChip 3	GeneChip 4	GeneChip 5	GeneChip 6	GeneChip 7

(b)

	Array 1	Array 2	Array 3	Array 4	Array 5	Array 6
Cy3 (Green)	Time 1	Time 2	Time 3	Time 4	Time 5	Time 6
Cy5 (Red)	Time 0	Time 0	Time 0	Time 0	Time 0	Time 0

(c)

	Array 1	Array 2	Array 3	Array 4	Array 5	Array 6	Array 7
Cy3 (Green)	Time 1	Time 2	Time 3	Time 4	Time 5	Time 6	Time 0
Cy5 (Red)	Time 0	Time 1	Time 2	Time 3	Time 4	Time 5	Time 6

Figure 10.4: Experimental design for time series study. Budding yeast is treated with sporulating medium; samples are taken at seven different time points and hybridised to microarrays. There are three common experimental designs: **(a)** Samples from each time point are hybridised to seven different Affymetrix GeneChips. **(b)** The sample from time zero is used as a reference sample, labelled with Cy5 and hybridised to all arrays. Samples from the other six time points are labelled with Cy3 and hybridised to six different two-colour arrays. **(c)** Samples from all seven time points are labelled twice: once with Cy3 and once with Cy5. These are hybridised to the arrays in the pattern shown. This is known as a loop design.

labelled samples on nylon filters. The stimulation of a cell culture can possibly lead to global changes in gene expression over the course of the time series. Suppose one particular array, corresponding to a particular time point, is "brighter" than the others. This could have two interpretations. First, it could be that this is an experimental artifact resulting from differential hybridisation; second, it could be that overall gene expression is higher at that time point (Figure 10.5).

If we use Affymetrix technology, or a different single-colour system, these two factors are confounded and no analysis can resolve the true situation. Although it is tempting to apply between-array normalisation (Section 5.4), this would be incorrect as it would remove all information about global changes in gene expression from the analysis (Figure 10.5b).

Experimental designs 2 and 3 resolve this problem. With experimental design 2, each sample is normalised relative to the sample at time zero, so that the measurements are log ratios of the time points relative to time zero. The presumption is that if an array is particularly bright, it will be bright for both samples (Figure 10.5a), and so the log ratio will be free from this artifact (Figure 10.5b).

Experimental design 3 has the advantage over experimental design 2 in that there are two independent measurements of each of the samples while using the same number of arrays. However, there are two disadvantages. This first is that it requires a more complex ANOVA analysis to be able to compare all the samples on all the arrays (Section 7.6), in contrast with experimental design 2, which can be readily analysed

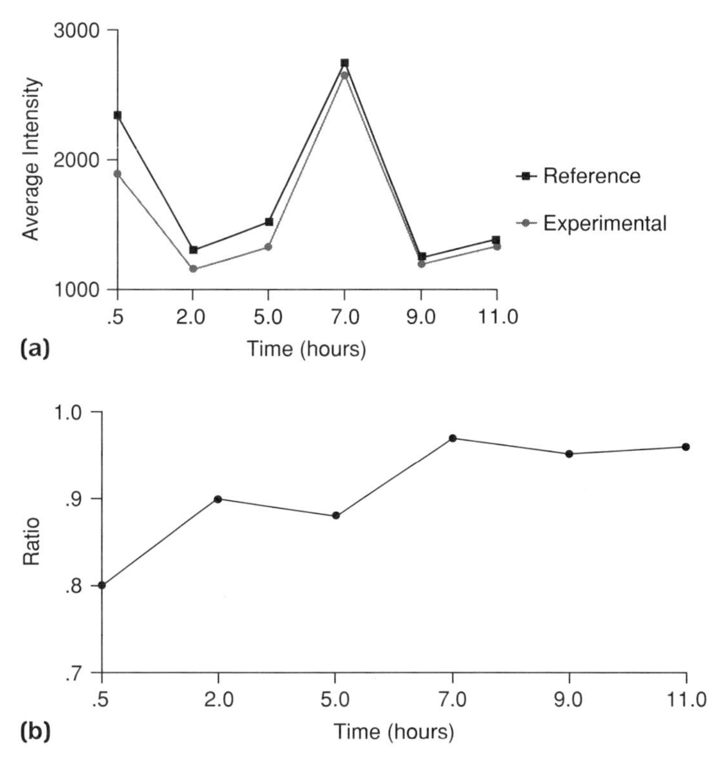

Figure 10.5: Normalisation problem in single-colour time-course experiments. Budding yeast is placed in sporulating medium and samples are taken at time 0, and after 30 minutes, 2, 5, 7, 9 and 11 hours. The experiment is designed according to experimental design 2, on two-colour arrays, with sample taken from time zero used as a reference sample. **(a)** Average intensity for all the genes on the array for both the experimental sample and the reference sample. The array at 7 hours is much brighter than at any other time point. Had this experiment been performed without a reference sample, for example on an Affymetrix platform, it would not be possible to determine whether this high signal represents maximum gene induction at 7 hours or is simply a bright array. In fact, the signal on the reference sample is also high, suggesting that this is a bright array. **(b)** Average log ratio of the all the genes on the array to the reference sample. The ratio increases with time until 7 hours, which indicates that gene expression over the whole genome is increasing over this time. Therefore, normalising by the average array signal would be incorrect because this information would be lost. It is therefore impossible to analyse this data without either a reference sample or a loop design.

without such an analysis. The second is that if a single array were to fail, it would affect the entire analysis. With experimental design 2, the data from a failed array can be omitted, and the other data points can continue to be used.

SECTION 10.4 HOW MANY REPLICATES?

This is the third consideration of experimental design and one of the most frequently asked questions about microarray experiments. The answer to this question depends on a number of factors: most critically, the type of experiment being performed, and the analysis to be applied to the data.

In this section, we will look at methods to estimate the number of replicates needed for experiments to detect differentially expressed genes (Chapter 7). For most practical purposes, the calculations of the number of replicates for these types of experiment suffice also for the types of experiment in which the data will be analysed with cluster analysis (Chapter 8) or classification analyses (Chapter 9). This is because the number of replicates can be used to determine the fold change we would want to detect in a gene before using it in a cluster or classification analysis.

The classical way to estimate the number of replicates needed in an experiment is with a **power analysis**. The concept of power is closely related to another concept, confidence, which we touched on in Chapter 7.

Confidence and Power

The **confidence** of a statistical test is the probability of not getting a false positive result. To put it another way, it is the probability of concluding that a gene is not differentially expressed when the gene is truly not differentially expressed. From a statistical perspective, it is the probability of accepting the null hypothesis when the null hypothesis is true. Statisticians sometimes refer to false positive results as Type I errors. Type I errors are generally controlled explicitly when we select a significance level for the statistical test. For example, when one performs a statistical test with a 1% significance threshold, one is selecting a confidence of 99%. In microarray experiments, the confidence has to be adjusted to take into account multiple testing (Section 7.5).

The **power** of a statistical test is the probability of not getting a false negative result. This is the probability of concluding that a gene is differentially expressed when the gene is truly differentially expressed. From a statistical perspective, it is the probability of rejecting the null hypothesis when the null hypothesis is false. Statisticians sometimes refer to false negative results as Type II errors. Type II errors cannot be controlled explicitly, but are controlled implicitly via the experimental design. In particular, the power of an experiment depends critically on the number of replicates used. Thus the choice of number of replicates is determined by the power you want to achieve in your analysis.

Type I and Type II errors are summarised in Table 10.1. When we choose a significance threshold, we have to make a choice about the balance between the confidence and power of the analysis: a more stringent significance threshold gives greater confidence but reduces power, and, conversely, a less stringent significance threshold gives less confidence but increases power.[2]

[2] A common question is which is worse: a Type I error or a Type II error. There is no statistical answer to that question: in different situations, Type I or Type II errors may be less preferred. Statistics give you the tools by which you can measure the rates of these errors, but it is then a matter of judgement as to how to balance the two types of error. For example, if the purpose of the microarray experiment is to identify novel targets for drug discovery, and a lot of money will be spent researching each chosen target, then it may be more important not to have Type I errors, as a false positive will result in an expensive failure. On the other hand, if the microarray is being used as a diagnostic tool for a public health screening program for a fatal cancer, such as breast cancer, then it might be more important not to have Type II errors. A false negative could result in a patient developing a fatal tumour that might have been curable had it been detected earlier.

TABLE 10.1: Type I and Type II Errors

	True Situation	
Our Decision	**Not Differentially Expressed**	**Differentially Expressed**
Not significant	Correct	Type II error
Significant	Type I error	Correct

Note: There are four possible outcomes of a statistical test. The two correct outcomes occur either if the gene is not differentially expressed and we say that it is not significant, or if the gene is differentially expressed and we say that it is significant. There are two possible wrong outcomes: a Type I error occurs when the gene is not differentially expressed and the analysis concludes that it is significant; a Type II error occurs when the gene is differentially expressed and the analysis concludes that it is not significant. The confidence of an analysis is the probability of not making a Type I error; the power of an analysis is the probability of not making a Type II error. The power is controlled by choosing an appropriate number of biological replicates.

The reason why a power analysis is used to determine the number of replicates in an experiment is because the power of a statistical test depends on the following factors:

- The number of replicates
- The type of analysis (paired or unpaired)
- The difference in mean that we are trying to detect (which is the log ratio)
- The standard deviation of the population variability
- The significance threshold of the test

Therefore, we estimate the number of replicates needed from knowledge of the other parameters and a predetermined desired power. Before we show how to perform such analyses, we will discuss each of the parameters in detail.

Types of Replicates

Microarray experiments can be replicated at many different levels. Fundamentally, there are two types of replicates: **biological** replicates and **technical** replicates.

Biological replicates are replicates taken at the level of the population being studied. In Example 10.2, where samples are taken from 20 patients suffering from hepatocellular carcinomas, each patient is a biological replicate. In power analyses, the replicates we refer to are always biological replicates. This is because the analysis we intend to perform is making a statistical inference to the population from which the replicates derive (Section 7.1). We need to include sufficient biological replicates to be certain that the effects we see can generalise to the population (e.g., the population of patients suffering from hepatocellular carcinomas).

Technical replicates, on the other hand, are taken at the level of experimental apparatus. The purpose of technical replicates is to account for variability in the experimental setup; if the experiment were of sufficiently high quality, then technical replicates would not be required at all. Technical replicates can be at many levels:

- Replicate features on the array, which can account for differential printing or hybridisation
- Replicate arrays hybridised with the same sample
- Replicate sample preparations; for example, two dye-reversed labellings

Technical replicates cannot count as different samples either in the power calculations or in the analysis of the data that is produced. Instead, it is common to take the average of technical replicates to provide a single measure for the individual. This does improve the reliability of the microarray data, because the experimental variability in an average of several technical replicates is decreased.

Some microarray users pool the samples from several individuals before hybridizing to arrays. For the purposes of power calculations and data analysis, pooled samples count as a single individual, because the information about variability between individuals has been lost in the pooling process. Therefore, in experiments where the variation between individuals is important (e.g., experiments on human disease), pooling should be avoided if at all possible.

Type of Analysis

In Chapter 7 we looked at data that was either paired or unpaired. The power is different for paired and unpaired data: in general, paired tests are more powerful because the differences between individuals are cancelled out via the pairing. In this section, we will discuss power analyses for these types of data. It is also possible to perform power analyses for more complex data types, such as those requiring ANOVA; this is more advanced statistics.

Difference in Mean

The power of a statistical test depends on the difference in mean we are trying to detect. In the case of paired data, this is the difference between the mean of the data and zero; in the case of unpaired data, this is the difference between the means of the two groups. In microarray analysis, where we are working with logged data, the difference in mean translates to an average log ratio.

It should be fairly intuitive that it is harder to detect smaller differences in means than larger ones. For example, it is more difficult to detect 1.5-fold differentially expressed genes than 3-fold differentially expressed genes. However, it is important to appreciate that we are not using the fold ratio as a threshold for detecting genes; we are simply stating that the power of the hypothesis tests described in Chapter 7 depends on the level of differential expression, as well as other parameters.

Standard Deviation

The power of a statistical test also depends on the level of variability in the population. In these calculations, we make the assumption that the errors in gene expression measurements are log-normally distributed (see Chapter 6). This assumption is only approximately true for most microarray experiments; for this reason, we do not recommend using *t*-tests to identify differentially expressed genes (Section 7.3). However, power analyses are only an approximate guide to estimating number of replicates and are not a precise measure; thus, deviations from the log-normal assumption are not a serious problem. If you prefer to perform power analyses without assuming that the data is log-normally distributed, it is possible to perform bootstrap power analyses; the interested reader is referred to the book on bootstrapping listed at the end of Chapter 7.

Power Analysis Calculation and Tables

The formulae for power analyses are complicated. However, most statistical packages have implementations of power analyses; in this section, we will show how to use the power analysis function in the R package. This has the advantage of allowing the user to select very stringent confidence levels. We need this flexibility because in microarray experiments we use very high levels of confidence (or low significance thresholds) in order to control the false-positive rate. The R function is

power.t.test(*n*, *delta*, *sd*, *sig.level*, *power*, *type*, *alternative*) (Eq. 10.2)

where

- *n* is the number of replicates (in a one-sample test) or group size (in a two-sample test).
- *delta* is the difference in mean that we are trying to detect (which is the log ratio).
- *sd* is the standard deviation of the population variability (calculated using Equation 10.2).
- *sig.level* is the significance threshold.
- *power* is the desired power.
- *type* can be *one.sample* or *two.sample*; *one.sample* is used for paired analyses, and *two.sample* is used for unpaired analyses.
- *alternative* can be *one.sided* or *two.sided*. (In microarray experiments we are almost always looking for both up- or down-regulated genes, so we generally use *two.sided*.)

To use the formula, one of the variables *n*, *delta*, *sd*, *sig.level* or *power* is omitted, and the function then calculates the value of the omitted variable. Usually, either we omit *n* and supply a desired *power* so that the formula returns the number of replicates we need, or we omit *power* and supply the number of replicates in the study, so the formula returns the power of our experiment.

In Table 10.2, we have used the R package to calculate powers of one-sample and two-sample analyses for detecting 2-fold differentially expressed genes, with a significance of 0.0001 (which would give 1 false positive on a 10,000-gene microarray), for a variety of levels of population variability and group sizes. The two examples that follow show how to perform a power analysis for specific experiments.

TABLE 10.2A: Power Analysis for Paired Test for 2-Fold Difference with $\alpha = 0.0001$

Num	Population Coefficient of Variation									
Reps	20%	25%	30%	35%	40%	45%	50%	60%	70%	80%
3	0.4%	0.2%	0.2%	0.1%	0.1%	0.1%	0.1%	0.1%	0.0%	0.0%
4	2.2%	1.2%	0.7%	0.5%	0.4%	0.3%	0.2%	0.1%	0.1%	0.1%
5	9.7%	4.8%	2.7%	1.6%	1.1%	0.8%	0.6%	0.3%	0.2%	0.2%
6	29.7%	14.8%	8.0%	4.7%	2.9%	2.0%	1.4%	0.8%	0.5%	0.4%
7	59.4%	33.7%	18.9%	11.0%	6.8%	4.4%	3.0%	1.6%	1.0%	0.7%
8	84.0%	57.4%	35.5%	21.6%	13.5%	8.8%	6.0%	3.1%	1.8%	1.2%
9	95.8%	78.1%	54.7%	35.9%	23.3%	15.4%	10.5%	5.3%	3.1%	2.0%
10	99.2%	91.0%	72.3%	51.7%	35.4%	24.1%	16.7%	8.6%	4.9%	3.1%
11	99.9%	97.0%	85.1%	66.7%	48.7%	34.6%	24.5%	12.9%	7.4%	4.6%
12	*	99.2%	93.0%	78.9%	61.6%	45.8%	33.5%	18.2%	10.5%	6.6%
13	*	99.8%	97.0%	87.7%	72.8%	56.9%	43.1%	24.4%	14.4%	9.0%
14	*	*	98.9%	93.3%	81.8%	67.1%	52.8%	31.3%	18.8%	11.9%
15	*	*	99.6%	96.6%	88.4%	75.8%	61.9%	38.6%	23.8%	15.3%
16	*	*	99.9%	98.4%	92.9%	82.8%	70.1%	46.0%	29.3%	19.1%
17	*	*	*	99.3%	95.9%	88.2%	77.2%	53.3%	35.0%	23.2%
18	*	*	*	99.7%	97.7%	92.2%	83.0%	60.3%	40.9%	27.7%
19	*	*	*	99.9%	98.8%	94.9%	87.6%	66.7%	46.8%	32.3%
20	*	*	*	*	99.4%	96.8%	91.2%	72.5%	52.6%	37.1%
25	*	*	*	*	*	99.8%	98.8%	91.4%	76.9%	60.7%
30	*	*	*	*	*	*	99.9%	97.9%	90.8%	78.9%
35	*	*	*	*	*	*	*	99.6%	96.9%	90.1%
40	*	*	*	*	*	*	*	99.9%	99.1%	95.8%
45	*	*	*	*	*	*	*	*	99.8%	98.4%
50	*	*	*	*	*	*	*	*	99.9%	99.4%

* > 99.9%.

Note: Power analysis for a paired test to detect a 2-fold difference (up- or down-regulated) in gene expression samples for a significance threshold of 0.0001; this would give approximately 1 false positive on a 10,000-gene array. The power depends critically on the coefficient of variability of the population. For example, when the population variability is 35%, we can achieve 95% power with 15 biological replicates. If the population variability were 50%, we would require 25 biological replicates to achieve similar power.

EXAMPLE 10.5 CALCULATING THE POWER OF A STUDY

Twenty breast cancer patients have been treated with a 16-week course of doxorubicin chemotherapy. Samples have been taken before and after treatment, and will be analysed for up- or down-regulated genes using a one-sample t-test. We are analysing 6,500 genes and want no more than one false positive. The coefficient of variability of the population is 50%. What is the power of the analysis for identifying 2-fold up-regulated genes? What fold regulation can we detect with 95% power?

Before applying the formula in Equation 10.2, we make some preliminary calculations:

Significance threshold. In order to have one false positive, we choose a significance threshold of 1/6500, which is approximately equal to 0.00015.

Standard deviation. We apply Equation 10.1 with $v = 0.5$ to obtain a standard deviation of 0.68 in log to base 2.

TABLE 10.2B: Power Analysis for Unpaired Test for 2-Fold Difference with $\alpha = 0.0001$

Group Size	Population Coefficient of Variation									
	20%	25%	30%	35%	40%	45%	50%	60%	70%	80%
3	1.3%	0.7%	0.4%	0.3%	0.2%	0.1%	0.1%	0.1%	0.1%	0.0%
4	7.9%	3.4%	1.7%	1.0%	0.6%	0.4%	0.3%	0.2%	0.1%	0.1%
5	25.5%	10.9%	5.2%	2.8%	1.7%	1.1%	0.8%	0.4%	0.3%	0.2%
6	50.9%	24.7%	12.2%	6.5%	3.7%	2.3%	1.6%	0.8%	0.5%	0.3%
7	74.1%	42.5%	22.5%	12.3%	7.1%	4.4%	2.9%	1.4%	0.8%	0.6%
8	88.8%	60.5%	35.4%	20.2%	11.8%	7.3%	4.8%	2.3%	1.3%	0.9%
9	95.9%	75.4%	49.0%	29.6%	17.9%	11.2%	7.3%	3.5%	2.0%	1.3%
10	98.7%	86.0%	61.9%	40.0%	25.1%	16.0%	10.5%	5.1%	2.8%	1.8%
11	99.6%	92.6%	72.9%	50.4%	33.0%	21.6%	14.4%	7.0%	3.9%	2.4%
12	99.9%	96.4%	81.6%	60.2%	41.3%	27.7%	18.8%	9.3%	5.1%	3.1%
13	*	98.3%	88.0%	69.0%	49.5%	34.3%	23.7%	11.9%	6.6%	4.0%
14	*	99.3%	92.4%	76.5%	57.4%	41.0%	28.9%	14.8%	8.3%	5.0%
15	*	99.7%	95.4%	82.6%	64.7%	47.7%	34.4%	18.1%	10.1%	6.2%
16	*	99.9%	97.3%	87.4%	71.2%	54.3%	40.0%	21.6%	12.3%	7.5%
17	*	*	98.5%	91.1%	76.9%	60.4%	45.6%	25.3%	14.5%	9.0%
18	*	*	99.1%	93.8%	81.7%	66.1%	51.1%	29.2%	17.0%	10.5%
19	*	*	99.5%	95.8%	85.8%	71.3%	56.4%	33.2%	19.7%	12.3%
20	*	*	99.8%	97.2%	89.0%	76.0%	61.4%	37.3%	22.5%	14.1%
25	*	*	*	99.7%	97.5%	91.4%	81.3%	57.3%	37.8%	24.9%
30	*	*	*	*	99.6%	97.4%	92.2%	73.7%	53.4%	37.4%
35	*	*	*	*	99.9%	99.3%	97.1%	85.2%	67.2%	50.2%
40	*	*	*	*	*	99.9%	99.1%	92.3%	78.2%	61.9%
45	*	*	*	*	*	*	99.7%	96.3%	86.2%	72.0%
50	*	*	*	*	*	*	99.9%	98.3%	91.7%	80.1%

* > 99.9%.

Note: Power analysis for an unpaired test to detect a 2-fold difference (up- or down-regulated) in gene expression samples for a significance threshold of 0.0001; this would give approximately 1 false positive on a 10,000-gene array. The group size is the number of biological replicates in each group: if this were a clinical study, the total number of patients would be twice the group size. This table also assumes that the two groups are of equal size (balanced design). If the groups are of unequal size, the power is decreased. The power depends critically on the coefficient of variability of the population. For example, when the population variability is 35%, we can achieve 95% power for a group size of 19. If the population variability were 50%, we would require a group size of 35 to achieve similar power.

Delta. A two-fold up- or down-regulation corresponds to a log ratio difference of 1 in log to base 2.

To calculate the power of detecting 2-fold regulated genes we use the formula

```
power.t.test(n=20, delta=1, sd=0.68, sig.level=0.00015,
type="one.sample", alternative="two.sided")
```

and obtain a *power* of 0.94, which is 94%. This means that applying a statistical analysis with a significance threshold sufficient to give approximately one false positive result will return 94% of the genes that are truly 2-fold differentially expressed.

To find out what fold regulation can be detected with 99% power, we use the formula

```
power.t.test(n=20, power=0.99, sd=0.68, sig.level=0.00015,
type="one.sample", alternative="two.sided")
```

and find *delta* equal to 1.16. The fold ratio is equal to $2^{1.16} = 2.23$, so we can detect 99% of the genes that are 2.23-fold regulated.

EXAMPLE 10.6 DETERMINING THE NUMBER OF PATIENTS NEEDED FOR A STUDY

In a new study of breast cancer chemotherapy, we want to identify genes that are two-fold up- or down-regulated following treatment with doxorubicin. We will be analysing 10,000 genes and want at most 1 false positive result. The coefficient of variability in the population is 50%. There are two possible experimental designs:

- Take samples from the same patient before and after therapy, and perform a paired analysis on the log ratio of the gene expression in the patients.
- Recruit two groups of patients, one to be treated and one to be untreated, and perform an unpaired analysis on the gene expression measurements from the patients in the two groups.

We want to identify how many patients we need in order to identify 95% of the 2-fold differentially expressed genes, and which experimental design requires fewer patients to be recruited to the experiment.

The preliminary calculations are similar to Example 10.5. The significance level is $1/10000 = 0.0001$; the standard deviation is 0.68 and delta is 1.

To find the number of patients needed for the first experimental design, we use the formula

```
power.t.test(power=0.95, delta=1, sd=0.68, sig.level=0.0001,
type="one.sample", alternative="two.sided")
```

and obtain $n = 21.48950$. A fractional number of patients is meaningless, so we round this number up and conclude that we would need 22 patients to achieve the desired power.

To find the number of patients needed for the second experimental design we use the formula

```
power.t.test(power=0.95, delta=1, sd=0.68, sig.level=0.0001,
type="two.sample", alternative="two.sided")
```

and obtain $n = 32.15861$. This is the group size, so we would need two groups of 33, which would mean a total of 66 patients.

We need fewer patients for the paired analysis than for the unpaired analysis, so the first experimental design is better. This illustrates a general principle that paired analyses are usually more powerful than unpaired analyses. With the paired analysis, the difference in gene expression is calculated with two measurements from the same patient, so individual variabilities are cancelled out. With the unpaired analysis, we compare the mean of gene expression in the two groups; the variabilities between individuals contribute to each of the means. Because of this, the unpaired analysis is less powerful than the paired analysis.

KEY POINTS SUMMARY

- Use blocking to remove problems of confounded variables.
- Use randomisation and blinding to remove the problem of bias.
- Avoid reference samples when comparing two samples from the same individual.
- Use reference samples for comparing many individuals or time series.
- Avoid single-channel technology for time series where there may be global changes in gene expression.
- Compute the number of biological replicates using power analyses.

USEFUL RESOURCES

Yang, Y.H. and Speed, T. 2002. Design issues for cDNA microarray experiments. *Nature Reviews Genetics* 3: 579–88.

Churchill, G.A. 2002. Fundamentals of experimental design for cDNA microarrays. *Nature Genetics* 32 Suppl: 490–95.

Two excellent reviews of experimental design for microarray experiments.

Witmer, J.A. and Samuels, M.L. 2002. *Statistics for the Life Sciences* (3rd Edition). Prentice-Hall: Englewood Cliffs, New Jersey.

An excellent introduction to statistics that I have found very helpful for teaching statistics to biology undergraduates. It covers experimental design as well as the analysis methods described here.

http://www.r-project.org/

The homepage for R, a free statistics package which is very similar to S. This is available for Unix, Windows and Macintosh, and has a wide range of statistics and graphing functionality. Unlike S+ or SPSS, it does not have a graphical user interface, and is operated via commands and scripts.

Papers from Which We Have Used Data or Experimental Design Ideas

Perou, C.M., Sorlie, T., Eisen, M.B., van de Rijn, M., Jeffrey, S.S., Rees, C.A., Pollack, J.R., Ross, D.T., Johnsen, H., Akslen, L.A., Fluge, O., Pergamenschikov, A., Williams, C., Zhu, S.X., Lonning, P.E., Borresen-Dale, A., Brown, P.O., and Botstein, D. 2000. Molecular portraits of human breast tumours. *Nature* 406: 747–52.

Paper for the breast cancer data.

Chu, S., DeRisi, J., Eisen, M., Mulholland, J., Botstein, D., Brown, P.O., and Herskowitz, I. 1998. The transcriptional program of sporulation in budding yeast. *Science* 282: 699–705.

The paper from which the yeast sporulation data are derived.

Okabe, H., Satoh, S., Kato, T., Kitahara, O., Yanagawa, R., Yamaoka, Y., Tsunoda, T., Furukawa, Y., and Nakamura, Y. 2001. Genome-wide analysis of gene expression in human hepatocellular carcinomas. *Cancer Research* 61: 2129–37.

A study of differential gene expression in hepatocellular carcinomas.

Alizadeh, A.A., Eisen, M.B., Davis, R.E., Ma, C., Lossos, I.S., Rosenwald, A., Boldrick, J.C., Sabet, H., Tran, C., Powell, J.I., Yang, L., Marti, G.E., Moore, T., Hudson, J. Jr., Lu, L., Lewis, D.B., Tibshirani, R., Sherlock, G., Chan, W.C., Greiner, T.C., Weisenberger, D.D., Armitage, J.O., Warnke, R., Levy, R., Wilson, W., Grever, M.R., Byrd, J.C., Botstein, D., Brown, P.O., and Staudt, L.M. 2000. Distinct types of diffuse large B-cell lymphoma identified by gene expression profiling. *Nature* 403: 503–11.

A study of diffuse large B-cell lymphomas.

CHAPTER ELEVEN

Data Standards, Storage and Sharing

SECTION 11.1 INTRODUCTION

In this book we have described many different types of microarray experiment. All these experiments generate large volumes of complex data. As scientists, we need to be able to communicate the results of our experiments with other scientists. There are many reasons why scientists seek to share data, including the following:

- **To verify the results of a published microarray experiment.** It is accepted that for scientific results to be published in a refereed journal, it is necessary to provide sufficient information so that others can reproduce the experiment.
- **To perform further experimental work based on the results.** Microarray data frequently generate hypotheses that require further experimental investigation; often, microarray experiments are performed precisely to generate such hypotheses.
- **To undertake further data analysis of the results.** Sometimes it is possible to perform further data analysis beyond the analyses carried out by the researchers in their original paper, which requires full access to the data.
- **To compare the results with other functional genomics data.** It is valuable to make comparisons either between different microarray experiments, or between microarray data and data from other sources (e.g., proteomics).
- **To develop novel data analysis methods.** Bioinformatics researchers developing novel data analysis methods need data sets for testing their methods.

Data sharing is not simply a function of scientists. A microarray laboratory will typically run a number of different computer applications to capture, store, publish and analyse microarray data (Figure 11.1). In order for the laboratory to operate successfully, each of these computer applications needs to be able to exchange data with the other.

This chapter looks at the mechanisms that allow microarray users and software to handle and share data. It is arranged into three further sections:

Section 11.2: Software and Standards, looks at the software components that are used in a microarray laboratory, the reasons for needing standards, and the Microarray Gene Expression Database group that is coordinating the standards.

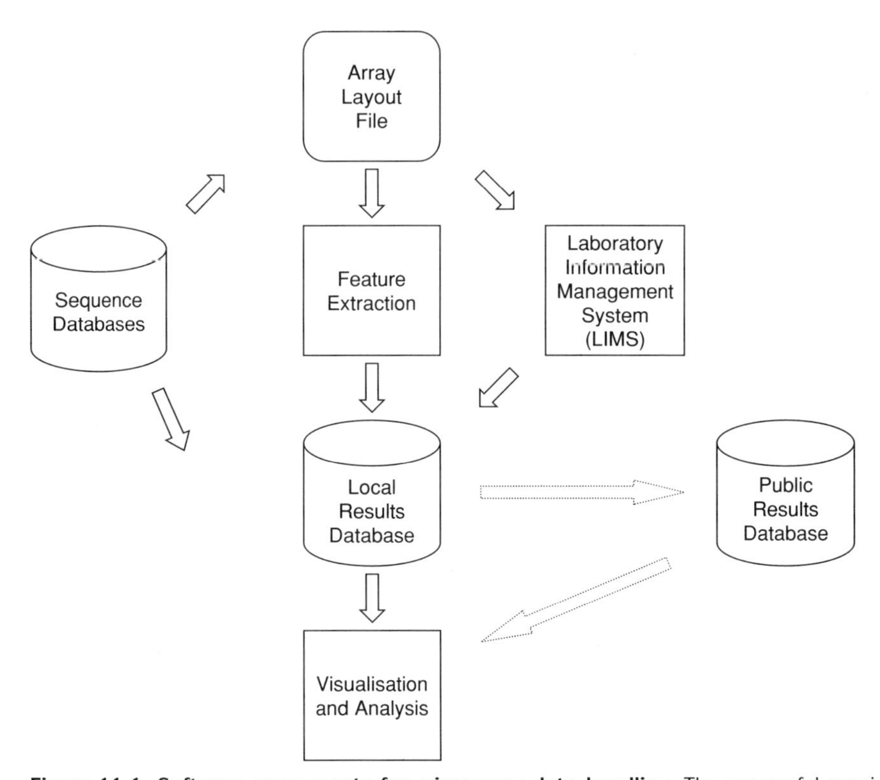

Figure 11.1: Software components for microarray data handling. The successful running of a microarray facility involves several software components that need to be integrated. Data should flow seamlessly between the different components, and ideally it should be possible to replace any component without affecting the other parts of the flow. **(a)** Array Layout File is a file containing details of what sequences and genes each feature represents. There are currently many formats for these files, depending on the platform you are using. **(b)** Sequence Databases contain information about the genes that the microarray is measuring and the sequences from which the sequences on the array derive. Accession numbers are included in the array layout so that it is possible to connect to these databases. Sequence databases are discussed in full in Chapter 2. **(c)** Feature Extraction software is discussed in Chapter 4. It converts the image of the microarray from the scanner into quantitative information about gene expression. It needs information from the array layout file to be able to annotate the features. **(d)** Laboratory Information Management System (LIMS) records all information about the laboratory methods and protocols used in microarray manufacture, sample preparation and labelling, hybridisation and washing. **(e)** Local Results Database contains results of experiments performed at the local institution. It can be in the form of a formal database or data warehouse, or the data might be stored as a series of files. The database needs to be able to link the feature extracted data with the experimental information stored in the LIMS. **(f)** Public Results Database contains results of microarray experiments that have been published in the public domain. If appropriate, data from the local results database might be transferred to a public database. **(g)** Visualisation and Analysis software allows the user to look at and interpret microarray data. The data could be from the local results database, public data, or a combination and comparison of the two.

Section 11.3: Minimal Information about a Microarray Experiment, describes the specifications drawn up for the Minimal Information About Microarray Experiments (MIAME), and the technologies to implement them in software.

Section 11.4: Ontologies, discusses the ontologies that are being developed to describe microarray experiments.

SECTION 11.2 SOFTWARE AND STANDARDS

In order to run a microarray facility, there are several software components that are needed: a laboratory information management system (LIMS), feature extraction software, and data analysis software. In addition, the running of microarray experiments also relies on databases, including sequence databases, local results databases and public microarray databases. The interrelationship between these components is shown in Figure 11.1.

Sequence databases and feature extraction software are discussed in Chapters 2 and 4, respectively. This section looks at LIMS, data warehouses and data analysis software. A selection of commonly used software packages is shown in Table 11.1. This list is by no means complete, and there are always new microarray software components becoming available both from academic groups and commercial organisations.

Laboratory Information Management System (LIMS)

A LIMS records all information about the laboratory experiment, including all procedures, protocols and methods in microarray manufacture, sample preparation and labelling, and hybridisation. At one level, a LIMS can be thought of as a computerised laboratory notebook, but with two key benefits:

- **Track data.** The LIMS can be used to record every step of the experimental process as it happens, including identity of experimenter, date, protocols used, and any experimental parameters. The advantages of tracking data are many:
  - **Quality control.** Any problems can be traced back to the source.
  - **Data reproducibility.** If the entire experimental process has been recorded, it is possible for other scientists to reproduce the experiment.
  - **Data comparison.** By knowing all parameters of the experiment, it is more meaningful to make comparisons between different microarray experiments and to know when comparisons are less meaningful.
  - **Data publication.** If the LIMS system is MIAME-compliant (see the following discussion), then it will record all the information necessary for publishing the data in a MIAME-compliant microarray database.
- **Standard protocols.** One feature of LIMS systems is that it is possible to include standard protocols as workflows that can help ensure that all staff in the laboratory (or group of collaborating laboratories) follow the same protocol. This helps to standardise microarray experiments performed by several people.

Local Data Warehousing

The results of microarray experiments performed in a laboratory or institution will usually be stored on local computers until such time as it might be relevant to store

TABLE 11.1: A Selection of Commonly Used Microarray Software Packages

Name	Institution	Functionality	Type	Availability
Array Informatics	Perkin Elmer	LIMS	Commercial	http://lifesciences.perkinelmer.com/areas/microarray/arrayinfo.asp
GeneDirector	BioDiscovery	LIMS	Commercial	http://www.biodiscovery.com/genedirector.asp
Limas	MRC	LIMS	Academic	http://www.mgu.har.mrc.ac.uk/microarray/limas/index.html
MIDAS	Sanger Institute	LIMS	Collaborators only	http://www.sanger.ac.uk/Projects/Microarrays/informatics/midas.shtml
GeNet	Silicon Genetics	Local database	Commercial	http://www.silicongenetics.com/cgi/SiG.cgi/index.smf
Resolver	Rosetta	Local database/ data analysis	Commercial	http://www.rosettabio.com/products/resolver/default.htm
GeneTraffic	Iobion Informatics	Local database/ data analysis	Commercial	http://www.iobion.com/
BASE	Lund University	Local database/ data analysis	Academic	http://base.thep.lu.se/
MIDAS/MEV/ MADAM	TIGR	Local database/ data analysis	Academic	http://www.tigr.org/software/#m
R	GNU	Data analysis	Free	http://www.r-project.org/
BioConductor	Open Source	Data analysis	Free	http://www.bioconductor.org/
Expression Profiler	EBI	Data analysis	Academic	http://www.ebi.ac.uk/microarray/ExpressionProfiler/ep.html
Cluster/TreeView	Berkeley	Data analysis	Academic	http://rana.lbl.gov/EisenSoftware.htm
GeneSpring	Silicon Genetics	Data analysis	Commercial	http://www.silicongenetics.com/cgi/SiG.cgi/index.smf
J-Express Pro	Molmine	Data analysis	Commercial	http://www.molmine.com
Excel	Microsoft	Data analysis	Commercial	http://www.microsoft.com/office/excel/default.asp
Spotfire	Spotfire	Data analysis	Commercial	http://www.spotfire.com/
Matlab	Mathworks	Data analysis	Commercial	http://www.mathworks.com/
ArrayExpress	EBI	Public microarray repository	Public access	http://www.ebi.ac.uk/microarray/ArrayExpress/arrayexpress.html
GEO	NCBI	Public microarray repository	Public access	http://www.ncbi.nlm.nih.gov/geo/
Stanford Microarray Database	Stanford University	Public microarray repository	Public access	http://genome-www5.stanford.edu/MicroArray/MDEV/index.shtml

the data in a public database. There are usually three types of data that would need to be stored:

- **Images.** The images of the arrays from the scanner. These are typically TIFF files and can be large: with 10-μm pixel sizes, a typical microarray image would be 7500×2200 pixels and 32 Mb in size.
- **Feature-extracted data.** The output from the feature extraction software (Table 4.2). These files can also be large: approximately 2.5 Mb for a 10,000-feature microarray.
- **Gene expression data table.** A summarised output of gene expression from all of the arrays in the experiment. The size of these files depends on the size and scope of the experiment.

There are two ways to store this data:

- **As files.** The data is stored directly on the laboratory's or institution's computer. This is easy to implement and does not require purchase of special software. However, it becomes difficult to track and query the data if larger numbers of experiments are being performed.
- **In a local database.** There are several products available both from commercial organisations and from academic groups that allow for local storage of microarray data. These enable good tracking and management of experimental data and integration with public microarray databases. However, it requires the purchase, installation and maintenance of a complex software product.

In general, smaller laboratories running a small number of microarray experiments can work effectively without a local data warehouse system. But larger institutions or companies working with a large number of microarrays will almost certainly need the data management facilities of a local data warehouse.

Data Analysis and Visualisation

It is the visualisation and analysis of your data that leads to the exciting results for which you have performed your microarray experiments. There is a wide range of software available for data analysis, some of which has been specifically written for microarrays. This software falls into three general categories (Table 11.2):

- **General graphical data analysis software packages.** These are software packages with graphical user interfaces that have been written for general use and which can be used for analysing microarray data. The most familiar example is Microsoft Excel, but other packages such as Spotfire also fall into this category.
- **General advanced statistics or data analysis software packages.** These are applications written to provide flexible and advanced data analysis functionality, with the ability to write and export scripts. The most commonly used examples are R and Matlab.
- **Graphical applications for microarray data analysis.** These are packages written specifically for the analysis of microarray data. There are many such packages, including GeneSpring, J-Express and Expression Profiler.

TABLE 11.2: Advantages and Disadvantages of Different Types of Data Analysis Software

General Graphical Data Analysis Packages	Advanced Statistics or Data Analysis Packages	Graphical Applications for Microarrays
Examples: Excel Spotfire ✓ Software is usually easy to use. ✓ Excel is familiar to many users because it is used widely for many purposes. ✓ Software contains useful range of functionality. × Not written specifically for microarrays so many data analysis methodologies are not included. × Restricted to the methods supplied in the software and it is difficult to implement new methods. × Not MIAME-compliant.	Examples: R Matlab ✓ Packages contain implementations of a wide range of statistical data analysis methods, e.g., Loess fitting or neural networks. ✓ Scripting language provides flexibility to implement new methodologies. ✓ Many groups have written data analysis libraries specifically for microarrays using both R and Matlab. × These packages use a command line and are difficult to use for people only used to graphical user interfaces. × Not MIAME-compliant.	Examples: GeneSpring J-Express Expression Profiler ✓ Easy-to-use graphical user interfaces. ✓ Software is written specifically for microarrays so it contains relevant data analysis methodologies. ✓ Link well with sequence databases and microarray data formats. ✓ Increasing MIAME-compliance. × Limited to the functionality of the methods that are implemented. × These are "black-box" systems that allow you to apply analysis methods without a full understanding of precisely what the software is doing.

Most microarray laboratories use a combination of all these types of software in order to benefit from the strengths of each of the platforms. If you are looking for a data analysis package to use, your best option is to evaluate as many software packages from as many sources as you can, and choose the ones that work best for you.

The Need for Standards

When sharing data, either between scientists or between computer applications, it is helpful if the data conform to standards. When communicating between individuals, standards are helpful because they can ensure that people understand what others are saying. For example, the terms *probe* and *target* have been used by different people interchangeably to mean the DNA on the array or the labelled DNA in solution that will be hybridised to the array.

Standards are essential for designing computer software that can integrate with other applications. If the people writing the software have a common set of standards for data representation and exchange, then it becomes possible for different academic groups or commercial organisations to write software applications or microarray databases that are able to work with applications developed elsewhere. It then becomes possible for an organisation using microarrays to obtain and connect software from several sources.

In order for standards to be successful, they need to have several qualities:

- **Useful.** The standards should be a genuine aid to storing and sharing microarray data.
- **Consensual.** The standards should be developed and agreed upon by microarray users.
- **Flexible.** The standards need to be designed to be able to accommodate all types of microarray experiments and data, including experiments that have not yet been thought of.
- **Comprehensible.** It should be easy for all microarray users to understand the standards.
- **Easy to implement.** It should be straightforward for programmers to implement the standards into software for microarray use.
- **Widely adopted.** In order to be of global benefit, the standards should be adopted by as many research groups and commercial organisations as possible. In order to achieve this, it is hoped that a requirement of publication of microarray results will be submission of data to a public-domain database that has adopted the standards.

Microarray data standards comprise three areas:

- **What to record.** Which aspects of the microarray experimental process and of the microarray data need to be recorded. This is the aim of MIAME: Minimal Information about a Microarray Experiment, discussed in Section 11.3.
- **How to describe it.** How to describe the experimental methods and microarray data. For this, we need **ontologies**: controlled vocabularies and relationships to describe genes, samples and data. Ontologies are discussed in Section 11.4.

TABLE 11.3: Organisations Represented on the MGED Board or MGED Advisory Board

Organisations
Research Institutes
DNA Data Bank of Japan (DDBJ)
European Bioinformatics Institute (EBI)
European Molecular Biology Laboratory (EMBL)
German Genome Resource Centre
Jensen Research Foundation
Max Plank Institute for Molecular Genetics
National Centre for Biotechnology Information (NCBI)
National Centre for Genome Resource (NCGR)
National Human Genome Research Institute (NHGRI)
RIKEN
The Institute of Genetics Research (TIGR)
The Jackson Laboratory
Vlaams Instituut voor Biotechnologie
Universities
Duke University
Imperial College, London
Stanford University
UMC Utrecht
Rockefeller University
Universite D'Aix-Marseille II
University of California at Berkeley
University of Colorado
University of Pennsylvania
University of Washington
Commercial Organizations
Affymetrix
Agilent
Clontech
Genelogic
Incyte
Iobion
Ipsogen
Lion Biosciences
Rosetta
Other
Open Informatics
Science Magazine

- **How to implement it.** How to implement MIAME and ontologies in computer software. This requires **object models**, **exchange languages** and language-specific modules, and is discussed briefly at the end of Section 11.3.

The Microarray Gene Expression Data Society (MGED)

The need for microarray data standards was recognised relatively early in the microarray community. In November 1999, the Microarray Gene Expression Data Society (MGED) was founded, with the intention of establishing standards for microarray data annotation and to enable the creation of public databases for microarray data. The MGED board of directors and advisory board now has representation from many of the major institutions involved with microarrays, including research institutes, universities, commercial organisations and journals (Table 11.3).

MGED has an annual meeting at which major developments are discussed and arranges regular workshops, tutorials and programming jamborees. MGED's work is arranged into four working groups:

- **MIAME.** Minimal Information About a Microarray Experiment formulates the information required to record about a microarray experiment in order to be able to describe and share the experiment.
- **Ontologies.** Determines ontologies for describing microarray experiments and the samples used with microarrays.
- **MAGE.** Formulates the object model (MAGE-OM), exchange language (MAGE-ML) and software modules (MAGE-stk) for implementing microarray software.
- **Transformations.** Determines recommendations for describing methods for transformations, normalisations and standardisations of microarray data.

SECTION 11.3 MINIMAL INFORMATION ABOUT A MICROARRAY EXPERIMENT

The aim of MIAME is to outline the minimum information that should be recorded about a microarray experiment so that the data can be fully understood and the experiment fully reproduced in another laboratory. It is intended to assist the exchange of microarray information between researchers, including doing so via the development of public microarray data repositories. It is not intended to be a formal specification, but a set of guidelines. However, it has become the standard for many microarray software packages and databases, so it is highly recommended that you record data from your experiments in a way that is compliant with MIAME.

MIAME is arranged into two broad areas (Figure 11.2):

- Array design description
- Experiment description

The reason for this distinction is that the array design is frequently independent of the experiment, with the same array design being used for many experiments. For example, the Affymetrix U133 GeneChip is used in many laboratories for many different types of experiment and can be described independently of any specific experiment using it.

Array Design Description

The aim of the array design description is to give a detailed description of the array, including physical factors (size and material), chemical factors (type of attachment) and logical factors (sequences). To describe the sequences on an array, MIAME introduced

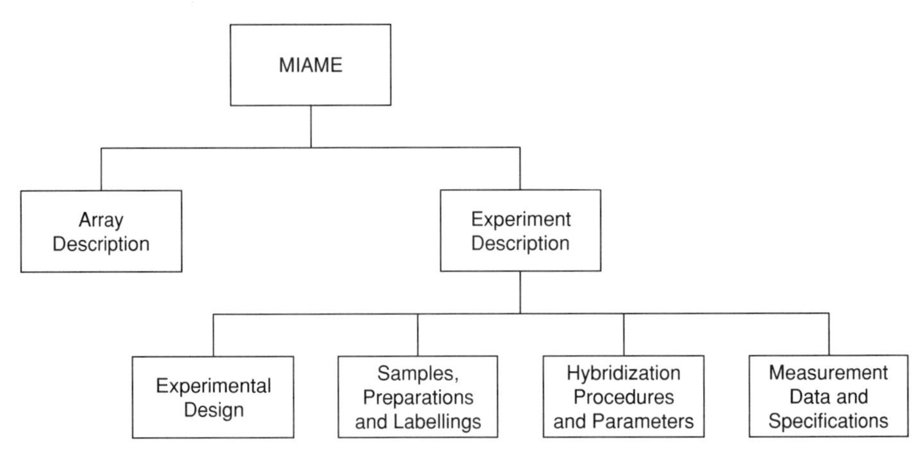

Figure 11.2: MIAME structure. The information in MIAME is arranged hierarchically. It is principally divided into two areas: information about arrays and information about experiments using arrays. The reason for this division is that microarrays are frequently manufactured independently of the experiments and then used for a wide range of experiments. For example, the Affymetrix U133 GeneChip is used by many laboratories for many different experiments, and thus can be described independently of any experiment. The information about experiments is further divided into four broad areas: the design of the experiment; the samples used in the experiment and the methods used to prepare and label them; the hybridisation steps and parameters from the hybridization; and the data itself, together with any information about normalization or transformation of the data.

three new terms:

- **Feature.** The location on the array containing the DNA sequence (also commonly referred to as *spots*).
- **Reporter.** The DNA sequence on a feature.
- **Composite sequence.** The (gene) sequence from which the reporter derives; there could be several different reporter sequences for the same gene.

The array design description is arranged into seven types of information.

1. Array-related information:
   - **Array design name.**
   - **Platform type.** Whether the array is in-situ synthesised, spotted or some other type of array.
   - **Surface and coating specification.** The physical composition of the array (nylon or glass), and description of any chemical derivitisation on the surface of the array.
   - **Physical dimensions of the array.**
   - **Number of features on the array.** Includes the number of features in both x and y, and details of any grids on the array.
   - **Availability.** Name of supplier and catalogue number for commercial arrays, or production protocol for custom-made arrays.
2. Reporter type information:
   - **Type of reporter.** Whether the reporters are synthetic oligonucleotides, PCR products, plasmids, colonies or other.
   - **Single- or double-stranded.**

3. For each reporter:
   - **Sequence or PCR information.** The sequence if known (e.g., oligonucleotides), sequence accession number or primer pairs (if relevant).
   - **Exact or approximate length of sequence.**
   - **Clone information.** If relevant, the clone ID, clone provider, date of provision and availability of the clone.
   - **Element generation protocol.** Sufficient information to reproduce the element on custom arrays that are not generally available.
4. Feature type information:
   - **Dimensions.** The physical size of the features.
   - **Attachment.** Covalent, ionic or other; if the feature is an oligonucleotide, whether attachment is from 3′ or 5′ end of oligonucleotide.
5. For each feature:
   - **Location on the array.** Both physical and logical coordinates.
   - **Which reporter.** Which reporter sequence is on the feature.
6. For each composite sequence:
   - **Which reporters it contains.**
   - **The reference sequence.**
   - **Gene or EST names.** Including links to appropriate databases (e.g., UniGene or RefSeq).
7. Control elements on the array:
   - **Position of the feature.** Logical coordinates.
   - **Control type.** Spiking, normalisation, negative or positive.
   - **Control qualifier.** Endogenous or exogenous.

Experiment Description

The aim of the experiment description is to give sufficient information that another laboratory would be able to repeat the experiment. An experiment may consist of one or more hybridisations to one or more types of array. The experimental description is broken into four main parts, each of which has several components:

1. Experimental design
   - **Authors, laboratory and contact information.**
   - **Type of experiment.** Typical experiments might be normal vs. disease comparison, treated vs. untreated comparison, time course or dose response.
   - **Experimental factors.** These are the parameters or conditions that are tested in the experiment; for example, treatment, time, dose or genetic variation.
   - **Number of hybridizations in the experiment.**
   - **Whether or not a common reference sample has been used.**
   - **Quality control steps.** These include replications at different levels, the use of dye reversal, or the inclusion of quality control features.
   - **Description of experiment and its goal.**
   - **Links to journal and/or web publication of the experiment.**
   - **Journal or URL citations.**

2. Samples used, extract preparation and labelling
   MIAME devised a hierarchical terminology for describing the samples that are hybridised to arrays.
   - **Biosource properties.** The biosource is the term used to describe the organism from which the sample that will by hybridised to the array is derived. It has the following properties:
     - **Organism.** Names are used from the NCBI taxonomy (e.g., *Homo sapiens*).
     - **Contact details.** Who to contact for information about the sample (e.g., dov@bius.co.uk).
     - **Descriptors relevant to the sample.**
       - **Sex**, e.g., male, female, hermaphrodite.
       - **Age.** Including relevant units (days, months, years), and whether from birth or embryolysis.
       - **Developmental stage.** An organism could develop at different rates depending on environmental conditions so this is included in addition to age.
       - **Organism part.** Tissue.
       - **Cell type.**
       - **Animal/plant strain or line.**
       - **Genetic variation**, e.g., wild-type, gene knockout or transgenic variation.
       - **Individual genetic characteristics.** Disease-associated alleles or polymorphisms.
       - **Additional clinical information.**
       - **Individual ID.**
   - **Biomaterial manipulations.** These are the laboratory processes carried out to the biosource as part of the experiment. They include
     - **Growth conditions.**
     - **In vivo treatments.**
     - **In vitro treatments**, including cell culture conditions.
     - **Treatment type**, e.g., small molecule (drug), heat shock, food deprivation.
     - **Separation technique**, e.g., none, microdissection, FACS.
   - **Hybridisation extract preparation protocol.** This is the nucleic acid that is extracted from the biomaterial that will be labelled:
     - **Extraction method**, e.g., URL of protocol.
     - **Extract type**, e.g., total RNA, mRNA or genomic DNA.
     - **Amplification**, e.g., RNA polymerases or PCR.
   - **Labeling protocol.** For each extract:
     - **Amount of nucleic acid labelled.**
     - **Label used**, e.g., A-Cy3, G-Cy5 or 33P.
     - **Label incorporation method**, e.g., URL of protocol.
   - **External controls added to hybridization extract.** These are spiking controls added for quality control purposes.
     - **Element on array expected to hybridise to spiking control.**

   - **Spike type**, e.g., oligonucleotide or bacterial DNA.
   - **Spike qualifier**, e.g., concentration, expected ratio or labelling methods.
3. Hybridisation procedures and parameters.
   - **Information about which labelled extracts have been hybridised to which arrays.** The labelled extracts relate to the sample, and the array will relate to array design information (see earlier discussion).
   - **Hybridisation protocol.** This would normally include
     - **The solution**, e.g., Na^+ concentration or formamide concentration.
     - **Blocking agent**, e.g., COT1.
     - **Wash procedure**, e.g., temperature and Na^+ concentration.
     - **Quantity of labelled target used.**
     - **Time, concentration, volume and temperature.**
     - **Hybridization instruments**, e.g., manufacturer and model.
4. Measurement data and specifications of data processing
   MIAME provides standards for describing the data from a microarray experiment at three levels. At the lowest level, the raw data is the image of the array. The second level is the image quantitation table, which contains the information produced by the feature extraction software such as mean pixel intensity, number of pixels and pixel standard deviation (Table 4.3). At the highest level, gene expression measurements from all the arrays in the experiment are normalised and combined to produce a gene expression measurement table for the experiment.
   - **Raw data description.** The protocols and settings for scanning including
     - **Scanning protocol**, including scanning hardware and software (e.g., make, model number or version), and scan parameters, including laser power, spatial resolution, pixel space and photomultiplier tube (PMT) voltage.
     - **Scanned images.** There is no consensus in MGED as to whether the images themselves should be provided. There are two advantages of providing images. First, they are the raw data, and thus provide better validation of results, particularly where features may be flagged. Second, advances in feature extraction software may mean that it would be desirable to revisit old images and obtain new quantitative data. However, images are large in size and so inclusion of images would be expensive and difficult for many laboratories.
   - **Image analysis and quantitation.**
     - **Image analysis software.** The specification and version of the feature extraction software, the algorithm and all parameters used.
     - **Image analysis output.** For each image, the complete output of the image analysis software. This is the image quantitation table.
   - **Normalized and summarized data.** This is the gene expression data matrix containing data from the whole experiment.
     - **Data processing protocol**, including details of any normalization algorithms used.

- **Gene expression data tables:**
  - **Derived measurement values.** These summarise the replicates (whether on the same or different arrays), or different elements (sequences) for the same gene.
  - **Reliability indicator for each data point**, e.g., a standard deviation or median absolute deviation from the median. The inclusion of a reliability indicator is strongly encouraged but not essential. (Much legacy data does not have reliability indicators.)

MAGE

MicroArray and Gene Expression (MAGE) is the technical implementation that allows software to be developed using MIAME. MAGE will be of interest to individuals seeking to develop microarray software that is fully supportive of MIAME. MAGE is under constant review, and it is recommended that if you are interested in using MAGE, you should visit the MGED web site and obtain the most current information. However, MAGE Version 1 is now set, and any new changes will appear in Version 2.

MAGE contains three subsections:

- **MAGE-OM.** The MAGE object model for MIAME. An object model is a design that represents structured complex information, such as the information in a microarray experiment. Object models are used as the design for databases, so a MIAME-compliant microarray LIMS or database would use MAGE-OM as its design.
- **MAGE-ML.** The XML implementation of MAGE-OM. XML (eXtensible Markup Language) is a data exchange language that is similar to HTML (the language via which web pages operate), except that the user can define tags and terms to be used. It is used to transfer information between databases and software components. Any microarray database that is MIAME-compliant will be able to import and export data in MAGE-ML format. Therefore, MAGE-ML operates as a standard around which any microarray databases or software can be integrated.
- **MAGE-stk.** MAGE software toolkit containing functionality to convert MAGE-OM into MAGE-ML in a number of languages, including Perl and Java.

SECTION 11.4 ONTOLOGIES

Section 11.3, MIAME, details *what* information is needed to be recorded from a microarray experiment in order to be able to reproduce the experiment. Ontologies provide a solution for *how* that information can be recorded. The aim of an ontology is to give the framework for a formal representation of a subject. An ontology consists of two parts:

- **Vocabulary.** The words and names of the items in the subject area that are to be described.

- **Relationships.** The ways in which the items in the subject area relate to one another.

There are three reasons for using ontologies:

- **Unambiguity.** The use of ontologies enables one scientist to understand precisely the terms that another scientist is using in an unambiguous way.
- **Conceptual framework.** Ontologies are built on a conceptual framework that can help in understanding the information about a subject.
- **Database design and querying.** The most important reason for using ontologies is that they help the development of computer databases that hold information about a subject. By introducing a controlled vocabulary, it is possible to query databases using the controlled terms and, in doing so, find all references that relate to the term of interest.

There are many applications of ontologies in information science. In the field of microarrays there are three sets of ontologies that are used: **taxonomic ontologies**, **gene ontologies** and **MGED ontologies**.

Taxonomic Ontologies

Taxanomic ontologies are the most familiar: every living organism is placed in a hierarchy of kingdom, phylum, class, order, family, genus and species. The genus and species together form the scientific name of the organism (e.g., *Homo sapiens*). There are controlled vocabularies for each of the terms, and the terms relate hierarchically.

Using scientific names for organisms makes good sense because it helps scientists to know exactly what species another scientist is referring to. For example, suppose you referred to an organism you had used for a microarray experiment as *yeast*. Another scientist might not know whether you were referring to *Saccharomyces cerevisiae* (baker's yeast), *Schizosaccharomyces pombe* (fission yeast), or possibly a different species commonly referred to as yeast (e.g., *Candida albicans*).

Gene Ontologies and the GO Consortium

Gene ontologies provide a set of terms for describing genes and their products. The Gene Ontology (GO) Consortium was set up in 1999 in order to provide a common framework for its members to be able to describe genes and gene products. The consortium members are shown in Table 11.4. GO has allowed its members to have a common set of terms for annotating genomes. There are three advantages to using GO:

- Unambiguous gene descriptions
- Simplified database querying
- Easier cross-species comparison

GO has organised ontologies for describing genes on three levels:

- **Molecular function.** The task performed by individual gene products (e.g., transcription factor or Serine/Threonine protein kinase).

TABLE 11.4: Members of the Gene Ontology (GO) Consortium

FlyBase	Database for the fruitfly *Drosophila melanogaster*
Berkeley *Drosophila* Genome Project	*Drosophila* informatics; GO database & software
Saccharomyces Genome Database (SGD)	Database for the budding yeast *Saccharomyces cerevisiae*
Mouse Genome Database (MGD) & Gene Expression Database (GXD)	Databases for the mouse *Mus musculus*
The *Arabidopsis* Information Resource (TAIR)	Database for the brassica family plant *Arabidopsis thaliana*
WormBase	Database for the nematode *Caenorhabditis elegans*
PomBase	Database for the fission yeast *Schizosaccharomyces pombe*
Rat Genome Database (RGD)	Database for the rat *Rattus norvegicus*
DictyBase	Informatics resource for the slime mold *Dictyostelium discoideum*
The Pathogen Sequencing Unit	The Wellcome Trust Sanger Institute
Genome Knowledge Base (GKB)	At Cold Spring Harbor Laboratory
EBI	InterPro – SWISS-PROT – TrEMBL groups
The Institute for Genomic Research (TIGR)	
Gramene	A Comparative Mapping Resource for Monocots
Compugen	(with its Internet Research Engine)

- **Biological process.** The broad biological goal of the gene product (e.g., mitosis or protein degredation).
- **Cellular component.** The subcellular organelle, location or macromolecular complex in which the gene product would operate (e.g., nucleus or telomere).

Each of these areas has a separate ontology defined for it, and any gene would have terms from all three ontologies.

EXAMPLE 11.1 GO ANNOTATION FOR ACT1 (ACTIN) IN *Saccharomyces cervisiae*

The GO terms for ACT1 in *S. cerevisiae* are shown in Box 11.1. There are GO terms for each of molecular function, biological process and cellular component. Each GO annotation has associated evidence of one or more citations and the type of evidence for the annotation (Table 11.5).

Each term in GO has several fields associated with it:

- **GO ID**, a unique numerical identifier
- **Synonyms**, alternative terms for the same term
- **Last updated**, date and time that the GO ID were updated
- **Parents**, more general classes to which the term belongs
- **Children**, more specific classes which derive from the term

BOX 11.1 GO Annotation for ACT1 in *Saccharomyces cerevisiae*

- Molecular Function
  - Histone acetyltransferase
  - Structural protein of cytoskeleton
- Process
  - Mitochondrian inheritance
  - Vacuole inheritance
  - Mitotic spindle orientation
  - Establishment of cell polarity (sensu *Saccharomyces*)
  - Regulation of transcription from Pol II promoter
  - Exocytosis
  - Endocytosis
  - Response to osmotic stress
  - Cell wall organization and biogenesis
  - Apical bud growth
  - Isotropic bud growth
  - Sporulation (sensu *Saccharomyces*)
  - Protein secretion
  - Cytokinesis
  - Histone acetylation
  - Cell cycle dependent actin filament reorganization
  - Vesicle transport along actin filament
- Component
  - Histone acetyltransferase complex
  - Actin cable (sensu *Saccharomyces*)
  - Contractile ring (sensu *Saccharomyces*)
  - Actin cortical patch (sensu *Saccharomyces*)
  - Actin filament

EXAMPLE 11.2 GO ENTRY FOR EXOCYTOSIS

The GO entry for the biological process Exocytosis is shown in Box 11.2. Each of the parent terms is a more general process of which exocytosis is one example; each of the child terms is a more specific type of exocytosis.

As we can see in Example 11.2, GO terms, and ontology terms in general, exist in a hierarchy of more general and more specific classes. In classical ontologies, each

TABLE 11.5: Evidence Annotation for the First Two GO Entries for ACT1

Histone acetyltransferase	Galarneau, L. et al., 2000. Multiple links between the NuA4 histone acetyltransferase complex and epigenetic control of transcription. *Mol Cell* 5(6):927–37.	IDA: Inferred from Direct Assay (*Last updated on 2002-09-18*)
Structural constituent of cytoskeleton	Botstein, D. et al., 1997. The yeast cytoskeleton in *The Molecular and Cellular Biology of the Yeast Saccharomyces: Cell Cycle and Cell Biology*, Vol. 3.	TAS: Traceable Author Statement (*Last updated on 2001-01-19*)

BOX 11.2 GO Entry for Exocytosis

- `Term Name: exocytosis`
- `Term ID: GO:0006887`
- `Synonyms: secretion, vesicle exocytosis`
- `Last updated: 2001-03-30 04:29:44.0`
- `Parent Terms: Process (3):`
  - `protein transport (GO:0015031)`
  - `vesicle-mediated transport (GO:0016192)`
  - `secretory pathway (GO:0045045)`
- `Child Terms: Process (4)`
  - `regulation of exocytosis (GO:0017157)`
  - `calcium ion dependent exocytosis (GO:0017156)`
  - `non-selective vesicle exocytosis (GO:0016194)`
  - `synaptic vesicle exocytosis (GO:0016079)`

term may only have one parent. However, due to the complexity of biological information, in GO each term can have more than one parent. It is sometimes useful to think of the terms as being in a tree; more precisely, the terms are organised in what is known as a **directed acyclic graph**.

EXAMPLE 11.3 DIRECTED ACYCLIC GRAPH

A directed acyclic graph showing a subset of GO terms for the molecular function ATP-dependent DNA helicase is shown in Figure 11.3. ATP-dependent DNA helicase has three parents, each of which connects to the root of the tree (molecular function) via several generations.

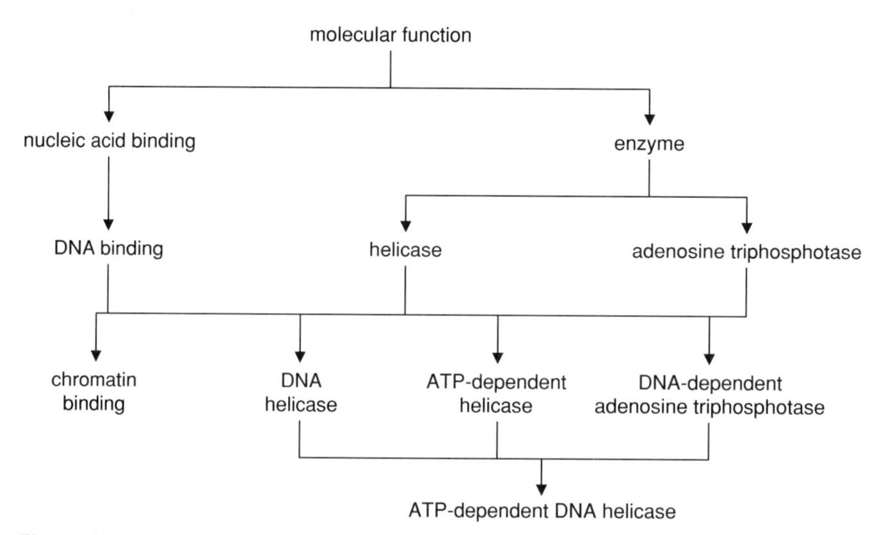

Figure 11.3: Directed acyclic graph. A directed acyclic graph showing the ancestry of the GO term ATP-dependent DNA helicase. This term has three parent terms, DNA helicase, ATP-dependent helicase and DNA-dependent adenosine triphosphotase. These terms in turn have parents, eventually leading back to the root of the ontology tree, molecular function. (Reproduced with Permission from the GO Consortium.)

Microarray Ontologies

The Ontologies Working Group at MGED has drawn up ontologies for microarray annotation with the aim of describing microarray data. The MGED ontology comprises three broad types of information:

- Classes
- Properties
- Individuals

Classes

Classes are the categories of information, for example, age, sex or protocol. Each class has a number of fields describing it:

- **Namespace.** A URL for the ontology.
- **Documentation.** A free text description of the class.
- **Type.** In the microarray ontologies, every class is of primitive type. This means that the class is not fully defined by its constraints.
- **Superclasses.** The parent classes of which this class is a special case.
- **Constraints.** These are rules by which any single instantiation of the class contains information. Each constraint is in the form of a property that the class may have.
- **Known subclasses.** These are child classes of the class which represent specialisations of the class.
- **Used in classes.** These are the classes that use this class as part of a constraint.

EXAMPLE 11.4 EXAMPLE MICROARRAY ONTOLOGY CLASS

The ontology entry for the class protocol is shown in Box 11.3. There is annotation for each of the fields. The superclass MGEDOntology is the root class from which all classes are derived. As protocols are widely used in microarray experiments, there are several constraints that can be used to describe the protocol and many subclasses or classes that contain protocols as a constraint.

Properties

Properties encapsulate information about classes. A class *has* properties; for example, the class *protocol* has the property *has_citation*. Each property is then linked to a class via the constraint in the class that contains the property. In the case of a protocol, *has_citation* will take a value in the class *BibliographicReference*.

EXAMPLE 11.5 EXAMPLE MICROARRAY ONTOLOGY PROPERTY

The microarray ontology entry for the property *has_citation* is shown in Box 11.4. Properties generally contain less information than classes: in this case, a link to the URL for the ontology and a list of classes and individuals which have the property.

BOX 11.3 Microarray Ontology Entry for the Class *Protocol*

```
class Protocol #1
namespace:
      http://www.cbil.upenn.edu/Ontology/MGEDontology.rdfs#
documentation:
      Documentation of the set of steps taken in a procedure.
type:
      primitive
superclasses:
      MGEDontology #3
constraints:
      restriction has_software #1 has-class Software #1
      restriction has_citation #1 has-class BibliographicReference #1
      restriction name #1 has-class thing
      restriction description #1 has-class thing
      restriction has_hardware #1 has-class Hardware #1
known subclasses:
      ArrayManufacture #5
      Hybridization #5
      ImageAcquisition #5
      ImageQuantification #5
      Labeling #5
used in classes:
      ArrayManufacture #5
      BiomaterialPreparation #1
      CompoundBasedTreatment #1
      ContaminantOrganism #1
      Hybridization #5
      ImageAcquisition #5
      ImageQuantification #5
      Labeling #5
      Preservation #1
      Treatment #1
      Water #1
```

Individuals

Individuals are instances of classes that are formally included in the ontology. For example, the class *Gender* has individuals *male, female, hermaphrodite, mixed_sex* and *unknown_sex* as instances of the class, which can be used to describe the gender of a biosource.

BOX 11.4 Microarray Ontology Entry for the Property *has_citation*

```
property has_citation #1
namespace:
      http://www.cbil.upenn.edu/Ontology/MGEDontology.rdfs#
used in classes:
      Protocol #1
      Study #3
```

BOX 11.5 Microarray Ontology Entry for the Individual *male*

```
individual male #1
namespace:
      http://www.cbil.upenn.edu/Ontology/MGEDontology.rdfs#
instance of:
      Gender #1
```

EXAMPLE 11.6 EXAMPLE MICROARRAY ONTOLOGY INDIVIDUAL

The microarray ontology entry for the individual *male* is shown in Box 11.5. Individuals have very little information associated with them: in this case, just a namespace and the class *Gender* of which *male* is an instance of.

KEY POINTS SUMMARY

- Microarray data is shared both by scientists and by computer software.
- Data sharing is enabled by data standards.
- MGED is responsible for data standards for microarrays.
- MIAME is the minimal information about a microarray experiment.
- MAGE enables software implementation of MIAME.
- Ontologies provide formal ways of describing species, genes and microarray experiments.

RESOURCES

http://www.mged.org/

The MGED web site, which contains information on MIAME, microarray ontologies and MAGE.

http://www.geneontology.org/

The GO Consortium web site containing information on gene ontologies and the laboratories that form part of the consortium.

http://protege.stanford.edu/publications/ontology_development/ontology101-noy-mcguinness.html

Ontology Development 101: A Guide to Creating Your First Ontology. A useful resource with an excellent description of ontologies.

Public Microarray Databases

http://genome-www5.stanford.edu/MicroArray/MDEV/index.shtml

The Stanford Microarray Database is the first microarray database available on the Internet. It contains many important data sets, including those from the pioneering papers that established microarrays as an important technology.

http://www.ebi.ac.uk/microarray/ArrayExpress/arrayexpress.html

Array Express is the first public implementation of a MIAME-compliant microarray database, which was developed at the European Bioinformatics Institute.

http://www.ncbi.nlm.nih.gov/geo/

The Gene Expression Omnibus at the NCBI contains data from microarray experiments, as well as gene expression experiments using other platforms such as SAGE.

Useful Papers

Brazma, A., Hingamp, P., Quackenbush, J., Sherlock, G., Spellman, P., Stoeckert, C., Aach, J., Ansorge, W., Ball, C.A., Causton, H.C., Gaasterland, T., Glenisson, P., Holstege, F.C.P., Kim, I.F., Markowitz, V., Matese, J.C., Parkinson, H., Robinson, A., Sarkans, U., Schulze-Kremer, S., Stewart, J., Taylor, R., Vilo, J., and Vingron, M. 2001. Minimum information about a microarray experiment – Towards standards for microarray data. *Nature Genetics* 29: 365–71.

Gardiner-Garden, M. and Littlejohn, T.G. 2001. A comparison of microarray databases. *Briefings in Bioinformatics* 2: 143–58.

APPENDIX

MIAME Glossary

This appendix has been reproduced in full with permission from MGED. It is also available from:

http://www.mged.org/Workgroups/MIAME/miame_glossary.html

Age The time period elapsed since an identifiable point in the life cycle of an organism. (If a developmental stage is specified, the identifiable point would be the beginning of that stage. Otherwise the identifiable point must be specified such as planting) [MGED Ontology Definition]

Amount of nucleic acid labeled The amount of nucleic acid labeled

Amplification method The method used to amplify the nucleic acid extracted

Array design The layout or conceptual description of array that can be implemented as one or more physical arrays. The array design specification consists of the description of the common features of the array as the whole, and the description of each array design elements (e.g., each spot). MIAME distinguishes between three levels of array design elements: feature (the location on the array), reporter (the nucleotide sequence present in a particular location on the array), and composite sequence (a set of reporters used collectively to measure an expression of a particular gene)

Array design name Given name for the array design, that helps to identify a design between others (e.g., EMBL yeast 12K ver1.1)

Array dimensions The physical dimension of the array support (e.g., of slide)

Array related information Description of the array as the whole

Attachment How the element (reporter) sequences are physically attached to the array (e.g., covalent, ionic)

Author, laboratory, and contact Person(s) and organization(s) names and details (address, phone, FAX, email, URL)

Biomaterial manipulation Information on the treatment applied to the biomaterial

Bio-source properties Information on the source of the sample

Cell line The identifier for the immortalized cell line if one was used to derive the BioMaterial [MGED Ontology Definition]

Cell type Cell type used in the experiment if non mixed. If mixed the targeted cell type should be used [MGED Ontology Definition]

Clone information For each reporter, the identity of the clone along with information on the clone provider, the date obtained, and availability

Common reference A hybridization to which all the other hybridisations have been compared

Composite sequence information The set of reporters contained in the composite sequence. The nucleotide sequence information for each composite element: number of oligonucleotides, oligonucleotide sequences (if given), and the reference sequence accession number (from relevant databases)

Composite sequence related information Information on the set of reporters used collectively to measure an expression of a particular gene

Compound A drug, solvent, chemical, etc., that can be measured [MGED Ontology Definition]

Contact details for sample The resource (e.g., company, hospital, geographical location) used to obtain or purchase the BioMaterial and the type of specimen [MGED Ontology Definition]

Control elements position The position of the control features on the array

Control elements related information Array elements that have an expected value and/or are used for normalization

Control type The type of control used for the normalization and their qualifier

Data processing protocol Documentation of the set of steps taken to process the data, including: the normalization strategy and the algorithm used to allow comparison of all data

Developmental stage The developmental stage of the organism's life cycle during which the BioMaterial was extracted [MGED Ontology Definition]

Disease state The name of the pathology diagnosed in the organism from which the BioMaterial was derived. The disease state is normal if no disease has been diagnosed [MGED Ontology Definition]

Element dimensions The physical dimensions of each features

Experiment description Free text description of the experiment and link to an electronic publication in a peer-reviewed journal

Experiment design Experiment is a lang=EN-US set of one or more hybridizations that are in some way related (e.g., related to the same publication MIAME distinguishes between lang=EN-US : the experiment design (the design, purpose common to all hybridisations performed in the experiment), the sample used (sample characteristics, the extract preparation and the labeling), the hybridisation (procedures and parameters) and the data (measurements and specifications)

Experiment type(s) A controlled vocabulary that classify an experiment

Experimental design Design and purpose common to all hybridisations performed in the experiment

Experimental factor(s) Parameter(s) or condition(s) tested in the experiment

Extraction method The protocol used to extract nucleic acids from the sample

Features related information Information on the location of the reporters on the array

Final gene expression table(s) Derived measurement value summarizing related elements and replicates, providing the type of reliability indicator used

Gene name The gene represented at each composite sequence: name and links to appropriate databases (e.g., SWISS-PTOR or organism specific database)

Genetic variation The genetic modification introduced into the organism from which the BioMaterial was derived. Examples of genetic variation include specification of a transgene or the gene knocked-out [MGED Ontology Definition]

Growth conditions A description of the isolated environment used to grow organisms or parts of the organism [MGED Ontology Definition]

Hybridization protocol Documentation of the set of steps taken in the hybridization, including: solution (e.g., concentration of solutes); blocking agent and concentration used; wash procedure; quantity of labelled target used; time; concentration; volume, temperature, and description of the hybridization instruments

Hybridization extract preparation Information on the extract preparation for each extract prepared from the sample

Hybridizations Procedures and parameters for each hybridization

Image analysis and quantitation Each image has a corresponding image quantitation table, where a row represents an array design element and a column to a different quantitation types (e.g., mean or median pixel intensity)

Image analysis output The complete image analysis output for each image

Image analysis protocol Documentation of the set of steps taken to quantify the image including: the image analysis software, the algorithm and all the parameters used

In vitro treatment The manipulation of the cell culture condition for the purposes of generating one of the variables under study and the documentation of the set of steps taken in the treatment

In vivo treatment The manipulation of the organism for the purposes of generating one of the variables under study and the documentation of the set of steps taken in the treatment

Individual genetic characteristics The genotype of the individual organism from which the BioMaterial was derived [MGED Ontology Definition]

Individual number Identifier or number of the individual organism from which the BioMaterial was derived. For patients, the identifier must be approved by Institutional Review Boards (IRBs, review and monitor biomedical research involving human subjects) or appropriate body [MGED Ontology Definition]

Label incorporation method The label incorporation method used

Label used The name of the label used

Measurements MIAME distinguishes between three levels of data processing: image (raw data), image analysis and quantitation, gene expression data matrix (normalized and summarized data)

Normalized and summarized data Several quantitation tables are combined using data processing metrics to obtain the 'final' gene expression measurement table (gene expression data matrix) associated with the experiment

Nucleic acid type The type of nucleic acid extracted (e.g., total RNA, mRNA)

Number of elements on the array The number of features on the array

Number of hybridisations Number of hybridizations performed in the experiment

Organism The genus and species (and subspecies) of the organism from which the BioMaterial is derived [MGED Ontology Definition]

Organism part The part or tissue of the organism's anatomy from which the BioMaterial was derived MGED Ontology Definition]

Platform type The technology type used to place the biological sequence on the array

Production protocol A description of how the array was manufactured

Provider The primary contact (manufacturer) for the information on the array design

Qualifier, value, source (may use more than once) Describe any further information about the array in a structured manner

Quality control steps Measures taken to ensure or measure quality: replicates (number and description), dye swap (for two channel platforms) or others (unspecific binding, low complexity regions, polyA tails)

Raw data Each hybridization has at least one image

Relationship between samples and arrays Relationship between the labelled extract (related to which sample which extract) and arrays (design, batch and serial number) in the experiment

Reporter and location The arrangement and the system used to specify the location of each features on the array (e.g., grid, row, column, zone)

Reporter approximate length The approximate length of the reporter's sequence

Reporter generation protocol A description of how the reporters were generated

Reporter related information Information on the nucleotide sequence present in a particular location on the array

Reporter sequence information The nucleotide sequence information for reporter: sequence accession number (from DDBJ/EMBL/GenBank), the sequence itself (if known) or a reference sequences (e.g., for oligonucleotides) and PCR primers pair information (if relevant)

Reporter type Physical nature of the reporter (e.g., PCR product, synthesized oligonucleotide)

Sample The biological material from which the nucleic acids have been extracted for subsequent labelling and hybridisation. MIAME distinguishes between: source of the sample (bio-source), its treatment, the extract preparation, and its labeling

Sample labeling Information on the labeling preparation for each labelled extract

Scanner image file The TIFF file including header

Scanning protocol Documentation of the set of steps taken for scanning the array and generating an image including: description of the scanning instruments and the parameter settings

Separation technique Technique to separate tissues or cells from a heterogenous sample (e.g., trimming, microdissection, FACS)

Sex Term applied to any organism able to undergo sexual reproduction in order to differentiate the individuals or type involved. Sexual reproduction is defined as the ability to exchange genetic material with the potential of recombinant progeny [MGED Ontology Definition]

Single or double stranded Whether the reporter sequences are single or double stranded

Spike type and qualifier The type of spike used (e.g. oligonucleotide, plasmid DNA, transcript) and its qualifier (e.g., concentration, expected ratio, labelling methods)

Spiking control External controls added to the hybridisation extract(s)

Spiking control feature Position of the feature (s) on the array expected to hybridise to the spiking control

Strain or line Animals or plants that have a single ancestral breeding pair or parent as a result of brother x sister or parent x offspring matings [MGED Ontology Definition]

Surface and coating specification Type of surface and name for the type of coating used

Targeted cell type The targeted cell type is the cell of primary interest. The BioMaterial may be derived from a mixed population of cells although only one cell type is of interest [MGED Ontology Definition]

Treatment type The type of manipulation applied to the BioMaterial for the purposes of generating one of the variables under study [MGED Ontology Definition]

INDEX